D0931223

WITHDRAWN

Nutrition of the dog and cat

portrait of the dog and cat.

Nutrition of the dog and cat

Waltham Symposium Number 7

Edited by

I.H. BURGER

Waltham Centre for Pet Nutrition,
Waltham-on-the-Wolds, Melton Mowbray, UK

and

J.P.W. RIVERS

Department of Human Nutrition,
London School of Hygiene and Tropical Medicine, London, UK

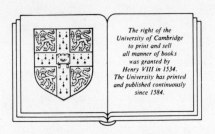

The right of the
University of Cambridge
to print and sell
all manner of books
was granted by
Henry VIII in 1534.
The University has printed
and published continuously
since 1584.

CAMBRIDGE UNIVERSITY PRESS

Cambridge

New York   Port Chester

Melbourne   Sydney

Published by the Press Syndicate of the University of Cambridge
The Pitt Building, Trumpington Street, Cambridge CB2 1RP
40 West 20th Street, New York, NY 10011, USA
10 Stamford Road, Oakleigh, Melbourne 3166, Australia

© Cambridge University Press 1989

First published 1989

Printed in Great Britain by Bath Press

British Library cataloguing in publication data

Nutrition of the dog and cat.
1. Pets : Dogs. Nutrition 2. Pets : Cats.
Nutrition
I. Burger, I.H. II. Rivers, J.P.W.
636.7'0852
ISBN 0 521 33019 X

Library of Congress cataloguing in publication data

Nutrition of the dog and cat / edited by I.H. Burger & J.P.W. Rivers.
    p.   cm.
Includes index.
ISBN 0-521-33019-X (U.S.)
1. Dogs–Nutrition. 2. Cats–Nutrition. 3. Dogs–Food. 4. Cats–
Food. I. Burger, I. H. II. Rivers, J. P. W.
SF427.4.N87 1989
636.7'0852–dc19   88-24619
                                                    CIP

ISBN 0 521 33019 X

SF
427.4
W36
1985

SE

Contents

v

Contributors

T.D. Adkins, University of Pennsylvania, School of Veterinary Medicine, 382 West Street Road, Kennett Square, PA 19348 USA

R.S. Anderson, University of Liverpool, Department of Animal Husbandry, Leahurst, Neston, South Wirral, L64 7TE UK

C.A. Banta, ALPO Petfoods Inc., PO Box 2187, Allentown, PA 18001, USA

J.E. Bauer, College of Veterinary Medicine, Dept. of Physiological Sciences, Box J-144 JHMHC, University of Florida, Gainesville 32610-0144, USA

C.H. Beauchamp, College of Veterinary Medicine, Dept. of Physiological Sciences, Box J-144 JHMHC, University of Florida, Gainesville 32610-0144, USA

S.E. Blaza, Pedigree Petfoods, National Office, Waltham-on-the-Wolds, Melton Mowbray, Leicestershire LE14 4RS UK

D. Booles, Waltham Centre for Pet Nutrition, Waltham-on-the-Wolds, Melton Mowbray, Leicestershire, LE14 4RT UK

J.E. Branam, Montevista Veterinary Hospital, Concord, California, USA

C.A. Buffington, Department of Internal Medicine, School of Veterinary Science, University of California, Davis, California 95616, USA

I.H. Burger, Waltham Centre for Pet Nutrition, Waltham-on-the-Wolds, Melton Mowbray, Leicestershire, LE14 4RT, UK

N.E. Cook, Department of Internal Medicine, School of Veterinary Science, University of California, Davis, California 95616, USA

R. De Wilde, Laboratory of Animal Nutrition, University of Ghent, Heidestraat 19, 9220 Merelbeke, Belgium

R.G. Downey, University of Pennsylvania, School of Veterinary Medicine, 382 West Street Road, Kennett Square, PA 19348, USA

G.C. Dunn, School of Veterinary Medicine, University of California, Davis, California 95616, USA

C.J. Gaskell, University of Liverpool, Small Animal Hospital, Crown Street, Liverpool, L7 7EX UK

R.G. Goring, Surgical Sciences College of Veterinary Medicine, University of Florida, Gainesville, FL USA

H.A.W. Hazewinkel, State University Utrecht, Small Animal Clinic, Yalelaan 8, De Uithof, PO Box 80 154, 3508 TD Utrecht, The Netherlands

T. Jansen, Laboratory of Animal Nutrition, University of Ghent, Heidestraat 19, 9220 Merelbeke, Belgium

J. Jones, Medical Sciences College of Veterinary Medicine, University of Florida, Gainesville, FL USA

E. Kane, Feline Research Centre, Carnation Research Farm, Carnation, Washington 98014-0500, USA

E. Kienzle, Institut für Tierernahrung, Tierarztliche Hochschule Hannover, Bischofsholer Damm 15, D-3000 Hannover 1, West Germany

D.S. Kronfeld, University of Pennsylvania, School of Veterinary Medicine, 382 West Street Road, Kennett Square, PA 19348, USA

R.B. Lavelle, University of Melbourne, School of Veterinary Science, Veterinary Clinical Centre, Princes Highway, Werribee, Victoria, Australia

G.G. Loveridge, Waltham Centre for Pet Nutrition, Waltham-on-the-Wolds, Melton Mowbray, Leicestershire LE 14, 4RT, UK

J.G. McLean, Swinburne Institute of Technology, PO Box 218, Hawthorn 3122, Australia

J.R. Mercer, Department of Animal Husbandry, The University of Sydney, NSW 2006, Australia

D.J. Meyer, Medical Sciences College of Veterinary Medicine, University of Florida, Gainesville, FL USA

H. Meyer, Institut für Tierernahrung, Tierarztliche Hochschule Hannover, Bischofsholer Damm 15, D-3000 Hannover 1, West Germany

A.R. Michell, Royal Veterinary College, Hawkshead House, Hawkshead Lane, North Mymms, Hatfield, Hertfordshire, AL9 7TA, UK

E.A. Monger, Department of Veterinary Preclinical Sciences, School of Veterinary Science, The University of Melbourne, Parkville, Victoria 3052, Australia

J.G. Morris, Department of Animal Science, University of California, Davis, California 95616, USA

H.-C. Mundt, Institut für Tierernahrung, Tierarztliche Hochschule Hannover, Bischofsholer Damm 15, D-3000 Hannover 1, West Germany

J.P.W. Rivers, Dept. of Human Nutrition, London School of Hygiene and Tropical Medicine, Keppel Street, London, WC1E 7HT, UK

Q.R. Rogers, School of Veterinary Medicine, Dept. of Physiological Sciences, University of California, Davis, California 95616, USA

M.C. Schaeffer, United States Department of Agriculture, Western Human Nutrition Research Centre, PO Box 29997, Presidio of San Francisco, California 94129, USA

B.E. Sheffy, James A. Baker Institute for Animal Health, Cornell University, Ithaca, New York 14853, USA

S.V.P.S. Silva, Department of Animal Husbandry, The University of Sydney, NSW 2006, Australia

Preface

The subject matter of this book is based on papers originally presented at the 7th Waltham Symposium on Recent Advances in Dog and Cat Nutrition held at Queens' College Cambridge in August 1985.

Although three years have elapsed between the conference and the appearance of the written record, this has been due principally to our desire, from the outset, to produce not just a conference proceedings, but a textbook on the current state of the art in dog and cat nutrition. For this reason we encouraged the authors to expand and update their manuscripts accordingly. It is also this aim which prompted us to eschew normal 'proceedings practice' of including a discussion section at the end of each paper and to combine the salient points into the text. We felt that the reader would appreciate a novel rather than a play.

The book is designed principally for nutritionists and practitioners and students of veterinary medicine, but we hope that food scientists in industry and academia will also find items of interest. We shall be more than happy if the book conveys the challenging nature and practical importance of this somewhat neglected area of animal nutrition.

In conclusion we must acknowledge the invaluable help given by many individuals during the various stages of the project but in particular Dorothy Howard and Christina Roberts (Waltham Centre for Pet Nutrition), Heather Pane and Fenella Jones (Gwynne-Hart and Associates) and the Cambridge University Press.

Acronyms

AAFCO American Association of Feed Control Officials
ACT Allometric Cancellation Technique
AP Activity Product

BMR Basal Metabolic Rate
BW Bodyweight

CCMP Carbohydrate Containing Medium Protein
CFMP Carbohydrate Free Medium Protein
CPE Crude Protein Equivalent
CPM Counts Per Minute
CT Critical Temperature
CV Coefficient of Variation

DCP Digestible Crude Protein
DE Digestible Energy
DM Dry Matter
DMI Dry Matter Intake

EAA Essential Amino Acid
ECF Extracellular Fluid Volume
EFA Essential Fatty Acid
ENL Endogenous Nitrogen Loss
EUNL Endogenous Urinary Nitrogen Loss

FCRD Feline Central Retinal Degeneration
FEDIAF European Pet Food Manufacturers' Federation
FSUS Feline Struvite Urolithiasis Syndrome
FU Feline Urolithiasis
FUNL Fasting Urinary Nitrogen Loss
FUS Feline Urological Syndrome

GE Gross Energy
GFR Glomerular Filtration Rate

HC High Calcium

x

ICS   Incremental Cost of Standing
IEMT   Institute for Interdisciplinary Research on the Human–Pet Relationship

LC   Low Calcium
LCP   Long Chain Polyunsaturated Fatty Acids
LDL   Low Density Lipoprotein
LH   Lateral Hypothalamus

ME   Metabolisable Energy
MFN   Metabolic Faecal Nitrogen

NAS   National Academy of Sciences (USA)
NC   No Calcium
ND-p   Net Dietary-protein
ND-p:E   Net Dietary-protein: Energy ratio
NPU   Net Protein Utilisation
NRC   National Research Council (USA)

OMR   Organ Metabolic Rate
ONL   Obligatory Nitrogen Loss

PCV   Packed Cell Volume
P:E   Protein: Energy ratio
PFMA   Pet Food Manufacturers' Association
PTH   Para-Thyroid Hormone

RNA   Ribonucleic Acid

SAA   Sulphur Amino Acids
SPF   Specific Pathogen Free

TEN   Total Endogenous Nitrogen
TONL   Total Obligatory Nitrogen Loss
TSAA   Total Sulphur Amino Acid
TWI   Total Water Intake

VFI   Voluntary Food Intake
VMH   Ventromedial Hypothalamus

WCPN   Waltham Centre for Pet Nutrition

1

Why dogs and cats?

R.S. ANDERSON

Why dogs and cats? Why not little pigs or monkeys? After all, little pigs and monkeys are as intelligent, expressive and probably have as long a history of domestication as dogs and cats. What denied them the intimacy with man that is accorded to dogs and cats? The keeping of monkeys as pets among the entourages of the upper classes in bygone days is well documented. Dogs and cats, however, predominate in present-day society as the animals with which we are most closely associated. There must be few, if any, individuals in the Western world who have not touched or been touched by a dog or a cat, whereas the numbers who have handled a living, breathing pig, monkey, or, for that matter, sheep, horse, cow or goat are, one suspects, very much smaller.

What is it therefore, about dogs and cats which have made them so pervasive? Twenty-four per cent of British households own a dog and 20 per cent own a cat, whereas the next most popular animal to be classed as a pet, the budgerigar, enjoys a presence in only 6 per cent of British households (PFMA 1986). In 1982 there were 26 million pet dogs and 23 million pet cats within the European Community (before it was joined by Spain) and, although millions of other animals are kept as pets, dogs and cats are more commonly kept than any other species (European Pet Food Manufacturers' Federation FEDIAF, 1982).

Archaeological evidence suggests that wolves were among the first of what are now classified as domestic animals to become associated with human settlements – and that was about 12,000 years ago. As in man's relationship with all other species, the relationship was encouraged to develop in the furtherance of his vested interest in his own survival and survivability; the particular benefits of a closer association with wolves probably included their role in removing the remnants of food – bones and other debris from the human 'table' which would otherwise have accumulated in the vicinity of the settlement. Serpell (1986) postulates that the initial entrance of dogs into our society was an accident of history; they were in the right place at the right time, that is, they were around when our

1

ancestors were starting to tame and domesticate other species. Whether dogs and cats were at that time valued for other than the strictly utilitarian reasons is a matter for speculation.

Though none of the individual characteristics of dogs or cats is unique, it would seem to be the particular spectrum into which these characteristics are combined which has singled them out for a special place in man's environment.

Size

Size is probably one factor in their favour. It would clearly be inconvenient to have anything much larger than an Irish Wolfhound in the average home while very small animals, like hamsters or mice, are inclined to get lost unless confined to their cages; their small size also renders them more vulnerable to a traumatic end half-way through the pantry door or under some careless foot. Although cats are of relatively constant size, the considerable and unique range in size of the various dog breeds nevertheless falls within the maximum and minimum which is compatible with the human establishments which they share. Despite the labile nature of the canine genes and man's predisposition to exploit this variability, it seems that we have reached the limits of their phenotypic potential at both large and small ends of the spectrum – no new and even bigger giant breeds, nor yet smaller miniature breeds have emerged in recent centuries, if the paintings in which they have so frequently featured are to be believed. It is also worth noting that, despite the range in size from which the dog owner has to choose, the most popular size is in the middle of the range and there are, therefore, no 'market forces' driving the breeder to extend the range. We must conclude, therefore, that their size, variable though it is, is a factor in their favour as household companions.

Predisposition to identify with individual people or localities

One of the characteristics of dogs and cats which brought them together with man was their predisposition to locate and remain within the area of human settlements. (Their cousin, the once-rural fox is also rapidly learning that there are advantages in living closer to man, as its natural environment is depleted.) This characteristic not only gives us the pleasure of another creature actually seeking our company, rather than shunning it as do other species, but it is convenient that we do not have the cost and trouble of putting up dog- and cat-proof fences nor do we (in general) find it necessary to tether, hobble or otherwise restrain them to keep them in the vicinity of our homes.

Greeting behaviour

Because of their predisposition to remain in the vicinity of our homes – a predisposition which we foster and exploit for our own purposes, their usual role is to await our return from our regular forays to the office or the shops. In the case of dogs, this return, whether after an absence of minutes or hours, is usually celebrated with an uninhibited enthusiasm which cannot fail to please, whereas in the case of the cat, it is usually marked by a more restrained, but none the less welcoming, behaviour. This behaviour is not only pleasurable but reassuring for older people who live alone and whose house is otherwise empty when they are out. The exuberant greeting behaviour of dogs is, moreover, usually reserved for the individual owner or other members of the household – which gives the added savour of a privilege conferred only on a favoured few.

Excretory behaviour

The ability to control excretory behaviour is a *sine qua non* of the inclusion of any creature in the household. It is socially and hygienically unacceptable to keep a creature of any size in a house unless it is house-trained or unless its excretory activities are confined to a very small purpose-built area. The ready trainability of dogs to excrete outside the house probably derives from the natural propensity of their wolf ancestors to remain within the boundaries of their own territory (and indeed 'mark' these boundaries with urine) and also to avoid soiling the immediate area of their dens or lairs with faeces. There is, therefore, a built-in predisposition for dogs to deposit urine and faeces within a defined area (the garden) but outside their lairs (the house) – a predisposition which is readily reinforced by house-training. The cat's habit of burying its faeces, and urine-marking its territory is also readily exploited to render it an acceptable house companion. Although less effort has probably been applied to house-training other animals, most of them in the wild are much more indiscriminate in their excretory habits and would, therefore, present significantly greater problems in being trained to a manageable standard of excretory control. The house-trained dog or cat may, therefore, share in a real sense the domestic environment and the intimacy with other members of the household which this allows, whereas other pet animals, regardless of how cuddly or appealing, must spend much of their time shut up in a cage or a pen as a penalty for their inability to adapt their excretory behaviour to the hygienic standards of the household.

Petting

Dogs and cats are by no means unique in their apparent enjoyment of contact with man. Horses, cattle, sheep, pigs, rats, rabbits, snakes and probably frogs and fish quickly learn to appreciate the pleasures of being scratched, stroked, tickled, rubbed or patted. Although we humans clearly enjoy doing it, partly because of the apparent enjoyment it gives to the animal, it is only recently that a more quantified explanation of this mutual enjoyment has been offered. Katcher and his co-workers, following an earlier observation that the pet-owning survivors of heart attacks had a significantly greater chance of surviving the following 12 months than the non-pet-owning (Friedman *et al.*, 1980), found that greeting and talking to their dogs actually helped their owners to relax (as indicated by blood pressure changes) in mildly stressful conditions (see Katcher 1981). A later study by the same team showed that even the presence of a dog in the same room as children required to carry out a mildly exacting task was sufficient to exert a calming effect as measured by blood pressure and heart rate changes (Friedman *et al.*, 1983). That this calming effect is not, however, a unique characteristic of dogs, is suggested by the observation that the contemplation of the fish in an aquarium was more effective than a blank wall in lowering the blood pressure of adult subjects (Katcher *et al.*, 1984). These apparent effects of contact with, or even the presence of, dogs must, however, depend on the cultural background of the subject and, as Serpell (1986) points out, an orthodox Muslim (or a citizen of a land in which rabies was endemic) might view the proximity of a dog with less equanimity than would, say, a European or a North American.

Life expectancy

The highest levels of pet ownership are in families with children (PFMA, 1986) and, although there are many interactions, e.g. type of housing, age of housewife, working status, floor space, and garden, the presence of children is an important predictor of pet ownership even after the data are adjusted or transformed to account for these interactions (Messent & Horsfield, 1985).

Although reliable data are not available, the longevity of dogs and cats is of the order of 10–15 years – and this appears to coincide with the period when the children in the family are between the ages of, say, 5 years and the late 'teens. Smaller pets, like hamsters, gerbils and mice have a much shorter life expectancy and are usually acquired when the children are quite young. They will, inevitably, die after two or three years, when the children are still quite young. Children have, therefore, much less of a length and continuity of association with these animals than with a dog or cat, which may be present throughout childhood and adolescence. This early childhood experience of dogs and cats appears to be an important predictor of adult attitudes to pet ownership (Serpell, 1981). (Whether the much shorter experience of the smaller caged rodents in childhood has an effect in later years on attitude to these or other pets does not seem to have been examined.) This 10–15 year period of 'imprinting' must, therefore, operate to the advantage of dogs and cats in perpetuating their popularity in households in which the parents retain positive memories of their own childhood association with these animals.

On a national basis, these generation-to-generation effects on the predisposition to own pets probably contribute to the marked differences in dog or cat ownership between different countries; the United States has, for example, five times as many dogs and eight times as many cats per head of the population as Japan, and, even countries as close geographically as France, Switzerland and West Germany, show marked differences in pet ownership (Messent & Horsfield, 1985). It is also likely that the tradition of pet owning which seems to be passed down through families will operate against any rapid increases or decreases in national pet populations and will thus help to maintain these marked national differences in pet ownership.

Dogs, cats and diet

The relationship between health, longevity and diet receives an increasing amount of attention in both the popular and scientific press. In the western world (though not unfortunately in countries such as Ethiopia) the interest in food has shifted from deficiences to excesses and to the effects of diet on the slowly progressive degenerative diseases of middle and old age.

Although not a prey to quite the same degenerative diseases as man, dogs and cats, nevertheless, share his relative immunity to the commoner infectious diseases conferred by increased hygiene and the development and use of vaccines. Their feeding has, furthermore, become much more controlled and scientific, as manufactured foods have gradually replaced the more haphazard feeding of the past. Because of this, and because of national and international controls and regulations, there is much more at stake for the manufacturer of pet foods in 'getting it right'. Millions of animals may now obtain a substantial proportion of their daily nutrient intake from food produced by a relatively small number of manufacturers. The consequences of failing to understand and implement the best current knowledge on nutrition are therefore potentially very much greater than those for any single manufacturer of human foods. Dog and cat owners, while probably sharing the growing awareness of the general public to the importance of balancing their own nutrient intake from the range of foods available, have come to expect that the pet foods they buy provide the whole range of nutrients in one can or pack. While this expectation is justifiable *to the best of our knowledge*, it would be foolish to suppose that our knowledge is so comprehensive that it cannot be improved. Within the last decade, the new knowledge on taurine, essential fatty acids and amino acids which has been generated by nutritional research is enough to dispel any notion that our present knowledge of canine and feline nutrition is complete enough for all eventualities.

The competitive search by the pet food industry for new raw materials and processes to improve value for money, palatability, texture and appearance of pet foods will always have nutritional implications for the dog and cat population. Unless the level of understanding of the digestion, absorption and metabolism of such materials by dogs and cats keeps pace with these technological developments, the possibility of nutritional problems arising cannot be discounted.

The continued support of nutritional research by the pet food industry, of which this book, and the symposium from which it arose are good examples, will continue to play an essential part in maintaining the health of the pet population.

REFERENCES

FEDIAF. (1982) *The European Pet Food Industry*.

Friedman, E., Katcher, A.H., Lynch, J.J. & Thomas, S.A. (1980). Animal companions and one-year survival of patients after discharge from a coronary care unit. *Public Health Reports* 95, 307–12.

Friedman, E., Katcher, A.H., Thomas, S.A., Lynch, J.J. & Messent, P.R. (1983). Interaction and blood pressure: influence of animal companions. *Journal of Nervous Mental Diseases*, 171, 461–5.

Katcher, A.H. (1981). Interactions between people and their pets: form and function. In *Interrelations between People and Pets*, ed. B. Fogle, pp. 41–67. Springfield, Ill.: Charles Thomas.

Katcher, A.H., Segal, D.D.S. & Beck, A.M. (1984). Contemplation of an aquarium for the reduction of anxiety. In *The Pet Connection* Eds. R.R. Anderson, B.L. Hart and L.A. Hart. pp. 171–8. Minneapolis: University of Minnesota.

Messent, P.R. & Horsfield, S. (1985), Pet Population and the Pet-Owner-Bond. In *The Human–Pet Relationship*. Proceedings of the International Symposium on the occasion of the 80th Birthday of the Nobel Prize Winner Prof. Dr. Konrad Lorenz. Vienna: Institute for Indisciplinary Research on the Human–Pet Relationship.

The Pet Food Manufacturers' Association. (1986), *Profile*.

Serpell, J.A. (1981). Childhood pets and their influence on adults' attitudes. *Psychological Report* 49, 651–4.

Serpell, J.A. (1986). *In the Company of Animals*. Oxford: Basil Blackwell.

2

The current consensus in dog and cat nutrition

J.P.W. RIVERS AND I.H. BURGER

The current consensus in nutritional topics tends to be illuminated, if not always precisely defined, by meetings of expert committees which gather together the central wisdom of a subject in an attempt to define nutritional needs, and such reports are the first port of call for the nutritionists seeking any information about a species. The symposium from which this book arose was held in the shadow of the publication of the definitive set of requirements for the dog (NAS/NRC 1985) and, during the gestation of this book, another NAS/NRC committee has produced a set of requirement estimates for the cat (NAS/NRC, 1986).

The fact that this flurry of activity exists is in itself evidence of an important consensus about the nutrition of these two species. It is an important matter, requiring serious attention. Traditionally, nutrition has tended to be divided between human nutrition, the object of which is self-evident, and animal nutrition, by which is meant the study of farm animals, producing food or work. Traditionally, also, an extensive literature exists on the nutrition of laboratory animals, primarily because they are used as models for other species, but a fairly extensive limbo exists when the nutrition of other animals is considered. The flurry of interest in producing requirement estimates for dogs and cats is evidence that they have moved out of that limbo, and establishes the first point of the consensus that the topic is important.

In a sense, dog and cat nutrition share attributes of factors that make both animal and human nutrition important. As Anderson shows in his chapter, the nutrition of dogs and cats is like the nutrition of farm animals, in that it is important financially. The pet food industry is now a major one, in the EEC in 1986 having an annual turnover of around £2,250 million and directly employing 18,000 people. That industry is justified by the fact that it makes diets which are safe, palatable and nutritionally balanced for the consuming animal, and, in this sense, is heavily dependent upon its nutritional bases. But, unlike farm animal nutrition and like the nutrition of the human species, the nutrition of dogs and cats is made difficult by the criteria for

7

success that are employed. In farm animal work, an animal is well fed if it does not have a deficiency disease, and is productive to an optimal level, often an economic not a physiological optimum. The owners of cats and dogs wish them to share the same health criteria as are used upon humans, that is the animal should not only avoid deficiency but should live a long and healthy life. Currently, the consensus in human nutrition is that human diets, as consumed in the West, do not meet these criteria and it is believed that much morbidity can be laid at the door of the diet. Though cat and dog nutritionists are aware that the problem may equally exist for these species, and are concerned with diets and health in a broader sense than requirements to avoid deficiency, they have at the present time every reason to be confident. Undoubtedly, cat and dog diets will be improved as knowledge advances, but it is often said that they are currently better than the diets the owners get!

It is not possible to say whether this new concept of nutrition is reflected in the pronouncements of the Committee on Dog Nutrition. But it is interesting in this context that the NAS committees abandoned the idea of estimating safe levels for dogs and cats, and interpreted their brief as providing estimates of minimal requirements, devoid of safety factors whether related to biological variability of the consumer, or bioavailability of the diet. This has proved to be a controversial move, a controversy which is aired in the chapters by Sheffy and Kronfeld & Banta. But it has the merits that it leaves uncertainty evident where uncertainty certainly exists.

The problems inherent in the extrapolation of results from studies with purified or semi-purified diets to practical feeding form the first of several themes of the book. These by no means cover all of the current questions and challenges in dog and cat nutrition but do serve as a focus for what we have called the current consensus. There are three other main points which emerge from the book.

The first of these is the special nutritional characteristics displayed by the cat. Most of the information has been gained in the last 10 to 15 years and our knowledge in this area is still expanding. The subject is explored extensively by Morris and Rogers in Chapter 5 but there are other papers devoted to specific aspects of this fascinating, but perhaps little known, area of animal nutrition. Feeding behaviour of the cat is also one of the topics discussed. It is clear from the data presented in these chapters that the cat possesses a wide portfolio of nutritional features which are indicative of an obligatory carnivorous lifestyle. It is probably not unique in this respect but has the distinction of being the animal that has been subjected to the most intensive study.

Most commercial pet foods comprise a complete diet, often for all life stages of the animal. The nutritional balance of these products then assumes a much more crucial position than it could ever do for a typical individual human food. The requirement for dietary carbohydrate is a good example of this question of nutritional balance which arises through provision of a whole diet, and it is addressed by a series of papers dealing primarily with dog requirements. It appears from the results of these studies that protein can substitute for carbohydrate but only if the concentration is sufficiently high to supply gluconeogenic amino acids in the required amounts.

Finally, we return to the 'healthy' diet theme. Can the diet, through nutritional imbalance, excessive nutrient intake or just unwise feeding practices, actually adversely affect the well-being of the animal? The last section of the book is devoted mainly to some of the contemporary problems in the area, in particular calcium and phosphorus balance in the dog and the Feline Urological Syndrome (FUS). The different views and theories presented are an excellent illustration of the challenges that confront all those who are interested in dog and cat nutrition, and who are involved in the formulation of the best possible diets for these animals.

REFERENCES

NRC (1985) *Nutrient Requirements of Dogs*. National Research Council. Washington D.C.: National Academy of Sciences.

NRC (1986) *Nutrient Requirements of Cats*. National Research Council. Washington D.C.: National Academy of Sciences.

3

The 1985 revision of the National Research Council nutrient requirements of dogs and its impact on the pet food industry

BEN E. SHEFFY

The discussion in this chapter concentrates on the latest update of the publication on 'Nutrient Requirements of Dogs' issued by the US National Academy of Sciences/National Research Council (NAS/NRC 1985). Such reports occupy a central or bridging role in the study of the subject. On the one hand, they collate the scientific advances in the nutrition of the dog, a subject which as other chapters in this book testify, is undergoing vigorous advance. On the other hand, the report is a document to provide guidelines for the veterinary profession, the petfood industry and Government, and it cannot be evaluated without considering that context.

After reviewing the salient points of difference in the 1985 update from the previous publications, this paper will conclude with a plea and a justification for the required updating of the industry and the profession as well as the regulatory officials so that dogs and the dog owning public may continue to enjoy the benefits of research scientists' concern and efforts on the dog's behalf. All that is required is understanding and a modicum of honest leadership. All are capable and worthy of the challenge.

The National Research Council

The National Research Council (NRC) is the working arm of the US National Academy of Sciences (NAS). The Academy was established in 1863 by proclamation of President Abraham Lincoln. The NRC was not established until 1916 but both were justified in response to the need for mobilisation of all scientific effort for war. Two important units operating under the NRC are the Commission on Natural Resources and the Board on Agriculture and Renewable Resources. Food is considered one of our most important natural renewable resources. To the extent that all life competes for or complements the available food supply it is imperative to utilise it effectively. Thus the Committee on Animal Nutrition is an important link, through its sub-committees, establishing minimum nutrient requirements of the various non-human species to support maximum productive ef-

11

ficiency with minimal food nutrient waste. Subcommittees are established in response to the need for updating or revising previous reports. Committees are charged to make an intensive review of all peer reviewed literature publications, make impartial evaluations of the studies and to integrate these into an updated nutrient requirement series for the species. The rationale for arriving at the requirements tabulated is to be discussed and documented in a clear, concise, scholarly manner. Prior to publication these documents are reviewed by both internal and external reviewers helping to assure accuracy and clarity.

To understand the environment in which the 1984–85 Committee worked it is necessary to review the 1974 publication (NRC/NAS, 1974) and its subsequent use by the pet food industry. While the documentary and literature review in the 1974 publication was discussed properly with weight given where required, the tabulated data on nutrient needs took a middle ground with a mixture of requirements and recommended allowances. Great liberties were taken in the recommendation for protein in particular and in the tacit implication that foods formulated to these guides would be complete and balanced to meet nutritional needs for all stages of the life cycle. This was done in spite of the fact that requirements for reproduction had not been sufficiently researched, and studies quantitating needs for more stressful situations did not exist. No statement was made that many values were really minimum requirements, rather the impression was given that these values were optimal. Many nutritionists and some regulatory officials were apprehensive about the application of these 1974 values by some pet food manufacturers and called for some form of feeding studies to document the adequacy of a food for any productive stage of the life cycle. The Industry responded and produced a series of test feeding protocols to justify nutritional claims made on labels. Criteria of adequacy included satisfactory performance in growth rate, development, clinical health and haematological responses and reproductive–lactation performance where it applied. These were adopted in 1974 by the American Association of Feed Control Officials (AAFCO) and are published annually in the AAFCO Manual (AAFCO, 1984). These protocols, which would have provided minimal standards of evaluation, were too restrictive for some sections of the Industry. Thus, an alternative procedure for claiming nutritional adequacy for all stages of the life cycle was approved. This was the statement that the food 'Meets or Exceeds' 1974 NRC on the basis of a chemical analysis called 'nutrient profiling'. The result was predictable: the bluebloods of the Industry, who exhaustively tested all their products by feeding to dogs in their own research kennels, produced foods in keeping with higher requirements for more demanding situations. They developed puppy foods, stress foods and even geriatric foods. Nutrient concentration was kept in concert with energy concentration, while palatability, digestibility and nutrient availability was monitored and kept high. Others, more cost- than quality-conscious, did not do feeding tests but based claims for adequacy ('Meets or Exceeds NRC'), purely on chemical analysis and profiling. Custom packers and petfood formula merchants abounded and the situation was not improved by the advent of 'least-cost' linear programming. In the wake of this development there were store brands and then generic brands. Many pet owners learned about nutrient inadequacies the hard way, while packers of canned meat products and producers of vitamin–mineral preparations and other supplements, thrived. It bordered on insanity to lecture to pet owners that 'all you need to add is water' – they saw the results and knew better.

As the largely untested products began making inroads into the sale of the well-tested products, an economic pinch was felt by the producers of the latter who began to fight back. For example, consider the efforts of a major dry dog food manufacturer, Ralston Purina, to educate the Veterinary Profession both on the limitations of 'Meets or Exceeds' by chemical analysis and the all important roles of palatability, digestibility and biological availability.

Initially, an independent laboratory was commissioned to evaluate, by chemical analysis, 78 generic brand dog foods purchased from supermarkets all over the United States, to determine whether these foods met their guaranteed label analysis claims. The results were hardly unexpected. Over 50% did not meet their own guaranteed analysis stated on the label for one or more of protein, fat and fibre. For the eight specific nutrients analysed, namely calcium, phosphorus, potassium, iron, copper, zinc and vitamins A and B_6, 65 brands (83%) failed to meet minimum values of NRC 1974 for one or more of the nutrients analysed. When Permitted Analytical Variations (AAFCO 1984) were considered, 65% were still in unacceptable ranges to meet NRC values. For specific nutrients, 63% failed to meet recommended levels for vitamin A and 19% had ratios of calcium to phosphorus outside the acceptable range of 1:1 to 2:1. For the trace elements, copper and zinc, the levels found varied widely from below the recommended level to 5–6 times that level. These results became even more disturbing when one considered that the inadequacies would be exacerbated by reduced nutrient availability and/or food palatability.

This aspect was further demonstrated in a study which involved feeding two similar products, both designated as complete and balanced nutritionally to meet requirements for all stages of the life cycle. The dog feeding study was conducted in accordance with the AAFCO protocol for a food designated for the growing dog. Two groups each of 15 weaned puppies were fed, either a dry dog food purchased from a major food chain (Store Brand R2), or Purina Dog Chow, designated R1. Dogs of three breeds were studied, representing littermates that were matched on the basis of breed, weight and sex. Both foods were chemically analysed and this demonstrated that the nutrients guaranteed on the labels were in fact present as stated. From their appearance, and by chemical analysis, both foods could be expected to be satisfactory for puppies as judged by the 1974 NRC. Feeding results indicated differently. Differences in growth performance were dramatic as judged by weight gain or body length. The average body weight gain of R2 puppies was less than half that of the R1 puppies, and the body length gain of R2 puppies was consistently less than their littermate R1 puppies. Puppies of the R2 group ate significantly less food (0.22 v. 0.34 kg/day) than did the R1 pups.

Clinically, the R1 puppies were in good health and had an attractive appearance while physical examination of R2 puppies indicated anorexia, increased illness, hair coat problems (including greying, now recognised as a clear sign of a zinc deficiency) and general lack of growth and development.

Serum chemistry and haematological values supported in the main the dogs performance and clinical signs. By the end of the test period, the R2 puppies showed significant deviations in both haemoglobin and haematocrit values. Similarly they had lower serum cholesterol, alkaline phosphatase, calcium and phosphorus levels.

In spite of the fact that chemical analysis of the two foods was similar, there were obvious differences in palatability of the foods, and it appeared the nutrients in the store brand were not as readily digested or as biologically available. In a subsequent test in adult dogs, food R1 was shown to be 18% higher in digestibility than the store brand. Higher digestibility of both protein (7.2%) and fat (13.6%) for Purina Chow was recorded. None the less, palatability would appear to have been the more important and overriding factor in the growth study comparison. This strong evidence reinforced what long was generally recognised by many veterinarians and dog nutritionists and clearly supported the need for compulsory feeding of all dog foods before a positive statement of nutritional value could be claimed on the food label. Without this the label should be required to state that the food has not been tested and feeding it as the sole source of nourishment may be injurious to the dog's health.

The 1985 NRC Revision (NRC 1985)

With the above as background, it becomes easier to understand the direction if not the intensity of the changes introduced in 1985. The major and significant changes were essentially three.

First, no requirement for protein as such specified, rather requirements for ten essential amino acids for growth and maintenance of dogs were listed on a per kilogram body weight per day basis (Table 1) and for pet foods, on a per 1,000 kilocalorie of metabolisable energy (ME) basis when fed for growth (Table 2). The development of this research data was made possible largely as a result of the interest and support of the petfood manufacturers. It was entirely from their own financial contributions to the Pet Food Institute coupled with the encouragement and scientific advice of their Nutrition

Table 1. *Minimum nutrient requirements of dogs for growth and maintenance (amounts per kg of body weight per day)[a]*

Nutrient	Unit	Growth[b]	Adult maintenance[c]
Fat	g	2.7	1.0
Linoleic acid	mg	540	200
Protein[d]			
Arginine	mg	274	21
Histidine	mg	98	22
Isoleucine	mg	196	48
Leucine	mg	318	84
Lysine	mg	280	50
Methionine–cystine	mg	212	30
Phenylalanine–tyrosine	mg	390	86
Threonine	mg	254	44
Tryptophan	mg	82	13
Valine	mg	210	60
Dispensable amino acids	mg	3414	1266
Minerals			
Calcium	mg	320	119
Phosphorus	mg	240	89
Potassium	mg	240	89
Sodium	mg	30	11
Chloride	mg	46	17
Magnesium	mg	22	8.2
Iron	mg	1.74	0.65
Copper	mg	0.16	0.06
Manganese	mg	0.28	0.10
Zinc	mg	1.94	0.72
Iodine	mg	0.032	0.012
Selenium	μg	6.0	2.2
Vitamins			
A	IU	202	75
D	IU	22	8
E[e]	IU	1.2	0.5
K[f]			
Thiamin	μg	54	20
Riboflavin	μg	100	50
Pantothenic acid	μg	400	200
Niacin	μg	450	225
Pyridoxine	μg	60	22
Folic acid	μg	8	4
Biotin[f]			
B_{12}	μg	1.0	0.5
Choline	mg	50	25

Notes (Table 1):
[a] Needs for other physiological states have not been determined.
[b] Average 3-kg-BW growing Beagle puppy consuming 600 kcal ME/day.
[c] Average 10-kg-BW adult dog consuming 742 kcal ME/day.
[d] Quantity sufficient to supply minimum amounts of available indispensable and dispensable amino acids specified.
[e] Requirement depends on intake of PUFA and other antioxidants. A fivefold increase may be required under conditions of high PUFA intake.
[f] Dogs have a metabolic requirement, but a dietary requirement was not demonstrated when natural ingredients were fed.

Table 2. *Required minimum concentrations of available nutrients in dog food formulated for growth*

Nutrient	per 1000 kcal ME	Dry basis (3.67 kcal ME/g)
Protein[a]		
Indispensable amino acids		
Arginine	1.37 g	0.50%
Histidine	0.49 g	0.18%
Isoleucine	0.98 g	0.36%
Leucine	1.59 g	0.58%
Lysine	1.40 g	0.51%
Methionine–cystine	1.06 g	0.39%
Phenylalanine–tyrosine	1.95 g	0.72%
Threonine	1.27 g	0.47%
Tryptophan	0.41 g	0.15%
Valine	1.05 g	0.39%
Dispensable amino acids	17.07 g	6.26%
Fat	13.6 g	5.0%
Linoleic acid	2.7 g	1.0%
Minerals		
Calcium	1.6 g	0.59%
Phosphorus	1.2 g	0.44%
Potassium	1.2 g	0.44%
Sodium	0.15 g	0.06%
Chloride	0.23 g	0.09%
Magnesium	0.11 g	0.04%
Iron	8.7 mg	31.9 mg/kg
Copper	0.8 mg	2.9 mg/kg
Manganese	1.4 mg	5.1 mg/kg
Zinc[b]	9.7 mg	35.6 mg/kg
Iodine	0.16 mg	0.59 mg/kg
Selenium	0.03 mg	0.11 mg/kg

Nutrient	per 1000 kcal ME	Dry basis (3.67 kcal ME/g)
Vitamins		
A	1011 IU	3710 IU/kg
D	110 IU	404 IU/kg
E[c]	6.1 IU	22 IU/kg
K[d]		
Thiamin[e]	0.27 mg	1.0 mg/kg
Riboflavin	0.68 mg	2.5 mg/kg
Pantothenic acid	2.7 mg	9.9 mg/kg
Niacin	3 mg	11.0 mg/kg
Pyridoxine	0.3 mg	1.1 mg/kg
Folic acid	0.054 mg	0.2 mg/kg
Biotin[d]	—	—
Vitamin B_{12}	7 μg	26 μg/kg
Choline	340 mg	1.25 g/kg

Notes:

[a] Quantities sufficient to supply the minimum amounts of available indispensable and dispensable amino acids as specified. Compounding practical foods from natural ingredients (protein digestibility 70%) may require quantities representing an increase of 40% or greater than the sum of the amino acids listed, depending upon ingredients used and processing procedures.

[b] In commercial foods with natural ingredients resulting in elevated calcium and phytate content, borderline deficiencies were reported from feeding foods with less than 90 mg zinc per kg (Sanecki et al., 1982).

[c] A fivefold increase may be required for foods of high PUFA content.

[d] Dogs have a metabolic requirement, but a dietary requirement was not demonstrated when foods from natural ingredients were fed.

[e] Overages must be considered to cover losses in processing and storage.

Task Force that studies at the University of Illinois over the past six years were made possible. These studies defined the amino acid requirements for growth. The committee therefore had good growth data as well as balance studies in Beagle puppies to consider in establishing requirements for growth. Data developed by Ward (1975) establishing amino acid needs for maintenance of dogs were the basis for the maintenance requirements.

The second important difference in 1985 from 1974 is that all of the values for the nutrients are listed as minimum *requirements*, and not recommended *allowances*. Additionally Table 2 of the NRC 1985 publication lists the requirements for nutrient content of dog foods in terms of available nutrients from the food when consumed. While the Committee could not predict nutrient bioavailability from respective ingredients or foods it did list

Table 3. *Factors for consideration in formulation of dog foods from natural ingredients*

Nutrient	Factors for consideration
Fat	Degree of unsaturation, antioxidants, vitamin E
Carbohydrate	Fibre, lactose, reducing sugars, processing, stage of life cycle
Protein	Energy content, digestibility, amino acid balance, processing, anti-nutrients, antitryptic factors
Amino acids	Availability; heat treatment in presence of reducing sugars reduces availability, especially of lysine; requirement for individual amino acids increases with increased dietary nitrogen.
Minerals	Ratios, source, availability
Calcium	Phytates, ligands, vitamin D
Phosphorus	Phytates, calcium, plant-animal
Sodium, potassium, chloride	High availability
Zinc	Phytates, calcium, plant-animal, fibre
Copper	Phytates, zinc
Iron	Source, availability, plant-animal
Vitamins	Processing, lipid content, source
A	Oxidation, toxicity
D	Toxicity, calcium level
E	PUFA, selenium
B_1	Losses in processing and storage, product pH, storage time and temperature, thiaminases
B_2	UV light
B_6 (Pyridoxine)	Protein level in diet
Niacin	Tryptophan, low availability of plant sources
Folate	Processing losses
B_{12}	Plant versus animal proteins
Choline	Methionine, folate, vitamin B_{12}, availability, fat

factors that might affect it and which should be considered in formulating foods from natural ingredients (Table 3).

Thirdly, requirements for dog foods for nutrients available to dogs are expressed as quantities present per 1000 kilocalories of ME content (Table 2). The latter is very helpful if not vital in specifying nutrient content of dog foods that can vary in moisture content from 8 to 80%. In addition, owing to the variety of ingredients used, and the direct addition of fats and oils, dog foods may vary from 5 to 40% fat content on a dry basis. Thus, expression of nutrient content on a unit energy basis makes nutritional comparison between foods easier and may improve assurance of nutritional adequacy.

Some will question the wisdom of expressing some nutrients on an available energy basis. Metabolisable energy none the less is the common denominator of choice in dog foods. We feed dogs quantities of food to meet their energy requirements. Thus it is prudent that this quantity of food, which is therefore dependent on ME content, should contain or provide the dog's daily need for the respective essential nutrients as well.

Many questions have been raised regarding two key points. Why specify requirements instead of recommended daily allowances? Why express the data as available nutrients rather than nutrients as fed?

The answer to the first is easy, the charge to the subcommittee from the parent Committee on Animal Nutrition stated that all publications in the series would be standardised to express *requirement values*. But available? Why? If swine nutritionists can predict availability why can't dog nutritionists do likewise? The answer is that swine rations are compounded from relatively few, mostly unprocessed, standard ingredients, e.g. corn, soybean meal. The digestibility and nutrient availability of these have been repeatedly demonstrated for swine and their analysis is consistent. Studies establishing nutrient requirements for swine were established using a combination of these ingredients. On the other hand, dog foods are formulated from a great variety of ingredients, many with different processing histories. In addition, after blending they are subjected to further cooking and processing all of which leads to a wide variation in nutrient availability. Livestock farmers are more astute evaluators of performance resulting from feeding different foods than pet owners, and will quickly detect differences in nutritional quality. Swine are fed exclusively one food as sole diet – dog foods are often supplemented. Swine foods are sold on merit while dog food is heavily promoted often by irrational advertising stressing human appeal. Quite simply, however, biological availability for many nutrients of many constituents of dog foods are unknown particularly in the various combinations and after processing.

Major changes in requirements

Since the 1985 report is one that stipulates requirements rather than allowances and since these are for available nutrients, generally the values listed either for foods (Table 2) or as intake per unit bodyweight (Table 1) have been lowered. The most dramatic has been for protein equivalent or the amino acids (6 g/kg bodyweight for growth, less than 2 g/kg for maintenance). However, the requirements for the major minerals calcium, phosphorus, potassium, sodium, chloride and iron have also been decreased, as they have for copper, zinc and for iodine. In recognition of these changes, caveats have been raised about the advisability of raising the level of amino acids by 40% or more when compounding foods from natural ingredients of low digestibility or those subjected to damaging processing procedures. Similar caveats were raised for zinc requiring three-fold increases (or more) where high calcium and/or phytate levels are present. The need for raising vitamin E levels as much as five-fold in foods containing high polyunsaturated fatty acid (PUFA) content and marginal selenium levels was also stated. Other vitamin levels were either decreased slightly or not at all.

Does it work?

There should be little reason for trepidation over the reductions made for the requirements of minerals and vitamins since the reductions are largely accounted for by the switch to minimum requirements for available nutrients. There might be reason for concern, however, about the requirements for the indispensable amino acids, since these were based in part on studies which were made in puppies that were over seven weeks of age, and which lasted only two weeks (Milner, 1979a, b, 1981, 1984). The criteria used in these studies were bodyweight gain and nitrogen balance only. Both are reliable parameters but some account of the composition of the gain would have been useful.

Therefore a feeding study was conducted to check the validity of the new NRC amino acid requirements for growth. A diet meeting these requirements (as shown in Table 2) provides the equivalent of 12% of the ME of the diet as protein, i.e. 12% protein kcal or P:E. This standard diet (Diet I – see Table 4) was compared with a diet (IV) in which the level of each amino acid was increased by one third to provide 16% P:E. Amino acids were substituted isocalorifically for carbohydrate, so that the ME of the diet remained constant at 4.02 kcal/g and the fat was constant at 10% of dry matter. Additionally, we undertook to investigate the effect of altering the level of fat in the diet, and hence the P:E ratio, on the adequacy of the diet for amino acids. The work of Milner (1981) on lysine suggested that the level of dietary fat in the standard diet had no effect on the requirement for this amino acid. Thus we fed two other diets, each containing the same level of amino acids per unit weight as Diet I but in which fat was increased to 20% (Diet II) and 30% of dry matter (Diet III), lowering P:E ratios to 10.9% and 9.9% respectively. Mineral and vitamin content was varied in concert with the energy content of respective diets. Four litters of Beagle puppies were used and distributed so that one animal from each litter was in each group. All puppies were fed once each day and offered quantities of food calculated on a metabolic body size basis to provide twice NRC maintenance, i.e. 264 kcal $W^{0.75}$. This amount was adjusted for changes in bodyweight every week.

The growth rates are shown in Figure 1. While there was no significant difference in growth rate of dogs fed diets with 12% and 16% P:E (Diets I and IV), clearly, feeding diets with P:E ratios lower than 12% (Diets II and III) was reflected in poorer growth. It is important to note that the dogs fed diets containing less than 12% P:E never reached normal adult weight even after being fed free choice a good commercial 25% protein expanded dog food after termination of the comparison study. Haematological values were lower for the lower P:E group of dogs but not significantly so.

Balance studies were performed at 4, 8, 12 and 16 weeks of age. Urine and faeces from five consecutive 24-hour periods were collected, composited separately and analysed for nitrogen content. The combined nitrogen content of urine and faeces was subtracted from the nitrogen intake to give the amount of nitrogen retained. The conversion factor of nitrogen to protein was 6.25. These nitrogen balance studies at each of the four periods during growth for each group of dogs reflected their respective amino acid intake (Table 4).

Dogs fed diet IV, 16% P:E, retained more nitrogen to a significant degree throughout the entire feeding regime than did dogs on the other three diets, as shown by the figures in Table 4. Additionally, the dogs on diet III retained

Fig. 1. Growth of Beagle pups fed diets providing different levels of protein and fat. Diets I, II and III contained amino acids at a level equivalent to the NRC protein requirement, diet IV contained 30% more than this. Diets I and IV contained 10% fat, diet II 20% fat, diet III 30% fat.

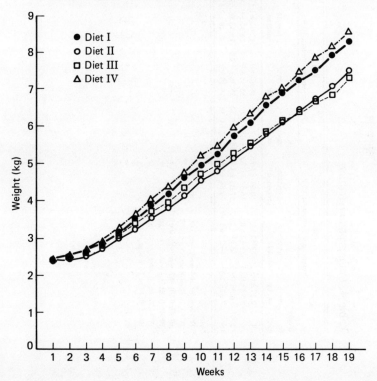

Table 4. *Daily protein retention*

Weeks on diet	Diet I Protein : energy content 12.2%			Diet II Protein : energy content 10.9%			Diet III Protein : energy content 9.9%			Diet IV Protein : energy content 16%		
	dog mean wt. kg	g protein fed	g protein retained	dog mean wt. kg	g protein fed	g protein retained	dog mean wt. kg	g protein fed	g protein retained	dog mean wt. kg	g protein fed	g protein retained
4	3.11±0.62	19.43±0	10.51±1.75	2.9±0.31	18.60±0	11.01±1.17	3.1±0.41	15.33±0	8.97±1.32	3.26±0.36	27.53±0	17.86±1.93[a]
8	4.53±0.61	25.42±0	16.54±0.62	4.08±0.27	23.54±0	16.47±0.15	4.31±0.45	19.41±0	12.25±2.79[c]	4.73±0.36	36.49±0	23.94±0.82[b]
12	6.0±0.87	30.91±2.85	17.03±3.14	5.42±0.32	28.88±1.48	15.33±2.62	5.5±0.45	24.14±4.78	12.11±4.78	6.31±0.49	46.26±0	27.79±3.83[d]
16	7.61±1.15	36.27±5.49	19.84±3.65	7.02±0.38	36.13±0	21.28±2.17	6.8±0.82	26.42±4.62	12.08±4.38[f]	8.16±0.71	55.67±2.17	28.01±3.72[e]

Notes:

[a] IV > I, II, III at 4 weeks $p < 0.05$
[b] IV > I, II, III at 8 weeks $p < 0.05$
[c] III < I, II, III at 8 weeks $p < 0.05$
[d] IV > I, II, III at 12 weeks $p < 0.05$
[e] IV > I, II, III at 16 weeks $p < 0.05$
[f] III < I, II, IV at 16 weeks $p < 0.05$

less protein than dogs on the other three diets, throughout the entire testing period, this being at a significant level during Balance Trials at 8 and 16 weeks.

At week 20 of the study, liver biopsies were taken from two dogs in each group. Since analysis of the biopsied liver (Table 5) revealed significantly elevated fat content in livers of dogs fed the diet of lowest P:E content and since choline content of all diets was equivalent on an ME content basis, other factors were responsible. Amino acid and/or protein deficiencies are commonly manifested by presence of fatty livers, therefore, the protein–amino acid content of all diets was increased by the addition of 5% by weight of casein to test this hypothesis. These diets were then fed to the remaining two dogs for 14 days prior to liver biopsy and assay for fat content (Table 5). Dogs fed less than 12% P:E had significantly greater liver fat content than dogs fed 12% and 16% P:E. However, there was no significant difference between the 12% and 16% P:E fed dogs. The addition of 5% casein (Table 5) did not significantly alter liver fat content of dogs on diet I and IV but did for dogs fed diet III. Liver fat of dogs on diet II was reduced by feeding of casein although this was not statistically significant.

Protein/energy malnutrition has repeatedly been found to alter immune function both in laboratory animals and in man (Suskind, 1977). Consistently cellular rather than humoral responses have been recorded (Gross & Newburn, 1980). To test cellular immune function, peripheral lymphocytes were taken and subjected to *in vitro* PHA–Con A stimulation, and mitogenesis was recorded at monthly intervals during growth.

Graded numbers of responder leukocytes were dispensed into 96 round-bottom well microtitre plates. The culture medium was RPMI 1640 supplemented with antibiotics (100 Units/ml of pencillin, 100 μg/ml of streptomycin), 10 mM of HEPES buffer and 10% heat-inactivated foetal calf serum. Mitogens, Con A (6 μg/ml) and PHA (3.7 μg/ml) were diluted in medium and added to the plate. The cultures were incubated for 96 h at 39 °C and labelled with 0.2 μCi/well of ^{3}H-thymidine during the last 18 h of incubation. Proliferation response was calculated as a Stimulation Index, i.e. ratio of counts per minutes (CPM) in culture with mitogen versus CPM in culture with medium only (Sheffy *et al.*, 1985).

Both studies showed superiority of the 30% + NRC AA (Group IV) over other groups, yet all dogs responded similarly to Parvo, CD and ICH vaccination. Thus if there was T-cell limitation, this was not reflected in the B-cell response. We concluded that the NRC AA level, 12% of energy is adequate but is indeed the minimum requirement. A level of 16% of energy protein equivalent AA, however, supported greater T-cell responses and resulted in greater nitrogen retention over 5-day study periods, and may be considered as a practical minimum.

Table 5. *Liver biopsies*

Diet		Liver DNA μg/g wet liver	Liver protein mg/g wet liver	Protein:DNA ratio	Liver fat* mg/g wet liver
I	A	65.72 ± 0.33	7.77 ± 0.76^d	119.53 ± 11.74	23.48 ± 3.87
	B	48.45 ± 14.79^a	5.20 ± 1.31^{b,c}	109.17 ± 4.93	16.64 ± 5.17
II	A	62.14 ± 33.24^a	5.67 ± 1.21^c	101.56 ± 35.75	72.60 ± 2.03
	B	120.86 ± 22.33^a	9.00 ± 2.19^{b,c}	74.30 ± 4.73	41.43 ± 3.47
III	A	29.66 ± 0.07	3.99 ± 0.38^d	134.67 ± 12.99	219.35 ± 168.36
	B	81.48 ± 60.21	7.58 ± 2.27^d	113.77 ± 56.08	59.49 ± 3.76
IV	A	76.26 ± 21.36	6.67 ± 0.29^d	90.75 ± 21.42	15.29 ± 4.20
	B	74.50 ± 7.60	8.02 ± 0.48^b	108.72 ± 17.77	19.76 ± 2.11

Notes:

A = diet fed throughout the trial

B = diet fed throughout the trial plus 5% casein added and fed for two weeks prior to biopsy

Each set of figures is the mean ± S.D. of values of two dogs on their respective diets

Significant figures: Liver DNA:

[a] IIB > IB, IIA $p < 0.05$

Liver protein:

[b] IB < IIB, IVb

[c] IIB > IIA, IB $p < 0.05$

[d] IIIA < IA, IIIB, IVA

*Liver fat:

IIIA > all other treatments $p < 0.05$

How does this translate to the real world of feeding dogs?

Quite naturally, one should not expect any products on the shelf to approach anything like the minimum amino acid contents discussed above (Sheffy, 1985). The reasons are multiple. Topping the list is that these are minimum values based on L-amino acids with essentially 100% availability, rather than protein diets. There were no antagonisms resulting from amino acid excesses or imbalances, the amino acid profiles were those of a perfectly utilisable protein. There were no necessary allowances for processing and/or storage losses. Furthermore, energy and other essential nutrient concentration was optimal. Palatability was not a problem and, perhaps most important, the subjects were selected healthy dogs, well adjusted to their environment and free of parasites and other infections. They were maintained in a low stress thermoneutral environment, in individual cages, restricted in exercise and not subjected to annoyances from interactions with each other or man. Differences or variations from some of the above factors will affect metabolic requirements while others would simply increase the quantities which need to be incorporated into complete and balanced foods. Either way it is not likely that commercial foods of 3.75 kcal of ME per gram with much less than 18% protein on a dry matter basis will be marketed in the near future except for maintenance purposes only. The reasons or limitations will be more to do with palatability and digestibility than of fulfilling required essential nutrient content. This will be a greater problem for the 'dry' dog foods than for those intermediate in water content or canned foods.

Protein from animal sources is a strong contributor to palatability of dog food. Judicious use of 'digests', particularly in coatings, can improve palatability but it is difficult to beat the real thing. It will be difficult to maintain high nutrient bioavailability when lower protein, poorer quality ingredients with more elevated fibre contents are used. Even if average digestibility is achieved, this does not necessarily indicate high bioavailability, although knowing apparent digestibility will be a big step forward.

However, if commercial foods with nutrient levels close to the requirements published in the 1985 NRC are to meet the challenge in the marketplace, thorough testing by feeding to dogs will be even more imperative than at present. Testing by feeding to dogs from weaning to maturity would appear to be the most direct evaluation of nutritional adequacy. It is one which is most relevant to the veterinarian and convincing to the dog owner. Simply ask the dog.

The industry's response

The response of the pet food industry to the change in the 1985 NRC report, to the all-important caveat that a simple 'meets or exceeds' is unacceptable and that dog feed protocols may be mandatory, ranged from disappointment to anger. The AAFCO regulatory officials have yet to make a firm decision on this point in spite of the fact that they had two years advance notice that these changes were coming. Until they do decide, all label claim regulations will continue to use the 1974 revision as their guide.

The pet food industry, through the Pet Food Institute has commissioned an independent laboratory to test, by fulfilling the AAFCO protocols for growth, three practical dry expanded foods. The ingredients of these foods are corn, soybean meal, meat and bone meal, wheat middlings, fat, vitamins and minerals. The foods have a calculated caloric density of about 3.3 kcal ME per gram. If nutrient availability is near 90%, one would approximate to NRC values, another would fall below and another would exceed. Before the start of the growth study, balance studies will be made to determine digestibility and an accurate estimate of ME content.

While any such study, properly executed, would yield important information it will be of limited value to the industry as a whole. It cannot infallibly be translated to foods formulated from other ingredients or subjected to different processing procedures.

What is needed?

In their review, the 1985 Committee noted an almost complete lack of published data on digestibility and bioavailability of specific nutrients from both common ingredients of US petfoods and those less commonly used. Such information would go a long way towards allowing the intelligent use of ingredients in compounding diets and offer a greater opportunity for prediction of nutritive values of the finished product, as well as their ME values. This, it is hoped, will be information available before the next NRC revision.

REFERENCES

Gross, R.L. & Newburn, P.M. (1980). Role of nutrition in immunologic function. *Physiological Review.* **60**, 188–302.

Milner, J.A. (1979a). Assessment of indispensable and dispensable amino acids for the immature dog. *Journal of Nutrition.* **109**, 1161.

Milner, J.A. (1979b). Assessment of the essentiality of methionine, threonine, tryptophan, histidine and isoleucine in immature dogs. *Journal of Nutrition.* **109**, 1351.

Milner, J.A. (1981). Lysine requirements of the immature dog. *Journal of Nutrition.* **111**, 40.

Milner, J.A., Garton R.L. & Burns R.A. (1984). Phenylalanine and tyrosine requirements of immature Beagle dogs. *Journal of Nutrition.* **114**, 2212.

National Research Council. (1974). Nutrient requirements of dogs. Washington, DC: National Academy of Sciences.

National Research Council. (1985). Nutrient requirement of dogs. Washington, DC: National Academy of Sciences.

Official Publication. Association of American Feed Control Officials Inc. (1984). Donald James, Charleston, W VA.

Sanecki, R.K., Corbin, J.E. and Forbes, R.M. (1982). Tissue changes in dogs fed a zinc-deficient ration. *American Journal of Veterinary Research,* **43**, 1642–6.

Sheffy, B.E. (1985). Pediatric nutrition in the 80's. *Proceedings of the Symposium Eastern States Veterinary Conference,* January 1985. Alpo Pet Center, Viewpoints in Veterinary Medicine, Canine Pediatrics, pp. 13–21.

Sheffy, B.E., Williams A.J., Zimmer J.F. & Ryan G.D. (1985). Nutritional metabolism of the geriatric dog. *Cornell Veterinarian.* **75**, 324.

Suskind, R.M. (ed). (1977). *Malnutrition and the immune response.* New York, Raven Press. pp. 111–15, 285–92.

Ward, J. (1975). Amino acid requirements of the mature dog. Ph.D. dissertation. Cambridge, England.

4

Optimal ranges of actual nutrients

D.S. KRONFELD AND C.A. BANTA

Introduction

Pertinent thoughts on nutritional standards – their history, uses and abuses, and especially the weaknesses in the recommendations for dogs and cats propounded by the US National Research Council – were presented in a Waltham-like symposium held in Oslo (Kronfeld, 1984). The critique suggested that the NRC recommendations for dogs and cats gave scant attention to energy density and less to interactions among ingredients. Also, it pointed out that the meaning of 'adequate' was obscure and appeared to change from near-minimum in 1953 to near-optimal in 1972 and 1974 (NRC, 1953, 1972, 1974).

The 1985 revision of *Nutrient Requirements of Dogs* has attended to those weaknesses by striving for minimal requirements of available nutrients expressed on the basis of metabolizable energy (NRC, 1985). Moreover, it has achieved a high and impartial level of scholarship.

The 1985 revision undoubtedly has made major advances in principle but, we submit, not in practice. It challenges the pet food industry: 'Users are advised to obtain evidence of nutritional adequacy by direct feeding to dogs' (NRC, 1985).

It probably should direct much the same challenge to nutritional science which could improve, we suggest, by addressing the needs of dog owners in several respects. One is the type of performance criteria appropriate for dogs; growth rate is less important than conformation, coat and behaviour. Another is the practical need for optimal ranges of nutrients rather than minimal requirements. A third is the influence of combinations of feedstuffs (diets) on nutrient availability, hence the actual contents of nutrients. We will try to illustrate these concerns with regard to protein and trace elements.

The optimal range for protein

The influence of a nutrient upon performance may be viewed theoretically as a curve that rises to a plateau then falls. The curve shown in

27

Figure 1 depicts 'Bertrand's rule' from 1912 (quoted by Mertz, 1981). If this concept merits experimental testing, it is necessary to describe the whole domain rather than only a zone somewhere near the left.

This left-zone, which hopefully defines minimum requirements, has drawn much attention, perhaps too much, from nutritional scientists. Even nutritionists sometimes find that these minima are too low for comfort, and that a 'recommended dietary allowance' with a safety margin is preferable. Also, minima are not more important to veterinarians than maxima, given the fashion of supplementation. Practical nutritional consultants need to know optimal ranges of nutrients, and whether narrower ranges are required for more demanding homeostatic situations in the animals or for more demanding and ambitious clients (see Table 1 in Kronfeld, 1984).

If we do address only the left-hand shoulder of the curve (Fig. 1), then we should seek a point of inflection to denote the minimum requirement. The simplest valid approach to this, we believe, is the broken-line analysis as shown, for example, in Figure 13 of the chapter by Shaeffer *et al.* (page 180). The break occurs at the intersecton of two regression lines that have significantly different slopes. In such an analysis, one needs at least three, preferably more, points to determine each of the two sections of the broken line.

Instead, most recent studies cited by the NRC for protein and amino acid requirements have only one or at the most two points (dietary nutrient levels) to determine the plateau. An example is a study of weight gain and nitrogen retention in growing Beagles fed 5, 7.5, 10, 15 or 20% lactalbumin (Burns, LeFaivre & Milner, 1982). In the data for animals from 13 to 17 weeks of age, nitrogen retention was significantly higher but bodyweight gain was not higher at the 20% level than at the 15% level. This can be clearly seen in Figure 7 of the chapter by Schaeffer *et al.* (page 173), which presents the data of Burns *et al.* (1982).

Burns *et al.* (1982) and the NRC (1985) concluded that 15% lactalbumin

Fig. 1. The relationship of biological performance to dietary concentration or rate of intake of an essential nutrient (Modified from Mertz, 1981.)

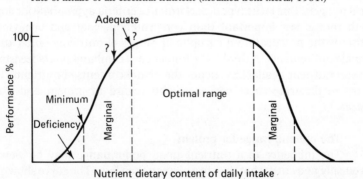

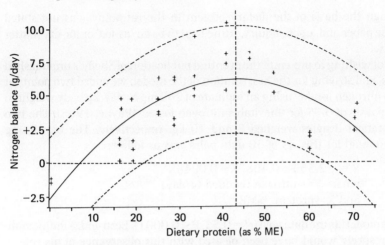

Fig. 2. A mathematical model of protein requirements in dogs, showing a parabolic relationship between nitrogen balance and the protein: energy ratio of the diet. The solid line shows the best fit polynomial.

Nitrogen balance (grams/day) $= -2.554 + 0.4057P - 0.0048P^2$ where P is the percentage of ME in the diet provided by protein. The dashed curves show the 95% confidence limits of the regression. (Modified from Sheffy & Banta, unpublished data.)

represents the minimum requirement of protein for growth at this age. Schaeffer *et al.* conclude from the same data that the requirement is at least 20% (see page 171 of this volume). Alternatively, one could conclude that the minimum requirement for growth was not established in 13–17-week old pups because no break or point of inflection is detectable in the best line representing the increase in nitrogen retention with increasing lactalbumin level. These data fit somewhere on the rising left-hand slope of the curve in Figure 1.

Does Bertrand's rule apply to the protein requirements of dogs? It is adumbrated by a hand-drawn parabola in a review (Sheffy, 1979). Dr Sheffy has kindly permitted us to use these unpublished data in a mathematical model (Fig. 2). Out of 39 nitrogen balances, 38 formed a coherent group and one was dropped from further statistical evaluation of the group because it was 8.4 standard deviations higher than the mean. (We suspect constipation!).

The data were from nitrogen balances on diets of different protein:energy ratios. The polynomial equation of best fit to these 38 data pairs was determined by means of the SAS General Linear Models programme on a VAX-11750 computor (Freund & Littell, 1981). The equation had an R-square of 0.45 (P < 0.0001) and a peak nitrogen retention at 43.1% protein. A similar analysis of data from rats yielded a peak also at 43.1% protein (Hartsook & Hershberger, 1971). That is close agreement, indeed, even though the basis of the dietary protein in the rat study was not stated in the paper and, upon inquiry, turned out to be an 'as-fed' or air-dry matter basis.

Not wishing to pre-empt the eventual publication of Sheffy's unpublished data, and striving for the best solution of the model, we added two points at zero nitrogen intake, using an estimate of 268 mg N/day/kg body weight to the power of 0.75 for 'inevitable nitrogen losses' (Payne, 1965). The two theoretical Beagles weighed 7 and 10 kg, respectively. The best fitting polynomial for this set of 30 data pairs was as follows:

$$Y = -2.554 + 0.4087X - 0.0048X^2,$$
where Y = nitrogen retained (g/day)
and X = protein, %ME.

This model fits the data well ($r^2 = 0.53$, $P < 0.0001$). Bertrand, a mathematician, surely would have been pleased with this observance of his rule.

This polynomial and its 95% confidence limits were drawn by SAS Graphics driving a Hewlett-Packard 7475A-Plotter (Figure 2). A simple polynomial equation describes the data well but has little physiological relevance, even though it has been used before for a similar purpose (Hartsook & Hershberger, 1971). More physiological equations for the description of nutrient–response relationships, for example, based on Michaelis-Menten kinetics, would be more suitable for the final analysis of these data (see Schultz, 1987).

Meanwhile, the implications of such a good-fitting model (Figure 2) merit examination. The most important point is the peak mean nitrogen retention of 6.1 g N/day at a dietary protein content of 43% of metabolisable energy. This 'peak' is fairly flat, however, and the scatter is relatively large (CV, 46%), as shown by the dashed lines that represent 95% confidence limits.

The peak point of the lower confidence limit is at 2.1 g N/day, and a horizontal line at this level intersects the mean curve at points that correspond to 14 and 72% protein. This indicates that among data of this kind, it would be difficult to find a statistically significant difference between nitrogen retentions of groups fed protein levels ranging from 14 to 72%. It is not surprising, therefore, that no significant differences were found in another study in which growing dogs were fed nine diets containing from 20 to 48% protein (Romsos et al., 1976). That range is none too wide in view of Figure 2.

Does the model suggest any narrower optimal range for growth that may be more suitable for the needs of competitive breeders? One SD below the peak mean is a nitrogen balance of 4.0 g N/day. This intersects the mean curve at points corresponding to 22% and 63% protein. It turns out that these are the same two protein concentrations where the lower 95% confidence limit intersects with zero nitrogen balance. Is this a coincidence or statistical tautology? The 22% value is familiar to readers of the NRC (1972, 1974).

An even tighter optimal range from 28% to 57% corresponds to 0.5 SD below the peak mean, i.e. 5.1 g N/day. The low end of this range agrees with the Wisconsin studies (Ontko *et al.*, 1957) and with much personal experience in the field. Following the usual interpretation of the classical production curve of economics, we would settle for an optimal growth range of 28–43% protein, leaving out the zone of diminishing returns. This zone is approximately the range for good business.

If we wish to partition factorially the protein requirement during growth into a component for maintenance plus a component above-maintenance for growth, then the equation may be used more precisely to predict a minimum requirement for maintenance of 6.8% protein in terms of metabolisable energy. This agrees well with data from many species (Payne, 1965).

Requirements of available trace elements

The 'generic dog food disease' (see previous chapter) might equally well be called the 'NRC disease' because previous versions of the *Nutrient Requirements of Dogs* failed to draw adequate attention to the adverse effects of vegetable fibre, starch, phytin and calcium on bioavailabilities of copper, iodine and especially zinc. In fact, the previous trace element requirements were suitable only for semi-purified experimental diets and practical diets based predominently on ingredients of animal origin. They did not have wide enough safety margins to be 'adequate' for diets based on cereal grains, milling byproducts and soybeans.

This hazard can be exacerbated by use of metabolisable energy (ME) as the basis for calculating nutrient contents. As ME/g decreases, so should concentrations of most nutrients on a weight basis. This situation is well recognised for protein, for example (Sheffy, 1979). Paradoxically, when energy density [kcalME/g] is decreased because the major ingredients are corn and soy, the contents of trace minerals on a weight basis should be increased, because, in these recipes, the decrease in bioavailabilities of trace minerals becomes relatively more important than the decrease in ME.

Such paradoxical situations are not unique to dogs and cats. Much of contemporary applied animal nutrition is concerned with finding ways of feeding today's least expensive form of food energy, cereal grains, to animals that have not become adapted to them during the course of evolution. Cereal grains are as alien to the nutritional heritage of cattle and horses as they are to dogs and cats.

Much of the 'research' conducted in this area by the pet food industry has been empirical, trial-and-error testing of different concentrations of trace elements in corn–soy diets. Bioavailabilities of these elements in these

Table 1. *Inferred bioavailabilities [%] of trace minerals calculated for five proprietary dry dog foods (A-E)*

Mineral	84NRC mg/kg	74NRC %	A %	B %	C %	D %	E %
Iron	31.9	53	23	15	32	18	13
Copper	2.9	40	21	15	26	19	20
Manganese	5.1	102	10	10	14	21	6
Zinc	35.6	71	65	61	32	190	23
Iodine	0.59	38	27	16	—	89	38

Note:
Bioavailabilities are inferred by comparing manufacturers' declared concentrations to the NRC (1985) minimal requirements. They are thus the bioavailability that would be necessary for the food just to meet requirement. The inferred bioavailability in the 1974 NRC requirement is also shown.

products may be inferred from comparisons of manufacturer's declared concentrations of trace elements in these products with the NRC (1985) minimum requirements. Such comparisons for five American dry dog foods are shown in Table 1.

These bioavailabilities refer to the whole animal concept (see below) and probably include safety margins. They are calculated as the reciprocal of the quotient of the product's content and the NRC's requirement. For example, if the iron content is 242 mg/kg and the NRC requirement for available iron is 31.9 mg/kg, the bioavailability of iron in this dog food is calculated as $31.9/242 = 0.13$ or 13% (Product E).

As can be seen in Table 1, the manufacturers appear to agree fairly well about the availability of iron, 13 to 23%, and copper, 15 to 26%, in dry dog foods. They diverge somewhat on manganese, with a range of bioavailability from 6 to 21%, and rather more on iodine which has a 5.5-fold range from 16 to 89%. The greatest apparent spread is shown for zinc, an 8.3-fold range from 23 to 190%; if the 190% for product D represents an error, then the upper value is 65%.

The iron content of product E might be too high for hard working dogs that consume twice maintenance or more. Such dogs are prone to haematochezia that may be partially attributable in some cases to irritation by excessive dietary iron.

Bioavailabilities implicit in the 1974 NRC requirements for trace elements also are shown in Table 1. They are much higher than those found in the proprietary dog foods. The 1974 NRC bioavailabilities are appropriate for semi-purified and meat-based diets, and the difference between these and those in the dry dog foods probably represents the adverse effects of relatively high contents of fibre, starch, phytin and calcium found in most dry dog foods.

Epilogue

The Waltham Symposium was a satellite of the XIII International Congress of Nutrition. At the latter, six symposium speakers on 'Assessment of Nutrient Availability' failed to agree on a common definition of availability. Some preferred a gastrointestinal concept (that is, efficiency of absorption), others a whole body concept that included urinary and other losses.

The speakers on nutrient availability also failed to agree on desirable methodologies. Indeed, the only good chance for matching methodologies with concepts was in studies of minerals using tracers. Even this prospect was confounded by the responses of bioavailabilities to changes in dietary levels not only of the mineral in question but also multiple interactive factors.

The bioavailability of iron in humans was determined indirectly by 'whole-animal' variables, such as ferritin saturation. Attempts to estimate bioavailabilities of vitamins were confounded by biosynthesis in the intestinal tract, by conversion in the digestive tract or intestinal mucosa, and by other interactive factors. When efficiency of absorption of a vitamin could be measured, as with vitamin B_6, it varied inversely with the dietary level.

The availability of amino acids generated the most heated discussion. There were proponents of the slope-ratio method and tracer techniques involving direct or indirect production of labelled carbon dioxide from labelled amino acids. These whole animal approaches were contrasted with a method for measuring the efficiency of absorption of individual amino acids, a gastrointestinal concept.

In the light of the numerous difficulties with the assessment of nutrient availability discussed at the Congress we should look again at the 1985 NRC *Nutrient Requirements for Dogs*. For it hinges on the concept of available nutrients, assuming that its readers know all that is meant by availability. And it issued a challenge to the pet food industry to determine availabilities, a challenge which might be extended providently to the whole science of nutrition.

REFERENCES

Burns, R.A., LeFaivre, M.H. & Milner, J.A. (1982). Effects of dietary protein and quality on the growth of dogs and rats. *Journal of Nutrition*, **112**, 1845–53.

Freund, R.J. & Littell, R.C. (1981). *SAS For Linear Models*. Cary, NC: SAS Institute.

Hartsook, E.W. & Hershberger, T.V. (1971). Interactions of major nutrients in whole-animal energy metabolism. *Federation Proceedings*, **30**, 1466–73.

Kronfeld, D.S. (1984). Optimal regimens based on recipes for cooking in home or hospital or on proprietary pet foods. In *Nutrition and Behaviour in Dogs and Cats*, ed. R.S. Anderson, pp. 43–53. Oxford: Pergamon Press.

Mertz, W. (1981). The essential trace elements. *Science*, **213**, 1332–8.

NRC (1953, 1972, 1974 and 1985). *Nutrient Requirements of Dogs*. National Research Council, Washington DC: National Academy of Sciences.

Ontko, J.A., Wuthier, R.E. & Phillips, P.H. (1957). The effect of increased dietary fat upon the protein requirement of the growing dog. *Journal of Nutrition*, **62**, 163–73.

Payne, P.R. (1965). Assessment of the protein value of diets in relation to the requirements of the growing dog. In *Canine and Feline Nutritional Requirements*, ed. O. Graham-Jones, pp. 19–31. Oxford: Pergamon Press.

Romsos, D.R., Belo, P.S., Bennink, M.R., Bergen, W.G. & Leveille, G.A. (1976). Effects of dietary carbohydrate, fat and protein on growth, body composition, and blood metabolite levels in the dog. *Journal of Nutrition*, **106**, 1452–64.

Schulz, A.R. (1987). Analysis of nutrient–response relationships. *Journal of Nutrition*, **117**, 1950–8.

Sheffy, B.E. (1979). Meeting energy–protein needs of dogs. *Compendium on Continuing Education for Small Animal Practitioners*, **1.5**, 345–54.

5

Comparative aspects of nutrition and metabolism of dogs and cats

JAMES G. MORRIS AND QUINTON R. ROGERS

Introduction

Cats and dogs are the most common companion animals kept by man. As they are both members of the biological order Carnivora, there is a tendency to assume that these two carnivores have similar nutritional requirements. However, cats and dogs are different, a fact succinctly stated by T.S. Eliot (1939):

> So first, your memory I'll jog,
> And say: A cat is not a dog

While all mammals have many similarities, there are important differences in the metabolism and nutritional requirements of cats and dogs.

Modern Carnivores are divided into two groups, the aquatic Pinnipeds and terrestrial Fissipeds. Divergence within the Fissipeds occurred in the late Eocene to early Oligocene period (Romer, 1974), or about 35 million years ago (Stokes, Judson & Picard, 1978), as illustrated in Figure 1. This divergence resulted in the emergence of three superfamilies, the *Canoidea* (or Arctoidea) which contains the dog *Canis familiaris*, the *Feloidea* (or Aeluroidea) which includes the cat *Felis catus* and the now extinct superfamily *Miacoidae*. Fossil records of the *Feloidea* indicate that there was a period of rapid evolution with the development of fully specialised forms, but subsequently there has been minimal change (Colbert, 1980). The rate of karyotypic evolution in the Carnivores is amongst the lowest of all mammals, with the average age of the Carnivora being over three times that of the Primates (Bush *et al.*, 1977, Table 1). Within the Carnivora, the Felidae have either 36 or 38 chromosomes (Matthey, 1973), whereas the Canids have chromosome numbers ranging from 38 to 78 (Fig. 2). If the chromosome number 36–38 is the primitive condition, it indicates that the Felidae have retained this number while the Canidae, through either progressive fission or other processes, have higher chromosome numbers.

Evolutionary events within the Fissiped Carnivores, which produced the Canoidea and Felidae resulted in animals with specialised dietary patterns.

35

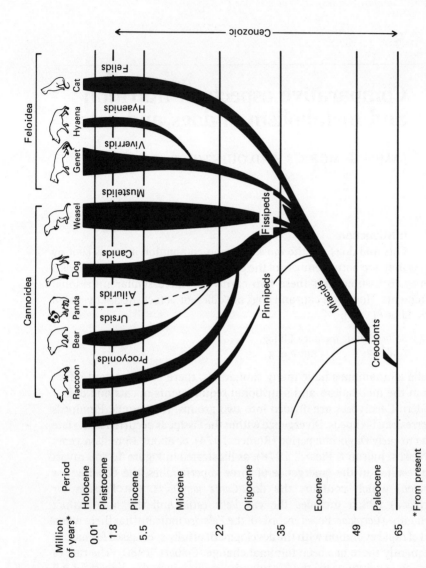

Fig. 1. Evolution of the Fissiped carnivores, showing the relationship of the dog to carnivores with omnivorous dietary habits and the cat with those which are strictly carnivorous. (Adapted from Romer, 1966; Colbert, 1980 and Morris & Rogers, 1983a.)

Table 1. *Rates of karyotype evolution in vertebrate mammals*

Mammal	Number of genera examined	Average age of genera (10^6 yr)	'Karyotypic charges'*/ lineage (10^6 yr)
Horses	1	3.5	1.395
Primates	13	3.8	0.746
Largomorphs	3	5.0	0.633
Rodents	50	6.0	0.431
Artiodactylas	15	4.2	0.561
Insectivores	7	8.1	0.187
Marsupials	15	5.6	0.176
Carnivores	10	12.9	0.078
Bats	15	9.0	0.059
Whales	2	6.5	0.025

Notes:
*Sum of chromosomes and arm numbers.
Adapted from Wilson *et al.* (1975) and Bush *et al.* (1977).

The Canoidea include strict herbivores (panda) and omnivores (racoons, bears and dogs), while the Feloidea are virtually strict flesh eaters (cats, lions, civets and hyenas). These dietary patterns may have been established early in evolution. There are marked differences in the dental and cranial anatomy of living members of the dog and cat families which can be traced back in the fossil records of their ancestors. A comparison of modern dogs with one of the first Canids (Hesperocyon) show there has been little loss of teeth or change in form and function of the dentition beyond the stage of the late Eocene or early Oligocene dogs (Colbert, 1980). Modern dogs and cats have the same number of incisor and canine teeth (six incisor and two canine teeth in the upper and lower jaws). However, dogs have 42 permanent teeth whereas cats have only 30 permanent teeth (Nickel, Schummer & Seiferle, 1979). Dogs have four premolar teeth on each side of the upper and lower jaw, and two molars on the upper, and three molars on the lower, jaw. In contrast, cats have only three premolars and one molar on the upper jaw, and two premolars and one molar on the lower jaw on each side (Nickel, Schummer & Seiferle, 1979). In modern dogs and cats the upper premolar (P4) or carnassiate tooth and the lower first molar (M1) are enlarged and, on occlusion, act as sectorial or cutting teeth. As dogs have crushing molar teeth which are associated with the capacity to utilise plant material, the dental formula of dogs suggests an omnivorous diet, whereas the dental formula of the cat is consistent with a strict carnivorous diet (Tedford, 1981).

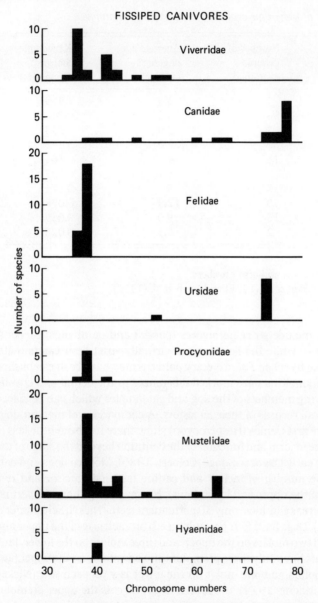

Fig. 2. Chromosome numbers of the Fissiped carnivores. The Felidae have either 36 or 38 chromosomes, whereas the Canidae range from 38 to 78 chromosomes.

Table 2. *Lengths of different segments of the intestine of pig, dog, cat and rabbit*

Animal	Segment of intestine	Relative length (%)	Ratio of intestine to to body length
Cat	Small	83	4:1
	Large	17	
Dog	Small	85	
	Caecum	2	6:1
	Colon	13	
Pig	Small	78	
	Caecum	1	14:1
	Colon	21	
Rabbit	Small	61	
	Caecum	11	10:1
	Colon	28	

Source: Adapted from Colin (1854).

Comparative digestive anatomy of dogs and cats

Morphological modifications of the gastrointestinal tract of mammals can be correlated with the diet. Mammals subsisting mainly on animal tissue (carnivorous) have a simple stomach, colon and small intestine, whereas folivorous species have a complex stomach and/or enlarged caecum and colon. Chivers & Hladik (1980) reported the coefficient of gut differentiation for a large number of mammals on the basis of the following index:

$$\frac{\text{(weight of stomach + caecum + colon)}}{\text{(weight of small intestine)}}$$

They reported that the domestic cat and lion have lower coefficients than the dog, which in turn has a lower coefficient than frugivorous and folivorous mammals.

The length of the intestine in proportion to body length is short in the cat compared to the dog. Colin (1854) reported the ratio to be 4:1 in cats; whereas for dogs, rabbits and pigs the ratios were 6:1, 10:1 and 14:1 respectively, (Table 2).

The large absorptive surface of the gut area of mammals compared to reptiles appears to be the main digestive adaptation which permits mammals to sustain digestive rates of an order of magnitude higher than those of reptiles (Karasov & Diamond, 1985). Most of the increase in transport capacity is achieved by increasing the surface area by a factor of about seven through folding of the mucosa and production of microvilli. The ratio of mucosal to serosal area of the intestines of the cat are: jejunum 15:1;

ileum 12:1; and colon 1:1 (Wood, 1944). Values for the rat were 6:1, 4:1 and 1:1 respectively. Despite the relatively short intestine of the cat compared to the rat, the ratio of mucosal surface area and bodyweight in both species are almost identical (mucosal surface area cat:rat, 6.3:1 and bodyweight of cat to rat, 6.6:1). High mucosal surface area to bodyweight ratios in small mammals have been associated with a carnivorous diet, and low ratios with a herbivorous diet (Barry, 1976). The dog and cat have a similar mucosal surface area/cm serosal length; (jejunum 54 v. 50 cm² and ileum 38 v. 36 cm², respectively), (Warren, 1939, Wood, 1944). However, the mucosal to serosal surface area ratio is lower in the dog than the cat.

The gastrointestinal tracts of cats and dogs, while similar, have several distinct features. The region of the proper gastric glands of the stomach of the cat is uniform. However, in dogs it is divided into a proximal zone which has a thinner mucous membrane and distinct foveolae and a distal zone of reddish brown colour with a thicker mucous membrane and less distinct foveolae (Nickel, Schummer & Seiferle, 1979). The second area of dissimilarity between dogs and cats is the caecum which is illustrated in Figure 3. In dogs, the terminal part of the small intestine, the ileum, communicates directly with the colon and the caecum exists as a diverticulum of the proximal portion of the colon (Evans & Christensen, 1979). In cats, the caecum is an unusually short comma-shaped diverticulum which connects with the ileum (Reighard & Jennings, 1966; Nickel, Schummer & Seiferle, 1979). Lions also have a simple conical caecum which Owen (1868)

Fig. 3. The caecum of the dog and cat. In the dog, the caecum exists as a diverticulum of the proximal colon and is more developed than that in the cat.

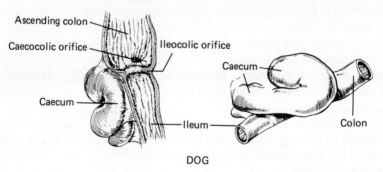

DOG

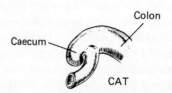

CAT

Table 3. *Comparative apparent digestion coefficients for the same commercial foods measured on dogs and cats*

Digestion coefficient for	Dog	Cat	SEM × 10²	Significance between species
Dry matter	0.85	0.78	0.9	xxx
Crude protein	0.87	0.82	1.0	xx
Acid ether extract	0.92	0.76	1.5	xxx
Nitrogen-free extract	0.70	0.67	2.5	x
Gross energy	0.89	0.79	1.0	xx

Source: Adapted from Kendall, Holme & Smith (1982).
x,xx,xxx $P < 0.05, 0.01, 0.001$, respectively.

reported as being 2 inches long in an adult. Mu & Lingham (1974) reported that, according to anatomical and histological criteria, cats have only a rudimentary appendix. The greater weight of caecal tissue in proportion to bodyweight in dogs than cats (Colin *et al.*, 1854), is consistent with the diet of dogs containing more plant material than the diet of cats.

Digestibility of diets by dogs and cats

While the overall digestive efficiencies of cats and dogs given a variety of diets have been measured, there is limited comparative data. James & McCay (1950) compared the apparent digestibility of three commercial diets when fed to dogs of three different breeds. Statistical analysis of their results indicated no effect of breed on digestion coefficient. No breed comparisons have been published for the cat and it is unlikely they exist. A comparison of the apparent digestibility of the same commercial foods by cats and dogs was undertaken by Kendall, Holme & Smith (1982). Dogs had significantly higher mean apparent digestion coefficients than cats absorbing about one-tenth more nutrients per unit food eaten. (Table 3). The difference between species in digestive efficiency was not constant, and decreased as the apparent digestibility of the food increased.

A comparison of the digestive efficiency of 13 felid species and a badger fed a high-meat diet was made by Morris, Fujimoto & Berry (1974). They reported similar apparent digestive efficiencies across species with a variance similar to that reported within species.

Carbohydrates

The apparent digestibility coefficients found for isolated sugars and starches (and proteins and fats) added to the diet has generally been higher than found for these nutrients in commercial diets. Early studies (Roseboom & Patton, 1929; Ivy, Schmidt & Beazell, 1936; Brown & Wood, 1936) qualitatively demonstrated extensive digestion by normal dogs of both raw and cooked starch. Later studies by Bennett & Coon (1966) also showed that dextrin-maltose, corn syrup or sucrose at 0.54 of the energy in the diet could be utilised with no signs of intolerance. However, the same fraction of the energy intake as lactose produced diarrhoea and malaise. Lower levels of lactose than this are reported to cause digestive disturbances in dogs. McCay (1949, p. 22) reported that diarrhoea often resulted if 0.05 of the dry feed was skim milk powder and additional milk was fed. Lactose contributes a higher proportion of the energy in cows' milk (0.3, USDA, 1976) than in bitches' milk (0.1, Oftedal, 1984). The apparent digestibility coefficients of glucose, sucrose, lactose, dextrin, starch and cellulose added to a meat-based diet and fed to adult cats were measured by Morris, Trudell & Pencovic (1977). They found that, with the exception of cellulose (which was indigestible), cats digested all the above carbohydrates with coefficients greater than 0.94. The inclusion of lactose in the diet at 0.15 of the energy produced diarrhoea.

These authors also reported a considerable proportion (0.79) of the raw starch in coarsely ground maize grain disappeared in the gut of cats. Fine grinding of the grain increased the apparent digestion coefficient to 0.94. Raw starch from coarsely ground wheat was less digestible than in finely ground wheat, but the differences due to grinding were not as great (0.93 and 0.97) as for maize grain. Starch from cooked coarsely ground wheat or maize grain was not digested to a greater extent than starch from these raw grains when finely ground.

The amylase activity of the pancreas of cats, although significant (McGeachin & Akin, 1979), on an equivalent tissue basis is only about one–one hundred and fiftieth of that for the pancreas of rats (McGeachin & Johnson, 1964). The saliva of cats has a low amylase activity which is probably derived from the plasma by diffusion, (McGeachin & Akin, 1979). The concentration of amylase in the salivary glands of cats and dogs is very low in comparison with rabbits, rats, mice, guinea pigs and humans in which definite synthesis in the gland has been reported (Rajasingham et al., 1971; Skude & Mardh, 1976).

The activity of disaccharidases of the small intestine of adult carnivores was investigated by Hore & Messer (1968). They found that the activities per g of mucosa of enzymes hydrolysing maltose and sucrose were greater in cats than in dogs. These results are consistent with the ability of cats to efficiently digest sucrose and starch. A finding of their study was that the cat and lion lacked the disaccharidase capable of hydrolysing trehalose, whereas this disaccharidase was present in the mucosa of the ferret, dog and polar bear. The lack of this enzyme would be of no practical significance to felids as the carbohydrate substrate alpha-trehalose occurs in haemolymph of insects (Gilmour, 1961) and some lower plants. The extent to which microbial fermentation in the large intestine contributes to the disappearance of carbohydrates in course of transit through the gut of cats is not known. The depressed apparent digestibility of protein when lactose was added to a meat-based diet of cats reported by Morris *et al.*, (1977) may have been due to the trapping of urea nitrogen by microbial fermentation. Alternatively, the depression in digestibility may have been due to an accelerated rate of passage of food along the tract. Phillipson (1947) reported the presence of (steam) volatile fatty acids (VFA) in the colon of dogs. However, of the simple stomached animals he examined, the quantity of VFA per unit body weight was least in dogs, and less in pigs, rabbits and rats than in a horse. Banta *et al.* (1979) measured the VFA and lactic acid concentrations in the ingesta of dogs given cereal-based and meat-based diets. They found that the concentration of VFA was not markedly affected by diet and concluded that the production of organic acid in the gastrointestinal tract of the dog was of little nutritional significance.

Fats

Dogs and cats can be maintained in good health on diets containing a wide range in level of fat, and fats of different sources provided adequate amounts of essential fatty acids are present. Appropriate adjustments have to be made to other essential nutrients in the diet to take account of the high energy value of high-fat diets.

Kendall, Holme & Smith (1982) measured the apparent digestibility coefficients of four commercial canned dog or cat foods, a dry cat food, and two experimental foods fed to cats and dogs. They found a mean apparent digestion coefficient of acid ether extract (AEE) of 0.92 for dogs and 0.76 for cats (see Table 3). Earlier reports of the apparent digestion coefficient of crude fat by dogs varied from approximately 0.80 to 0.95 when mixtures of glycerides from plant and animal sources were fed (James & McCay, 1950).

Apparent digestion coefficients for ether extract fat in experimental diets fed to cats have been comparable to the AEE values reported for dogs by Kendall, Holme & Smith (1982). Morris, Trudell & Pencovic (1977) reported a value of 0.99 for a mixture of beef and mutton fats in fresh mince and Kane, Morris & Rogers (1981) found apparent digestibility coefficients ranging from 0.97 to 0.99 for five different fats using purified diets containing 25% fat in the dry matter. When yellow grease (a yellow coloured low melting-point fat rendered from restaurant used fat) was fed at 10% of the dry matter the apparent digestibility coefficient declined to 0.90 but remained high (0.98) when the diet contained 50% yellow grease in the dry matter. This observation is consistent with the decrease in apparent digestibility of fat as the level in the diet decreases (Schneider *et al.*, 1951).

The low mean apparent digestion coefficient for AEE reported by Kendall *et al.* (1982) for cats is in the main due to the very low digestion coefficients found for one of the canned dog foods and the dry cat food (0.52 and 0.56

respectively). As the value they reported for the digesion coefficient of AEE in fresh mince was 0.96, it appears that processing of fat may reduce its ability to be digested by cats.

Protein

Digestion coefficients for proteins depend on the source of the protein and its processing treatment. Kendall *et al.* (1982) reported mean apparent digestion coefficients of 0.87 for dogs and 0.82 for cats given four commercial dog or cat foods, two experimental diets and fresh mince (see Table 3). Mince had a significantly higher digestion coefficient for crude protein than the commercial diets for both cats and dogs. The digestion coefficient for fresh mince (0.96) by cats was similar to that reported (0.94) by Morris *et al.* (1977).

Meal patterns of cats and dogs

Most wild felids with the exception of the lion are solitary hunters (Ewer, 1973) which restricts the size of their prey to that which can be killed by an individual. In contrast, many of the Canids such as the cape hunting dog and wolf (Mech, 1975) hunt in groups which enables them to prey on swifter and larger species than could be captured by a solitary hunter.

Measurement of cumulative food intake of cats and dogs over 24-hour periods by Mugford (1977) showed a distinct diurnal feed intake pattern by dogs, whereas cats distributed their meals evenly throughout the 24-hour period. Kane *et al.* (1981) recorded meal patterns of cats given four types of diets: commercial canned and dry expanded, and two purified diets based on casein or amino acids. Meals were defined as a period of eating with a minimal inter-meal interval of 3 minutes. They found that the mean number of meals per day for each diet varied over a very narrow range (15.7 to 17.4). The number of meals in the light and dark cycle of the day was similar, showing that the cat lacks the diurnal eating patterns of many other mammals. In another experiment in which cats were fed high- and low-fat diets, the mean number of meals per day was 11.4 and 10.4 respectively (Kane, Leung, Rogers & Morris unpublished data). The reason for the difference between experiments in number of meals per day is not apparent.

Small rodents are a staple prey of all species of the genus Felis that have been investigated (Ewer, 1973, p. 214). Studies on the diet of feral domestic cats in Victoria (Australia) showed that mammals, including rabbits (presumably young ones), constituted 88% of the diet (Coman & Brunner, 1972). In order for cats to meet their energy needs from small mammals, frequent predation is necessary. As an example, the energy supplied by a 26 g mouse is approximately 140 kJ (calculated from Bailey *et al.*, 1960). An active adult cat needs about 335 kJ/kg body weight for maintenance. Thus, for a 4 kg cat 1.34 MJ/day would be required, which could be met from ten 26 g mice/day.

The meal pattern of domestic cats contrast with those of large Felids such as lions and tigers, which are capable of taking prey exceeding their own bodyweight and therefore are adapted to be infrequent eaters. Statements in reference to the domestic cat such as 'In the wild, cats consume as much food as possible or as is available in a relatively short period of time. It may be several days before they eat again' (Lewis & Morris, 1984) are without foundation. They appear to have arisen from the assumption that domestic cats have a behavioural pattern similar to large Felids.

The meal patterns of three dogs given dry food were measured by Ardisson *et al.* (1981). They found no consistent effect of a 16:8, light:dark cycle on eating pattern. Only one dog ate significantly more per meal in the light cycle. The mean number of meals per day was 3.96, with a mean meal size of 115 g which indicates dogs are less frequent meal eaters than cats, and eat a greater amount of food per meal. These data contrast with the 10.4 meals per day reported by Mugford (1977) for dogs and 13.6 meals per day for cats. However, part of the disparity could arise from definition of a meal. Mugford's graphs show long periods in which dogs did not eat. Mugford (1977) also reported that video-tape recordings of dogs and cats in conspecific feeding situations indicated that 12% of the meals for cats and 20% for dogs were associated with another animal feeding.

Drinking patterns of dogs and cats

The water content of the diet has a marked effect on the amount of water drunk by cats. Kane *et al.* (1981) reported that cats given a canned food (78% moisture) did not drink any water after the first day, whereas cats given a dry expanded diet had a mean ($\pm$ SE) of 16.0) $\pm$ 1.3 drinking bouts/day, for two purified diets the mean ($\pm$ SE) number of drinking bouts were 12.5 $\pm$ 1.8 and 12.4 $\pm$ 2.1/day. Although these authors did not correlate drinking with eating, the number of drinking bouts was similar to the number of meals. Ardisson *et al.* (1981) reported that, for about half the periods they measured, there was a correlation between the amount of dry matter consumed and the quantity of water ingested between one meal and the next meal.

Food preference of dogs and cats

Most mammals show preference for foods or liquids containing sucrose over those without. Houpt *et al.* (1979) reported that both male and female dogs prefer diets containing sucrose over the same diet without sucrose. However, female dogs exhibit a significantly greater preference than male dogs for diets containing 1% sucrose. In contrast, for aqueous solutions, Carpenter (1956) reported that cats failed to discriminate between water and sucrose at any concentration investigated up to 1.0 M (3.42%) sucrose. This concentration of sucrose caused vomiting and diarrhoea and death in one cat. Bartoshuk, Harned & Parks (1971) also showed that cats did not show a preference for aqueous solutions of sucrose to water, but, if the sucrose was dissolved in 0.03 M (0.176%) sodium chloride, they preferred the sucrose solution to plain saline. They suggested, on the basis of electrical recordings, that the taste fibres of the cat discharge in response to water, but not to dilute solutions of sodium chloride.

This finding appeared to offer explanations for the observations of Frings (1951) that cats preferred solutions of sucrose in dilute milk over dilute milk alone and of Pfaffmann (1955) on the low responsiveness of the gustatory nerve of the cat to sucrose solutions. However, subsequent studies on flavour preferences of domestic cats and species of Panthera (lions, tigers, leopards and jaguars) by Beauchamp, Maller & Rogers (1977), failed to substantiate any preference for sucrose, whether it was dissolved in water or 0.03 M (0.176%) sodium chloride. They were able to show preference for sucrose and lactose when they were dissolved in diluted milk over diluted milk alone. But they suggested that this preference may be a consequence of sensory cues other than sweetness, as solutions of sucrose and lactose in dilute milk were equally acceptable to cats but are perceived as having different sweetness by humans.

Other sugars (glucose, mannose, fructose and galactose) were also tested by Beauchamp et al., (1977) in domestic cats without an indication of preference. Cats are, however, indifferent to, or reject, artificial sweeteners such as saccharin, cyclamates (sodium and calcium salts) and dulcin (Bartoshuk et al. 1975; Beauchamp et al. 1977).

Beauchamp et al. (1977) showed that solutions containing hydrolysate of proteins (casein, lactalbumin and soy) were preferred by cats to water. Panthera species also preferred a 3% casein hydrolysate and a 3% soy hydrolysate to water. However, the casein solution was preferred to soy solution. Amino acids, peptides and other nitrogenous compounds probably present in these hydrolysates, gave rise to increased acceptance of these solutions. White & Boudreau (1975), showed that cats preferred solutions of the L-amino acids lysine, histidine and proline in 50 mM (0.29%) sodium chloride to plain saline, but rejected solutions of tryptophan, isoleucine and adenine in favour of 50 mM (0.29%) sodium chloride.

Because of the higher protein requirement of the cat compared to other mammals (Rogers & Morris, 1980; 1982; 1983), it could be anticipated that cats would select more of a high protein than a low protein diet. Cook et al. (1985), using two-bowl choice conditions, gave kittens casein-based diets containing 18, 36, or 54% protein in all three possible combinations. Kittens consistently avoided the higher casein diet and significantly decreased food intake when only the 36% and 54% protein combinations were offered. In another protein choice study, either 16, or 31 or 63% soy-based protein diets were offered along with a protein-free diet. Kittens consumed comparable amounts of protein-containing and protein-free diet at each test when the diets were of similar physical consistency. Thus, when kittens were given the

16% protein diet along with the protein-free diet, they chose a combination considerably less than their requirement for protein (Smalley *et al.*, 1985). This behaviour of the cat is markedly different from that of the rat which, when prefed 'normal' levels of protein, will avoid a protein-free diet although it too selects low or moderate levels of protein in preference to high-protein diets (Leung, Gamble & Rogers, 1981).

It is apparent from these choice studies that dietary selection by cats is not based on protein *per se* but may be influenced by compounds associated with proteins; such as amino acids, small peptides and other water soluble constituents. Mugford (1977) reported that the duration and size of meals of cats was influenced by odours; cooked rabbit odour being a very effective stimulus for the cat. We consistently find that, when the 'juice' from high quality commercial canned cats foods is added to a purified diet, it increases the acceptance of the diet for kittens.

Fat may also contribute to the acceptance of a food as a number of organic compounds contributing to flavour and aroma are lipid soluble. Beauchamp *et al.* (1977) showed that emulsions of butter fat, and to a lesser extent corn oil, were preferred by domestic cats to water or water containing the emulsifying agent. Kane *et al.* (1981) also examined the effect of fat on the acceptability of diets by cats. When 25% fat was added, diets based on bleached tallow were significantly more acceptable than those based on butter or chicken fat. The majority of cats selected unbleached tallow and yellow grease over bleached tallow diets, but the differences were not statistically significant.

The physical characteristics of a diet are affected by the level of fat present. Fat levels in the diet could lead to preferences based on physical consistency. In the studies of Kane *et al.* a purified diet containing 25% fat were more acceptable than ones containing either 10% fat (powdery) or 50% fat (oily).

Cats do not appear to discriminate between beef tallow and hydrogenated beef tallow (HBT), but avoid diets based only on hydrogenated coconut oil (HCO), (MacDonald *et al.*, 1985). The basis for this choice appears to be the medium chain fatty acids in HCO. Replacing 5% of HBT in a diet with a mixture of medium chain triglycerides (designated MCT8 and containing 3% 6:0, 60% 8:0, 36% 10:0, 1% 12:0) causes a significant aversion to the diet. Further replacement of HBT with 10% MCT8 gave a greater discrimination against the diet. When a longer medium chain triglyceride, designated MCT12 (99% 12:0, 1% 10:0) replaced up to 15% of HBT, no significant discrimination occurred.

The fatty acid which appeared responsible for aversion of HCO is caprylic acid (8:0); (HCO contains 6% 8:0, 6% 10:0, and 50% 12:0). This is supported by the finding that adding either 0.1 or 1.0% caprylic acid to a HBT diet caused marked avoidance of the diet. As caprylic acid is present in HCO as triglyceride it might be expected that it would be ineffective until it is hydrolysed to the free fatty acid. However, the aversion is so rapid in its onset that it appears that either the HCO is hydrolysed by a lingual or salivary lipase such as is present in other species, or that MCT *per se* have an aversive taste.

We were unable to find reports of the reaction of dogs to various sources of fats and fatty acids. From behavioural traits one may predict that dogs are tolerant to, or prefer at least some, free fatty acids. Many Canids including the fox, jackals, and the domestic dog, have the habit of burying surplus food remaining after a meal (Ewer, 1973, p. 162). It is recovered in an advanced stage of decomposition and eaten when there would be high levels of free fatty acids present.

Essential nutrients

Carbohydrates

All animals have a metabolic requirement for glucose, but this can generally be met from glucose precursors such as amino acids and glycerol. The gluconeogenic capacity of the liver and kidney is adequate to supply body needs of growing animals for glucose. Romsos *et al.* (1976) fed Beagle puppies diets with either 0 or 62% of the metabolisable energy from carbohydrate. Bodyweight gain of puppies fed the carbohydrate-free diet (24% energy from protein, 76% of the energy from fat), from 2 months to 10 months of age was comparable to puppies fed diets containing 20 to 62% of the energy from carbohydrates.

For puppies fed the carbohydrate-free diet, the concentration of glucose in plasma and utilisation rate (measured by the disappearance of $2 - {}^3H$ glucose) were similar to puppies fed a diet containing 27% of the energy from carbohydrate (Belo *et al.*, 1976).

The carbohydrate content of the diet had a marked effect on the activity of phosphoenolpyruvate carboxykinase (PEPCK), a key gluconeogenic enzyme. In rat liver and kidney, this enzyme occurs only in the cytosol, but in the dog it occurs both in the cytosol and in the mitochondrion. Activity of mitochondrial PEPCK in dog liver and kidney increased when carbohydrate-free diets and low-protein diets were fed. Mitochondrial pyruvate kinase in liver and kidney also increased with the low carbohydrate diet. The concentration of protein in these diets exceeded the minimal requirements for growth (National Research Council, 1986) so would have provided excess amino acids for gluconeogenesis.

There are several reports by Resnick (1974*a, b*; 1978) of low survivability of puppies fed carbohydrate-free diets. While the problem appeared to be associated more with young (7–10 weeks of age) than older puppies (greater than 16 weeks), the cause of death was not determined. These observations are dealt with in chapter 12 of this book.

Low-carbohydrate, high-protein, high-fat diets have been suggested by Kronfeld *et al.* (1977) and Downey *et al.* (1980) as being advantageous to dogs undergoing heavy exercise stress, again an issue discussed elsewhere in this volume (Chapter 8).

The metabolism of carbohydrates and control of gluconeogenesis of cats shows divergence from the pattern of other simple stomached animals. The cat has a normal concentration of hexokinase in the liver, but glucokinase is virtually absent (Ballard, 1965). Therefore, the cat might not be expected to be well adapted for the ingestion of high carbohydrate meals.

An example of the inability of cats to metabolise high intakes of carbohydrate is the fructosuria observed by Drochner & Muller-Schlosser (1980) when cats were given high intakes of sucrose. Whether the fructosuria is due to a relative lack of phosphofructokinase in cats does not appear to have been examined. The virtual absence of glucokinase (Ballard, 1965) in cat liver and the higher K_m of hexokinase for fructose than glucose (1.5 $v.$ 0.15 mM) (Lehninger, 1975) would result in the preferential catabolism of glucose over fructose.

The limited data available on carnivores suggested that gluconeogenesis is active at all times. Kettlehut *et al.* (1978) investigated the effect of food deprivation in adult cats previously fed high-protein, low-carbohydrate diets and found that glucose concentration of their blood was maintained for over 72 hours. In contrast, cats previously fed the high-carbohydrate diet exhibited an initial decline in blood glucose over the first 24 hours of food deprivation, then maintained a steady state. Migliorini *et al.* (1973) showed that carnivorous birds also had the capacity to maintain blood glucose concentration during food deprivation.

Cats, food-deprived for 5 days, have a significant increase in liver PEPCK activity over fed values (Rogers *et al.*, 1977), but no significant difference was found when cats were previously fed 17.5 or 70% protein diets. PEPCK is distributed in both the cytosol and mitochondria fractions of the liver of cats (Kettlehut *et al.*, 1978). These investigators also found that cats fed a high-protein diet exhibit little, if any, adaptation in PEPCK activity.

Rowsell *et al.* (1979) reported that the cat and other flesh eaters have a high activity of hepatic serine–pyruvate aminotransferase and Beliveau *et al.* (1981) measured the comparative rates of gluconeogenesis and oxidation of ^{14}C-labelled glycine, serine, and lactate by isolated hepatocytes from cats and rats. They found two inhibitors of cytosolic PEPCK activity (quinolinic acid and mercaptopicolinic acid) depressed gluconeogenesis from serine in isolated rat hepatocytes, but not in cat hepatocytes. Furthermore, threonine in excess did not inhibit conversion of serine to glucose in cat hepatocytes. This suggested that serine in the cat liver is converted to glucose by a route not involving pyruvate. Serine and threonine are metabolised by a common enzyme (serine dehydratase) in the rat. The evidence of Beliveau *et al.* (1981) plus the low activity reported for serine dehydratase in the cat (Rogers *et al.*, 1977; Rowsell *et al.*, 1979) suggests that serine is metabolised by the cat through hydroxypyruvate, D-glycerate and 2-phosphoglycerate to produce glucose rather than via pyruvate, oxaloacetate and phosphoenolpyruvate.

The pancreas of cats releases insulin in a biphasic pattern to a continuous glucose or amino acid stimulus. However, the pattern of release is dissimilar to that observed in most other species (Curry *et al.*, 1982). The beta cells of the cat's pancreas appear to be less responsive to glucose as a stimulus than the cells of the rat's pancreas. Amino acids such as arginine and its analogues are more potent insulin secretogogues relative to glucose in cats than in dogs or rats, (Kanazawa *et al.*, 1966; Fajans *et al.*, 1974, Curry & Bennett, 1973, Curry *et al.*, 1982). A comparison of the insulin biphasic response of the cat's pancreas to amino acids and to glucose shows a higher proportional response in the second phase to amino acids than to glucose. Also the insulin response continues for a longer period after withdrawal of the stimulus with amino acids than with glucose.

Complete mixtures of amino acids and arginine alone in the presence of glucose are potent stimuli for the release of glucagon from the alpha cells of the cat pancreas. Glucose, even at high concentrations (330 mg/dl), in the presence of arginine or a complete amino acid mixture, does not markedly suppress glucagon release by the pancreas (Curry *et al.*, 1982). This observation suggests that, in cats, glucose does not serve as an important modulator of glycogen secretion in the presence of amino acids. It appears that amino acids are more important modulators of pancreatic hormone release in cats than in rats or other omnivores.

Protein

Essential amino acids

The minimal essential amino acids requirements of the growing puppy and of the kitten are presented in Table 4. When compared on an isoenergetic basis, the requirement of the kitten significantly exceeds that of the puppy for arginine and leucine and sulphur amino acids; and the puppy has an apparent higher requirement for tryptophan than the kitten. A recent study (Morris & Rogers, unpublished) indicates that the requirement of the growing kitten for tryptophan is greater when the diet is limited in nitrogen. As the requirements of the kitten for essential amino acids were established using diets with considerable excess of nitrogen, and those of the puppy with diets without an excess of nitrogen (Schaeffer *et al.*, 1989), the apparent differences in tryptophan requirements may be an artefact.

No explanation can be advanced for the higher requirement of the kitten than the puppy for leucine. The slope of the relationships of weight gain and nitrogen retention to level of dietary leucine for the kitten were relatively flat without definite 'break points'. However, the higher value for N retention rather than weight gain was taken as the requirement. The high requirement of the cat for sulphur amino acids compared to other species has been

Table 4. *Minimal essential amino acid requirements of kittens and puppies for growth.*

Amino acid	Kitten[a]	Puppy[b]	Ratio kitten:puppy
	mg amino acid/MJ diet		
Arginine	478	327	1.5
Histidine	144	117	1.2
Isoleucine	239	234	1.0
Leucine	574	380	1.5
Lysine	383	335	1.1
Methionine + cysteine	359	253	1.4
Phenylalanine + tyrosine	407	466	0.9
Threonine	335	304	1.1
Tryptophan	72	98	0.7
Valine	287	251	1.1

Notes:
[a] National Research Council (1986).
[b] National Research Council (1985).

discussed by MacDonald *et al.* (1984*b*) and National Research Council (1986). Although the cat has differences in its sulphur amino acid metabolism, e.g. taurine synthesis and felinine production, these do not account for the high requirement.

Arginine metabolism

A number of mammals, e.g. humans (Rose, Haines & Warner, 1954), swine (Easter, Katz & Baker, 1974), and rats (Wolf & Corley, 1939; Burroughs, Burroughs & Mitchell, 1940) can maintain nitrogen balance (adult) or growth (children), (Nakagawa *et al.*, 1963) without a dietary source of arginine. Rose & Rice (1939) suggested, on the basis of nitrogen balance, that the adult dog was independent of a dietary source of arginine. However, Burns, Milner & Corbin (1981) demonstrated that emesis occurred in adult dogs given an arginine-free diet.

A comparison of the clinical signs in cats, dogs, ferrets and rats following consumption of an arginine-free diet, shows that cats and ferrets are most sensitive to a deficiency of dietary arginase, dogs are intermediate among these animals, while growing rats exhibit only a depression of food intake (Morris, 1985). The clinical signs following consumption of a diet devoid of arginine can be explained by the hyperammonaemia which occurs following ingestion of an arginine-free meal. The addition of ornithine (at about twice the minimal molar concentration of arginine for growth) to an arginine-free diet prevents hyperammonaemia in kittens, but does not permit bodyweight gain (Morris & Rogers, 1978). Aparently, sufficient dietary ornithine enters the liver for the Krebs–Henseleit cycle to dispose of the ammonia arising from catabolism of amino acids. When citrulline was added to an arginine-free diet, the growth rate of kittens was comparable to that obtained when the same diet with added arginine was fed (Morris *et al.*, 1979). However, when arginine was replaced by citrulline on an equimolar basis, the efficiency of utilisation of citrulline was less than that of arginine (Johannsen, C., Morris, J.G. & Rogers, Q.R., unpublished data).

Czarnecki & Baker (1984) examined the ability of ornithine and citrulline to replace arginine on an equimolar basis in the diet of young (7–10 weeks old) and older (17–20 weeks old) puppies at concentrations approximating the minimal arginine requirement (0.4% of the diet) for maximal body weight gain. They found that ornithine somewhat reduced the number of emetic episodes which followed consumption of an arginine-free diet. Also ornithine promoted either a greater rate of bodyweight gain (young puppies) or a reduced rate of bodyweight loss (older puppies) in comparison to an arginine-free diet. In one experiment with young puppies, the addition of citrulline to an arginine-free diet resulted in comparable weight gain to that obtained with a diet containing arginine. However, in another experiment with older puppies, significantly lower body weight gains followed consumption of the citrulline diet than that from a diet containing arginine.

The ability of the intestinal mucosa of rats to metabolise glutamine and to synthesise citrulline was demonstrated by Windmueller & coworkers (see Windmueller & Spaeth, 1981). The resulting citrulline is converted to arginine in the kidney and so a significant proportion of the metabolic arginine requirement in growing rats is met by *de novo* synthesis. In cats, a key enzyme in the synthetic pathway (pyrroline-5-carboxylate synthase) has an extremely low activity. Cats have only about 5% of the activity of rats on a bodyweight basis (Rogers & Phang, 1985). In addition, the activity of ornithine aminotransferase in cat mucosa is lower than in rat mucosa (see Morris, 1985). The activities of the other enzymes, e.g. carbamyl phosphate synthase, in the pathway may also be low but, to our knowledge, have not been measured. These lower enzyme activities result in the cat being more sensitive to a dietary deficiency of arginine than other mammals such as rats. The dog appears to be intermediate between these two species in sensitivity to arginine deficiency.

Excess dietary lysine has been reported to antagonise arginine in the chicken (Jones, 1964; O'Dell & Savage, 1966), rat (Jones, Wolters & Burnett, 1966) and guinea pig (O'Dell & Regan, 1962). Czarnecki, Hirakawa & Baker (1985) reported a depression in growth rate and gain/feed ratio in puppies when the lysine concentration of an amino acid-based diet was increased from 0.9 to 4.9%. Smaller increments of lysine 1% and 2% added to the basal (0.9% lysine) diet had no effect on performance. Doubling the dietary concentration of arginine (i.e. 0.4 to 0.8%) of the diet part overcame the antagonism. Excess dietary lysine neither induced or inhibited hepatic arginase. Induction of kidney arginase, inhibited renal reabsorption and gut uptake of arginine have been proposed to explain the amino acid antagonisms in other species (Larsen, Ross & Tapley, 1964; Jones, Petersberg & Burnett, 1967; Nesheim, 1968; Austic & Nesheim, 1970; Harper, Benevenga & Wohlhueter, 1970; Boorman, 1971; Robbins & Baker, 1981).

Adverse arginine–lysine interactions have not been reported in the cat. If it occurs, its mode of action would not likely be based on induction of liver arginase which appears to behave more as a constitutive than an adaptive enzyme in the cat (Rogers *et al.*, 1977).

Tryptophan metabolism

Tryptophan metabolism in the cat has several points of divergence from that of other mammals. The activity of the enzyme tryptophan dioxygenase is increased several fold in the liver of the rat following administration of tryptophan or glucocorticoids (Knox, 1951; Knox & Auerbach, 1955). In contrast, in the cat liver the response of tryptophan dioxygenase to tryptophan loading is much less than in the rat (Leklem, Woodford & Brown, 1969). Also these workers showed that the enzyme is not induced by cortisol administration.

In most mammals, degradation of tryptophan leads to synthesis of nicotinic acid. The efficiency of conversion varies with species, and other factors, but, for rats and humans, 33–40 and 60 mg tryptophan respectively have been calculated to be required per mg niacin synthesised (Hankes *et al.*, 1948; Horwitt *et al.*, 1956). In contrast, the synthesis of niacin from tryptophan catabolism in cats is nutritionally insignificant (Da Silva, Fried & De Angelis, 1952; Leklem *et al.*, 1969) despite cats apparently having all the enzymes of the pyridine nucleotide pathway (De Castro, Brown & Price, 1957; Sukadolnik *et al.*, 1957; Ikeda *et al.*, 1965). High activity of the enzyme picolinic carboxylase is probably the reason for the inefficient conversion of tryptophan to niacin. This enzyme would remove amino–carboxyl–muconic semialdehyde, the common intermediate for both niacin synthesis and the degradation pathway (Ikeda *et al.*, 1965; Sukadolnik *et al.*, 1957).

Cats also exhibit a lower urinary excretion of tryptophan and its metabolites than rats, following loading doses of these compounds. The low urinary excretion of these metabolites of tryptophan which have carcinogenic properties may offer an explanation for the low incidence of urinary bladder neoplasms in the cat.

Micronutrient requirements

Vitamin requirements of cats and dogs

A comparison of the vitamin requirements of cats and dogs on an equivalent dietary energy basis is presented in Table 5. This method of

Table 5. *Comparative dietary vitamin concentrations recommended for growing cats and dogs per MJ of metabolisable energy*

Vitamin	Units	Cat[a]	Dog[b]
Vitamin A (retinol)	μg	48	72
Vit. D (cholecalciferol)	μg	0.60	0.66
Vit. E (-tocopherol)	mg	1.4	1.5
K	μg	4.7	NS
B_1	μg	239	65
B_2	μg	191	163
Pantothenic acid	μg	239	645
Niacin	μg	1912	717
Pyridoxine	μg	191	71.7
Folic acid	μg	38.2	12.9
Vitamin B_{12}	μg	1.0	1.7
Choline	mg	115	81

Source:
[a] National Research Council, 1986.
[b] National Research Council, 1985.
NS: not stated.

presentation was used since most cat diets have a higher metabolisable energy value than dog diets. For the fat soluble vitamins, given the uncertainty of the estimates, the requirements for dogs and cats are broadly equivalent. The similarity in the requirements for vitamin A of cats and dogs is interesting as cats are unable to utilise carotene as a source of vitamin A. The cat does not have the dioxygenase enzyme which is needed to cleave the carotene molecule to produce vitamin A (Gershoff *et al.*, 1957). Despite this difference the utilisation of preformed vitamin A *per se* is similar in both species.

Among the water soluble vitamins, the cat appears to have higher requirements for thiamin, niacin, pyridoxine, and folic acid, whereas the dog has a higher requirement for pantothenic acid than the cat. The different specific estimates for the latter two vitamins are based on limited data and may be artefacts.

The stated requirement for thiamine is approximately four times greater for cats than dogs. This difference appears to be a real species difference, but its metabolic basis is not apparent.

The cat has a higher stated requirement for niacin than the dog. Possibly this higher requirement may arise from the cat's virtual inability to convert tryptophan to niacin because of the high activity of the enzyme picolinic carboxylase (Suhadolnik *et al.*, 1957; Ikeda *et al.*, 1965). The higher requirement of the cat for pyridoxine could be a consequence of the higher protein requirement of the cat than the dog since pyridoxal phosphate is an important co-enzyme in amino acid catabolism.

Taurine

No dietary requirement for taurine has been demonstrated in the dog. However, the cat, especially when given a diet low in sulphur amino acid, requires a source of dietary taurine for the maintenance of tissue concentrations at a level to sustain the integrity of the retina (Schmidt *et al.*, 1976, 1977). New-born kittens require dietary taurine for maintenance of normal growth (K.C. Hayes personal communication). The feeding of a taurine-free diet to queens during gestation produced kittens of below normal birth weight which had low survivability, cerebellar dystrophy and abnormalities of the skeleton. Growth of kittens nursed by queens given the taurine-free diet through lactation, was subnormal (Sturman *et al.*, 1985 *a, b*).

Circulating taurine concentrations in the blood of the cat are directly related to the dietary concentration of sulphur amino acids, which are the precursors of taurine (See Chapter 10). However, the efficiency of hepatic synthesis of taurine by the cat is lower than that of other species, such as the rat, due to a presumed lower activity of the enzyme cysteine sulphinic acid (CSA) decarboxylase (Knopf *et al.*, 1978; Hardison *et al.*, 1977). In the rat, the percentage of oxidised cysteine converted to taurine is 83% (Stipanuk & Rotter, 1984). The percentage of the oxidised cysteine converted to taurine does not appear to have been measured in cats. However, per unit weight of liver tissue, about six times as much CSA is decarboxylated and four times as much CSA is converted to taurine in the liver of rats than in the liver of cats (Hardison *et al.*, 1977).

Cats, unlike dogs, use taurine as an obligatory conjugator of bile salts which results in a greater loss of taurine by bile salts not being absorbed by the enterohepatic circulation.

Essential fatty acids

Dogs in common with most other mammals have a dietary requirement for polyunsaturated fatty acids of the $n-6$ series. A dietary deficiency of these fatty acids in young animals results in failure of growth and the appearance of dermatitis (Hansen *et al.*, 1948, 1954; Hansen & Wiese, 1951; Wiese *et al.*, 1965, 1966). The dog is capable of synthesising the other metabolically essential fatty acids from dietary linoleic acid by alternate desaturation and chain elongation. This process requires the delta-6 and delta-5 desaturases. Rivers *et al.* (1975, 1976 *a, b, c*) and Hassam *et al.* (1977) presented evidence that cats and a lion failed to convert linoleic acid to gamma-linolenic or dihomo-gamma-linolenic acid (which are intermediates in the synthesis of arachidonic acid) and suggested that the desaturase activity of the cat was absent or very slow.

Subsequent work by MacDonald *et al.* (1983, 1984 *a, b, c, d*) demonstrated the essentiality of linoleate for maintenance of skin resistance to water loss and the failure of the cat to convert linoleate $18:2n-6$, to $20:3n-6$ or

Table 6. *Comparative dietary mineral concentrations recommended for growing cats and dogs per MJ of metabolisable energy*

Mineral	Units	Cat[a]	Dog[b]
Calcium	mg	382	382
Phosphorus	mg	287	287
Potassium	mg	191	287
Sodium	mg	24	36
Chloride	mg	91	55
Magnesium	mg	19	26
Iron	mg	3.8	2.1
Zinc	mg	2.4	2.3
Copper	μg	239	191
Manganese	μg	239	335
Iodine	μg	17	38
Selenium	μg	4.8	7.2

Notes:
[a] National Research Council, 1986.
[b] National Research Council, 1985.

$20:4n-6$. It was further demonstrated by these workers that arachidonate was required for female cats to deliver viable young. Reproduction in males was apparently supported by dietary linoleate as there was some conversion of linoleate to arachidonate in the testis.

There is no evidence to suggest that dogs require a dietary source of arachidonate to maintain reproduction if adequate linoleate is present.

Minerals

The comparative dietary mineral requirements of cats and dogs on the same energy basis is presented in Table 6. Some variation occurs in the estimates across these species, but, in general, the requirements are very similar.

Despite this similarity, dogs appear (from the number of papers in the literature) to suffer a higher incidence of skeletal deformities than cats (See Chapter 17). One can speculate that this difference might be a consequence of the greater selection pressures imposed on dogs by humans for the breeding of animals of widely different physical proportions. It is known that at least some of the skeletal abnormalities have a heritable component, i.e. canine hip dysplasia, (See Chapter 18).

Domesticated cats exhibit a higher evidence of urinary calculi of the struvite type than dogs. The basis for these observed differences may be the type of commercial foods available to cats and dogs rather than differences between the species in mineral metabolism.

Zinc deficiency has been reported more frequently in dogs than cats. This may also be related to the higher proportion of vegetable than animal products in commercial dog foods which would affect the availability of zinc (Schugel, 1982).

Secondary plant compounds

Cats are sensitive to some of the secondary compounds which occur in plants (e.g. benzoic acid derivatives) if they are included in the diet. These compounds are conjugated with glycine and glucuronic acid in most animals, including dogs, before excretion in the urine. The glucuronic pathway (which has the greater capacity) is defective in cats, so they must rely entirely on the glycine pathway, (Williams, 1967). In contrast to cats, herbivorous animals have the capacity to dispose of large quantities of secondary plant compounds. One of the main evolutionary advantages conferred by ruminant digestion may have been the catabolism of secondary plant compounds rather than utilisation of plant cell walls (Morris & Rogers, 1983*b*).

Onions contain a compound (or compounds, presumably disulphides) which cause Heinz body formation and haemolytic anaemia in cats (Kobayashi, 1981). Normal cat blood generally has a higher percentage of erythrocytes with Heinz bodies than the blood of other mammals. Dogs can also exhibit a haemolytic anaemia following ingestion of onions (Sebrell, 1930; Spice, 1976).

Summary

This review refers to some of the more important differences in the nutrition and metabolism of cats and dogs. These differences in nutritional requirements correlate with the evolution of the two species; dogs in common with many other members of the Canoidea evolved as omnivores, while cats and other Feloidea evolved as strict carnivores. This course of evolution rendered some of the metabolic pathways present in the dog and other omnivores redundant for the cat. The nutrition of the cat therefore differs from that of the dog precisely because 'a cat is not a dog'.

REFERENCES

Ardisson, J.L., Dolisi, C., Ozon, C. & Crenesse, D. (1981). Caracteristiques des prises d'eau et d'aliments spontanees chez des chiens en situation *ad lib. Physiology & Behavior* **26**, 361–70.

Austic, R.E. & Nesheim, M.C. (1979). Role of kidney arginase in variation of the arginine requirement of chicks. *Journal of Nutrition* **100**, 855–67.

Bailey, C.B., Kitts, W.D. & Wood, A.J. (1960). Changes in the gross chemical composition of the mouse during growth in relation to the assessment of physiological age. *Canadian Journal of Animal Science* **40**, 143–55.

Ballard, F.J. (1965). Glucose utilization in mammalian liver. *Comparative Biochemistry & Physiology* 14, 437–43.

Banta, C.A., Clemens, E.T., Krinsky, M.M. & Sheffy, B.E. (1979). Sites of organic acid production and patterns of digesta movement in the gastrointestinal tract of dogs. *Journal of Nutrition* 109, 1592–600.

Barry, R.E. (1976). Mucosal surface areas and villous morphology of the small intestine of small mammals: Functional interpretations. *Journal of Mammalogy* 57, 273–90.

Bartoshuk, L.M., Harned, M.A. & Parks, L.H. (1971). Taste of water in the cat: Effects on sucrose preference. *Science* 171, 699–701.

Bartoshuk, L.M., Jacobs, H.L., Nichols, T.L., Hoff, L.A. & Ryckman, J.J. (1975). Taste rejection of nonnutritive sweeteners in cats. *Journal of Comparative and Physiological Psychology* 89, 971–5.

Beauchamp, G.K., Maller, O. & Rogers, J.G. (1977). Flavor preferences in cats (*Felis catus* and *Panthera* sp.). *Journal of Comparative and Physiological Psychology* 91, 1118–27.

Beliveau, G.P., Morris, J.G., Rogers, Q.R. & Freedland, R.A. (1981). Metabolism of serine, threonine and glycine in isolated cat hepatocytes. *Federation Proceedings* 40, 807, Abstract Number 3504.

Belo, P.S., Romsos, D.R. & Leveille, G.A. (1976). Influence of diet on glucose tolerance, on the rate of glucose utilization and on gluconeogenic enzyme activities in the dog. *Journal of Nutrition* 106, 1465–74.

Bennett, M.J. & Coon, E. (1966). Mellituria and postprandial blood sugar curves in dogs after the ingestion of various carbohydrates with the diet. *Journal of Nutrition* 88, 163–8.

Boorman, K.N. (1971). The renal reabsorption of arginine, lysine and ornithine in the young cockerel (*Gallus domesticus*). *Comparative Biochemistry and Physiology* 39A, 29–38.

Brown, R.L. & Wood M. (1936). A study of starch digestion in hospitalized dogs. *North American Veterinarian* 17, 46–49.

Burns, R.A., Milner, J.A. & Corbin, J.E. (1981). Arginine: an indispensable amino acid for mature dogs. *Journal of Nutrition* 111, 1020–4.

Burroughs, E.W., Burroughs, H.S. & Mitchell, H.H. (1940). The amino acids required for the complete replacement of endogenous losses in the adult rat. *Journal of Nutrition* 19, 363–84.

Bush, G.L., Case, S.M., Wilson, A.C. & Patton, J.L. (1977). Rapid speciation and chromosomal evolution in mammals. *Proceedings of the National Academy of Science USA* 74, 3942–6.

Carpenter, J.A. (1956). Species differences in taste preferences. *Journal of Comparative & Physiological Psychology* 49, 139–44.

Chivers, D.J. & Hladik, C.M. (1980). Morphology of the gastrointestinal tract in primates: Comparisons with other mammals in relation to diet. *Journal of Morphology* 166, 337–86.

Colbert, E.H. (1980). Adaptation of the carnivorous mammals. In: *Evolution of the Vertebrates: A History of the Backboned Animals Through Time.* (3rd edition), pp. 334–9. New York: John Wiley & Sons.

Colin, G. (1854). *Traite de Physiologie Comparee des Animaux Domestiques*. Tome premier, pp. 408–10. Paris: J.-B. Bailliere.

Coman, B.J. & Brunner, H. (1972). Food habits of the feral house cat in Victoria. *Journal of Wildlife Management* 36, 848–53.

Cook, N.E., Kane, E., Rogers, Q.R. & Morris, J.G. (1985). Self-selection of dietary casein and soy-protein by the cat. *Physiology & Behaviour* 34, 583–94.

Curry, D.L. & Bennett, L.L. (1973). Dynamics of insulin release by perfused rat pancreas: Effects of hypophysectomy, growth hormone, adrenocorticotrophic hormone, and hydrocortisone. *Endocrinology* 93, 602–9.

Curry, D.L., Morris, J.G., Rogers, Q.R. & Stern, J.S. (1982). Dynamics of insulin and glucagon secretion by the isolated cat pancreas. *Comparative Biochemistry & Physiology* 72A, 333–8.

Czarnecki, G.L. & Baker, D.H. (1984). Urea cycle function in the dog with emphasis on the role of arginine. *Journal of Nutrition* 114, 581–90.

Czarnecki, G.L., Hirakawa, D.A. & Baker, D.H. (1985). Antagonism of arginine by excess dietary lysine in the growing dog. *Journal of Nutrition* 115, 743–52.

Da Silva, A.C., Fried, R. & De Angelis, R.C. (1952). The domestic cat as a laboratory animal for experimental nutrition studies. *Journal of Nutrition* 46, 399–409.

De Castro, F.T., Brown, R.R. & Price, J.M. (1957). The intermediary metabolism of tryptophan by cat and rat tissue preparations. *Journal of Biological Chemistry* 228, 777–84.

Downey, R.L., Kronfeld, D.S. & Banta, C.A. (1980). Diet of beagles affects stamina. *Journal of American Animal Hospital Association*, 16:273–7.

Drochner, W. & Muller-Schlosser, S. (1980). Digestibility and tolerance of various sugars in cats. In: *Nutrition of the Dog and Cat*. Editor R.S. Anderson, pp. 101–11. Oxford: Pergamon Press.

Easter, R.A., Katz, R.S. & Baker, D.H. (1974). Arginine: A dispensable amino acid for postpubertal growth and pregnancy swine. *Journal of Animal Science* 39, 1123–8.

Eliot, T.S. (1939). *Old Possum's Book of Practical Cats*. New York: Harcourt, Brace & Jovanovich.

Evans, H.E. & Christensen, G.C. (1979). Alimentary canal. In: *Miller's Anatomy of the Dog*. 2nd edition, Chapter 7, pp. 455–506. Philadelphia: W.B. Saunders Company.

Ewer, R.F. (1973). *The Carnivores*. Ithaca, New York: Cornell University Press.

Fajans, S.S., Christensen, H.N., Floyd, J.C. & Pek, S. (1974). Stimulation of insulin and glucagon release in the dog by a nonmetabolizable arginine analog. *Endocrinology* 94, 230–3.

Frings, H. (1951). Sweet taste in the cat and the taste-spectrum. *Experimentia* 7, 424–6.

Gershoff, S.N., Andrus, S.B., Hegsted, D.M. & Lentini, E.A. (1957). Vitamin A deficiency in cats. *Laboratory Investigation* 6, 227–40.

Gilmour, D. (1961). *The Biochemistry of Insects.* New York: Academic Press.

Hankes, L.V., Henderson, L.M., Brickson, W.L. & Elvehjem, C.A. (1948). Effect of amino acids on growth of rats on niacin-tryptophan deficient rations. *Journal of Biological Chemistry* **174**, 873–81.

Hansen, A.E. & Wiese, H.F. (1951). Fat in the diet in relation to nutrition of the dog. I. Characteristic appearance and gross changes of animals fed diets with and without fat. *Texas Reports of Biological Medicine* **9**, 491.

Hansen, A.E., Wiese, H.F. & Beck, O. (1948). Susceptibility to infection manifested by dogs on a low-fat diet. *Federation Proceedings* **7**, 289.

Hansen, A.E., Sinclair, J.G. & Wiese, H.F. (1954). Sequence of histologic change in skin of dogs in relation to dietary fat. *Journal of Nutrition* **52**, 541.

Hardison, W.G.M., Wood, C.A. & Proffitt, J.H. (1977). Quantification of taurine synthesis in the intact rat and cat liver. *Proceeding Society of Experimental Biological Medicine* **155**, 55.

Harper, A.E., Benevenga, N.J. & Wohlhueter, R.M. (1970). Effects of ingestion of disproportionate amounts of amino acids. *Physiological Reviews* **50**, 428–558.

Hassam, A.G., Rivers, J.P.W. & Crawford, M.A. (1977). The failure of the cat to desaturate linoleic acid; Its nutritional implications. *Nutrition and Metabolism* **21**, 321.

Hore, P. & Messer, M. (1968). Studies on disaccaridase activities of the small intestine of the domestic cat and other carnivorous mammals. *Comparative Biochemistry and Physiology* **24**, 717–25.

Horwitt, M.K., Harvey, C.C., Rothwell, W.S., Culter, J.L. & Haffron, D. (1956). Tryptophan-niacin relationships in man. *Journal of Nutrition* **60**, Supplement 1.

Houpt, K.A., Coren, B., Hintz, H.F. & Hilderbrant, J.E. (1979). Effect of sex and reproductive status on sucrose preference, food intake, and body weight of dogs. *Journal of the American Veterinary Medical Association* **174**, 1083–5.

Ikeda, M. Tsuji, H., Nakamura, S., Ichiyama, A., Nichizuka, Y. & Hayaisi, O. (1965). Studies on the biosynthesis of nicotinamide adenine dinucleotide. II. A role of picolinic carboxylase in the biosynthesis of nicotinamide adenine dinucleotide from tryptophan in mammals. *Journal of Biological Chemistry* **240**, 1395–401.

Ivy, A.C., Schmidt, C.R. & Beazell, J. (1936). Starch digestion in the dog. *North America Veterinary* **17**, 44–6.

James, W.T. & McCay, C.M. (1950). A study of food intake, activity, and digestive efficiency in different type dogs. *American Journal of Veterinary Research* **11**, 412–13.

Jones, J.D. (1964). Lysine–arginine antagonism in the chick. *Journal of Nutrition* **84**, 313–21.

Jones, J.D., Wolters, R. & Burnett, P.C. (1966). Lysine–arginine–electrolyte relationships in the rat. *Journal of Nutrition* **89**, 171–88.

Jones, J.D., Petersburg, S.J. & Burnett, P.C. (1967). The mechanism of the lysine-arginine antagonism in the chick: Effect of lysine on digestion, kidney arginase, and liver transamidinase. *Journal of Nutrition* **93**, 103–16.

Kanazawa, Y., Kuzuya, T., Ide, T. & Kosaka, K. (1966). Plasma insulin responses to glucose in femoral, hepatic, and pancreatic veins in dogs. *American Journal of Physiology* **2**, 442–8.

Kane, E., Morris, J.G. & Rogers, Q.R. (1981). Acceptability and digestibility by adult cats of diets made with various sources and levels of fat. *Journal of Animal Science* **53**, 1516–23.

Kane, E., Rogers, Q.R., Morris, J.G. & Leung, P.M.B. (1981). Feeding behaviour of the cat fed laboratory and commercial diets. *Nutrition Research* **1**, 499–507.

Karasov, W.H. & Diamond, J.M. (1985). Digestive adaptations for fueling the cost of endothermy. *Science* **228**, 202–4.

Kendall, P.T., Holme, D.W. & Smith, P.M. (1982). Comparative evaluation of net digestive and absorptive efficiency in dogs and cats fed a variety of contrasting diet types. *Journal of Small Animal Practice* **23**, 577–87.

Kettlehut, I.C., Foss, M.C. & Migliorini, R.H. (1978). Glucose homeostasis in a carnivorous animal (cat) and in rats fed a high-protein diet. *American Journal of Physiology* **239**. (Regulatory Integrative Comp. Physiol. 3) R115–R121.

Knopf, K., Sturman, J.A., Armstrong, M. & Hayes, K.C. (1978). Taurine: an essential nutrient for the cat. *Journal of Nutrition* **108**, 773.

Knox, W.E. (1951). Two mechanism which increase *in vitro* the liver tryptophan peroxidase activity: Specific enzyme adaptation and stimulation of the pituitary–adrenal system. *British Journal of Experimental Pathology* **32**, 462–9.

Knox, W.E. & Auerbach, V.H. (1955). The hormonal control of tryptophan peroxidase in the rat. *Journal of Biological Chemistry* **214**, 307–13.

Kobayashi, K. (1981). Onion poisoning in the cat. *Feline Pract.* **11**, 22.

Kronfeld, D.S., Hammel, E.P., Ramberg, C.F. & Dunlap, H.L. (1977). Hematological and metabolic responses to training in racing sled dogs fed diets containing medium, low, or zero carbohydrate. *American Journal of Clinical Nutrition* **30**, 419–30.

Larsen, P.R., Ross, J.E. & Tapley, D.F. (1964). Transport of neutral, dibasic and N-methyl-substituted amino acids by rat intestine. *Biochimica et Biophysica Acta* **88**, 570–7.

Lehninger, A.L. (1975). *Biochemistry* (2nd ed.). p. 193. New York: Worth Publishers Inc.

Leklem, J.E., Woodford, J. & Brown, R.R. (1969). Comparative tryptophan metabolism in cats and rats: Differences in adaptation of tryptophan oxygenase and *in vivo* metabolism of tryptophan, kynurenine and hydroxykynurenine. *Comparative Biochemistry and Physiology* **31**, 95–109.

Leung, P.M.B., Gamble, M.A. Rogers, Q.R. (1981). Effects of prior protein ingestion on dietary choice of proteins and energy. *Nutritional Reports International* 24, 257–66.

Lewis, L.D. & Morris, M.L. (1984). Diet as a causative factor of feline urolithiasis. In: *The Veterinary Clinics of North America: Small Animal Practice*, Vol. 14, No. 3, Symposium on Disorders of the Feline Lower Urinary Tract. pp. 513–27. Philadelphia: W.B. Saunders Company.

MacDonald, M.L., Rogers, Q.R., & Morris, J.G. (1983). Role of linoleate as an essential fatty acid for the cat independent of arachidonate synthesis. *Journal of Nutrition*, 113, 1422–33.

MacDonald, M.L., Rogers, Q.R., Morris, J.G., & Cupps, P.T. (1984a). Effects of linoleate and arachidonate deficiencies on reproduction and spermatogenesis in the cat. *Journal of Nutrition* 114, 719–26.

MacDonald, M.L., Rogers, Q.R., & Morris, J.G. (1984b). Nutrition of the domestic cat, a mammalian carnivore. *Annual Review of Nutrition*, 4, 521–62.

MacDonald, M.L., Rogers, Q.R., & Morris, J.G. (1984c). Effects of dietary arachidonate deficiency on the aggregation of cat platelets. *Comparative Biochemistry and Physiology*, 78C, 123–6.

MacDonald, M.L., Anderson, B.C., Rogers, Q.R., Buffington, C.A., & Morris, J.G. (1984d). Essential fatty acid requirements of cats: Pathology of essential fatty acid deficiency. *American Journal of Veterinary Research*, 45, 1310–17.

MacDonald, M.L., Rogers, Q.R. & Morris, J.G. (1985). Aversion of the cat to dietary medium-chain triglycerides and caprylic acid. *Physiology and Behavior* 34, 1–5.

Matthey, R. (1973). The chromosome formulae of eutherian mammals. In: *Cytotaxonomy and Vertebrate Evolution*, Eds. A.B. Chiarelli & E. Capanna, Chapter 15, pp. 531–616. New York: Academic Press.

McCay, C.M. (1949). *Nutrition of the Dog*. 2nd edition. New York: Comstock Publishing Company, Inc.

McGeachin, R.L. & Akin, J.R. (1979). Amylase levels in the tissues and body fluids of the domestic cat (*Felis catus*) *Comparative Biochemistry & Physiology* 63B, 437–9.

McGeachin, R.L. & Johnson W.D. (1964). The *in vivo* effects of puromycin on amylase levels in the serum, liver, salivary glands and pancreas of the rat. *Arch. Biochem. Biophys.* 107, 534–6.

Mech, L.D. (1975). Hunting behavior in two similar species of social canids. In: *The Wild Canids: Their Systematics, Behavioral Ecology and Evolution*. Ed. M.W. Fox, Ch. 24, pp. 363–79. New York: Van Nostrand Reinhold Company.

Migliorini, R.H., Linder, C., Moura, J.L. & Veiga, J.A.S. (1973). Gluconeogenesis in a carnivorous bird (black vulture), *American Journal of Physiology* 225, 1389–92.

Morris, J.G. (1985). Nutritional and metabolic responses to arginine deficiency in carnivores. *Journal of Nutrition* 115, 524–31.

Morris, J.G., Fujimoto, J. & Berry, S.C. (1974). The comparative digestibility of a zoo diet fed to 13 species of felid and badger. *International Zoo Yearbook*, 14, 169–71.

Morris, J.G. & Rogers, Q.R. (1978). Arginine: An essential amino acid for the cat. *Journal of Nutrition* 108, 1944–53.

Morris, J.G. & Rogers, Q.R. (1983a). Nutritional implications of some metabolic anomalies of the cat. *American Animal Hospital Association's 50th Annual Meeting Proceeding*, pp. 325–31.

Morris, J.G. & Rogers, Q.R. (1983b). Nutritionally related metabolic adaptations of carnivores and ruminants. In *Plant, Animal, and Microbial Adaptations to Terrestrial Environment*, eds. N.S. Margaris, M. Arianoutsou-Faraggitaki & R.J. Reiter, pp. 165–79. New York: Plenum Publishing Corporation.

Morris, J.G., Rogers, Q.R., Winterrowd, D.L. & Kamikawa, E.M. (1979). The utilization of ornithine and citrulline by the growing kitten. *Journal of Nutrition* 109, 724–9.

Morris, J.G., Trudell, J. & Pencovic, T. (1977). Carbohydrate digestion by the domestic cat *(Felis catus)*. *British Journal of Nutrition* 37, 365–73.

Mu, M.M. & Lingham, R. (1974). The rudimentary appendix of the cat *(Felis domesticus)*. *Acta Anatomica* 87, 119–23.

Mugford, R.A. (1977). External influences on the feeding of carnivores. In: *The Chemical Senses and Nutrition*. Eds. M.R. Kare & O. Maller, Ch. 2, pp. 25–50. New York: Academic Press.

Nakagawa, I., Takahashi, T., Suzuki, T. & Kobayashi, K. (1963). Amino acid requirements of children. Minimal needs for tryptophan, arginine and histidine based on nitrogen balance methods. *Journal of Nutrition* 80, 305–10.

National Research Council (1985). *Nutrient Requirement of Dogs*. Revised edition. Washington, D.C.: National Academy Press.

National Research Council (1986). *Nutrient Requirement of Cats*, Revised edition, Washington, D.C.: National Academy Press.

Nesheim, M.C. (1968). Kidney arginase activity and lysine tolerance in strains of chickens selected for a high or low requirement of arginine. *Journal of Nutrition* 95, 79–87.

Nickel, R., Schummer, A. & Seiferle, E. (1979). *The Viscera of the Domestic Mammals*, 2nd edition, Revised by Schummer, A., Nickel, R. & Sack, W.O. New York: Springer-Verlag.

O'Dell, B.L. & Regan, W.O. (1962). Effect of lysine and glycine upon arginine requirements of guinea pigs. In: *Proceedings Society of Experimental Biology and Medicine*, Vol. 112, pp. 336–7.

O'Dell, B.L. & Savage, J.E. (1966). Arginine-lysine antagonism in the chick and its relationship to dietary cations. *Journal of Nutrition* 90, 364–70.

Oftedal, O.T. (1984). Lactation in the dog: Milk composition and intake by puppies. *Journal of Nutrition* 114, 803–12.

Owen, R. (1868). Mammals. In: *Anatomy of Vertebrates*, Vol. 3, pp. 442–3. London: Longman, Green, & Co.

Pfaffmann, C. (1955). Gustatory nerve impulses in rat, cat and rabbit. *Journal of Neurophysiology* **18**, 429–40.

Phillipson, A.T. (1947). Fermentation in the alimentary tract and the metabolism of the derived fatty acids. *Nutrition Abstracts and Reviews* **17**, 12–18.

Rajasingham, R. Bell, J.L. & Baron, D.N. (1971). A comparative study of the isoenzymes of mammalian α-amylase. *Enzyme* **12**, 180–6.

Reighard, J.S. & Jennings, H.S. (1966). The alimentary canal. Apparatus digestorius. In: *Anatomy of the Cat.* 3rd edition, pp. 221–42. New York: Holt, Reinhart & Winston.

Resnick, S. (1974a). Effects of feeding a high-protein diet and an all-meat diet on the hemogram of the dog. *Veterinary Medicine, Small Animal Clinician* **69**, 70–4.

Resnick, S. (1974b). Results of feeding high-carbohydrate and all-meat diets to pups at various ages. *Veterinary Medicine, Small Animal Clinician* **69**, 585–92.

Resnick, S. (1978). Effect of age on survivability of pups eating a carbohydrate-free diet. *Journal of the American Veterinary Medical Association* **172**, 145–8.

Rivers, J.P.W., Sinclair, A.J. & Crawford, M.A. (1975). Inability of the cat to desaturate essential fatty acids. *Nature* **258**, 171.

Rivers, J.P.W., Sinclair, A.J., Moore, D.P. & Crawford, M.A. (1976a). The abnormal metabolism of essential fatty acids in the cat. *Proceedings of the Nutrition Society* **35**, 66a.

Rivers, J.P.W., Hassam, A.G. & Alderson, C. (1976b). The absence of 6-desaturase activity in the cat. *Proceedings of the Nutrition Society* **35**, 67a.

Rivers, J.P.W., Hassam, A.G., Crawford, M.A. & Brambell, M.R. (1976c). The inability of the lion, *Panthera leo L.*, to desaturate linoleic acid. *FEBS Letters* **67**, 269.

Robbins, K.R. & Baker, D.H. (1981). Kidney arginase activity in chicks fed diets containing deficient or excessive concentrations of lysine, arginine, histidine, or total nitrogen. *Poultry Science* **60**, 829–34.

Rogers, Q.R. & Morris, J.G. (1980). Why does the cat require a high protein diet? In: *Nutrition of the Dog and Cat.* Ed. R.S. Anderson, pp. 45–66. Pergamon Press.

Rogers, Q.R. & Morris, J.G. (1982). Do cats really need more protein? *Journal of Small Animal Practice* **23**, 521–32.

Rogers, Q.R. & Morris, J.G. (1983). Protein and amino acid nutrition of the cat. *American Animal Hospital Association's 50th Annual Meeting Proceedings*, pp. 333–6.

Rogers, Q.R., Morris, J.G. & Freedland, R.A. (1977). Lack of hepatic enzymatic adaptation to low and high levels of dietary protein in the adult cat. *Enzyme* **22**, 348–56.

Rogers, Q.R. & Phang, J.M. (1985). Deficiency of pyrroline-5-carboxylate synthase in the intestinal mucosa of the cat. *Journal of Nutrition* **115**, 146–50.

Romer, A.S., (1966). *Vertebrate Paleontology.* (3rd ed.). pp. 468 Chicago: The University of Chicago Press.

Romer, A.S. (1974). Carnivorous mammals. In: *Paleontology*, (3rd ed.). 4th impression, Ch. 19, pp. 229–38. Chicago: The University of Chicago Press.

Romsos, D.R., Belo, P.S., Bennink, M.R., Bergen, W.G. & Leveille, G.A. (1976). Effects of dietary carbohydrate, fat and protein on growth, body composition and blood metabolite levels in the dog. *Journal of Nutrition* 106, 1452–64.

Rose, W.C., Haines, W.J. & Warner, D.T. (1954). The amino acid requirements of man. V. The role of lysine, arginine and tryptophan. *Journal of Biological Chemistry* 206, 421–30.

Rose, W.C. & Rice, E.E. (1939). The significance of the amino acids in canine nutrition. *Science* 90, 186–7.

Roseboom, B.B. & Patton, J.W. (1929). Starch digestion in the dog. *Journal of the American Veterinary Medical Association* 74, 768–72.

Rowsell, E.V., Carnie, J.A., Wahbi, S.D., Al-Tai, A.H. & Rowsell, K.V. (1979). L-serine dehydratase and L-serine-pyruvate aminotransferase activities in different animal species. *Comparative Biochemistry & Physiology* 637, 543–55.

Schaeffer, M.C., Rogers, Q.R. & Morris, J.G. (1989). Protein in the nutrition of dogs and cats. (This symposium).

Schmidt, S.Y., Berson, E.L. & Hayes, K.C. (1976). Retinal degeneration in cats fed casein. I. Taurine deficiency. *Investigative Ophthalmology* 15, 47.

Schmidt, S.Y., Berson, E.L., Watson, G. & Huang, C. (1977). Retinal degeneration in cats fed casein. III. Taurine deficiency and ERG amplitudes. *Investigative Ophthalmology* 16, 673.

Schneider, B.H., Lucas, H.L., Pavlech, H.M. & Cipolloni, M.A. (1951). Estimation of the digestibility of feeds from their proximate composition. *Journal of Animal Science* 10, 706.

Schugel, L.M., (1982). Zinc. Its role in skin condition of dogs. *Petfood Industry* 24(2), 24–7.

Sebrell, W.H., (1930). Anemia of dogs produced by feeding onions. *Public Health Reports* 45, 1175–91.

Skude, G. & Mardh P.-A. (1976). Isoamylase in blood, urine, and tissue homogenates from some experimental mammals. *Scandinavian Journal of Gastroenterology* 11, 21–26.

Smalley, K.A., Rogers, Q.R., Morris, J.G. & Eslinger, L.L. (1985). The nitrogen requirement of the weanling kitten. *British Journal of Nutrition* 53, 501–12.

Spice, R.N. (1976). Hemolytic anemia associated with ingestion of onions in a dog. *Canadian Veterinary Journal* 17, 181–3.

Stipanuk, M.H. & Rotter, M.A. (1984). Metabolism of cysteine, cysteinesulfinate and cysteinesulfonate in rats fed adequate and excess levels of sulfur-containing amino acids. *Journal of Nutrition* 114, 1426–37.

Stokes, W.L., Judson, S. & Picard, M.D. (1978). *Introduction to Geology: Physical and Historical*, Second Edition. Englewood Cliffs, New Jersey: Prentice-Hall, Inc.

Sturman, J.A., Moretz, R.C., French, J.H. & Wisniewski, H.M. (1985a). Taurine deficiency in the developing cat: Persistence of the cerebellar external granule cell layer. *Journal of Neuroscience Research* 13, 405.

Sturman, J.A., Moretz, R.C., French, J.H. & Wisniewski, H.M. (1985b). Postnatal taurine deficiency in the kitten results in persistence of the cerebellar external granule cell layer: Correction by taurine feeding. *Journal of Neuroscience Research* 13, 521.

Sukadolnik, R.J., Stevens, C.O., Decher, R.H., Henderson, L.M. & Hankes, L.V. (1957). Species variation in the metabolism of 3-hydroxyanthranilate to pyridinecarboxylic acids. *Journal of Biological Chemistry* 228, 973–82.

Tedford, R.H. (1981) History of Dogs and Cats. In: *Nutrition and Management of Dogs and Cats*. St. Louis, Missouri: Ralston Purina Company.

USDA (1976). Composition of foods: Dairy and egg products, raw, processed, prepared. *Agriculture Handbook No. 8–1*. United States Department of Agriculture, Agriculture Research Service.

Warren, R. (1939). Serosal and mucosal dimensions at different levels of the dog's small intestine. *Anatomical Record* 75, 427–37.

White, T.D. & Boudreau, J.C. (1975). Taste preferences of the cat for neurophysiologically active compounds. *Physiological Psychology* 3, 405–10.

Wiese, H.F., Bennett, M.J., Coon, E. & Yamanaka, W. (1965). Lipid metabolism of puppies as affected by kind and amount of fat and of dietary carbohydrate. *Journal of Nutrition* 86, 271.

Wiese, H.F., Yamanaka, W., Coon, E. & Barber, S. (1966). Skin lipids of puppies as affected by kind and amount of dietary fat. *Journal of Nutrition* 89, 113.

Williams, R.T. (1967). Comparative patterns of drug metabolism. *Federation Proceedings* 26, 1029.

Wilson, A.C., Bush, G.L., Case, S.M. & King, M.-C. (1975). Social structuring of mammalian populations and rate of chromosomal evolution. *Proceedings National Academy of Sciences USA* 72, 5061–5.

Windmueller, H.G. & Spaeth, A.E. (1981). Source and fate of circulating citrulline. *American Journal of Physiology* 241 (*Endocrinology Metabolism*5), 473–80.

Wolf, P.A. & Corley, R.C. (1939). Significance of amino acids for the maintenance of nitrogen balance in the adult white rat. *American Journal of Physiology* 127, 589–96.

Wood, H.O. (1944). The surface area of the intestinal mucosa in the rat and in the cat. *Journal of Anatomy* 78, 103–5.

6

Allometric considerations in the nutrition of dogs

J.P.W. RIVERS AND I.H. BURGER

The nature of allometry

Allometry is the relationship between size and form or function, and given the nature of animals it is not surprising that it is a subject of which biologists have become increasingly aware. It is not, however, purely a manifestation of biological rules, and the constraints placed upon function by size are part of the nature of the world.

Size indeed imposes so many constraints that it is difficult not to see allometric phenomena everywhere. In architecture, for example, the need to put windows in buildings to provide illumination has traditionally meant that, as the size of buildings increases above a certain level, so their shape must also change to increase wall, and hence window, area. So small houses can be like 'little boxes' but larger ones must become increasingly rectangular boxes, and eventually quite un-boxlike in shape. As Stephen Jay Gould (1978) has pointed out, whatever its symbolic value, the cruciform shape of a Western church provides more wall for windows, and it may not be a coincidence that early Saxon churches were both small and rectangular. The college quadrangle (like the hole in a doughnut), is another device for increasing the surface area.

All these are architects' strategies which cope with the simple allometric constraint that, for a given shape with increasing size, volume increases faster than surface, so the extent of window illumination decreases. The nature of the problem is obvious if dimensional analysis is performed. With any similarly shaped objects, surface area increases as the square of the linear dimensions $[l^2]$, while volume rises as the cube of linear dimensions $[l^3]$. Thus a set of cubes of side l, will have a surface area $6l^2$, while their volume will be l^3. Thus the ratio of surface to volume will be $6l^2/l^3$, that is $6/l$. Thus, as l rises, the surface to volume ratio declines; as volume rises, the ratio of surface area to volume declines as $V^{\frac{2}{3}}$.

A brief history of allometry

This simple analysis can be traced back to Archimedes, but in the application of such principles to biology, a variety of names deserve

67

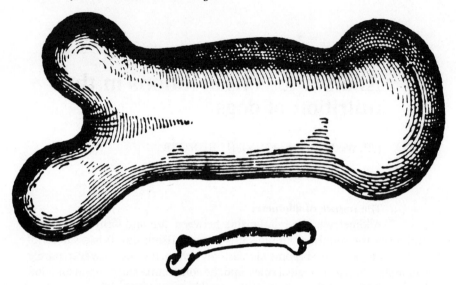

Fig. 1. Galileo's (1638) illustration of the effect of size upon form. An anatomically somewhat imprecise drawing shows the fact that the femur of a large animal must be thicker than that of a small one to provide increased mechanical strength.

mention. The first is Galileo (1638) who, under house arrest by order of the Inquisition for his heretical espousal of the Copernican heliocentric cosmology, was the first to write about the biological aspects of this phenomenon, which he called his Principle of Similitude. Thus, as his famous diagrams eloquently show, a large column is proportionately thicker for its length than a small one because the increase in deforming torque (weight) is proportional to the cube of length, while the strength to resist it is proportional to cross sectional area which, if shape were preserved, would only rise with l^2. For stability, cross sectional area must rise as l^3, so the relative diameter must rise in direct proportion to the square root of increasing length. Animals' legs must get thicker with increasing body size on the same principle, because, as Galileo pointed out, they need thicker bones to support their increasing weight (Fig. 1). Shape is therefore a function of size.

The term 'allometric' is a neologism that has been applied to such relationships for 50 years now, since it was coined by Sir Julian Huxley in his studies of relative organ size (Huxley, 1932). Huxley noted that the part (Y) was often related to the whole (X) by an equation of the form:

$$Y = aX^b \qquad \text{[Equation 1]}$$

If the size of the part is always directly proportional to that of the whole, b has the value 1, and Equation 1 is a simple proportion. If, however, the ratio of part/whole changes, b has a value greater or less than 1. Huxley originally borrowed the term isometric to describe situations where $b = 1$, and termed the others allometric, using positive and negative allometry to describe situations where $b > 1.0$ and < 1.0 respectively (Huxley & Tessier, 1936). His original nomenclature has been modified by subsequent authors, and allometric has come to be used of all relationships described by Equation 1, and allometry has taken on the looser meaning defined at the outset.

It follows from Equation 1 that

$$\log Y = \log a + b \log X \qquad \text{[Equation 2]}$$

and thus that a graph of log X against log Y will be a straight line with slope b (the allometric exponent) and an intercept which is the logarithm of a, the allometric coefficient. The basic process of determining allometry has come to involve regression analysis of the logarithms of the variables X and Y. Generally, least squares regression is used, though this can be theoretically objected to since the squares of the logarithms of X and Y are very different from the logarithms of the squares of X and Y (see Calder, 1987). A theoretically more satisfying approach is to use the major axis transformation technique (Martin, 1981), though, when values of the correlation coefficient, r, are high, the results from the two approaches are not significantly different.

Although allometric relationships were originally seen by Huxley and his school as anatomical relationships, function is implicit in anatomy as Galileo made clear. Functional allometry was already flourishing at the time that Huxley wrote, though reading his book gives the distinct impression that he wrote as an anatomist apparently ignorant or uninterested in allometry of physiological function. Interestingly enough for the present context this functional work was both comparative and quintessentially nutritional, originating in attempts to understand the variation in heat production and hence food needs of mammals.

It must have always been evident that the bigger animal needed more food, but the exact relationship only began to fascinate scientists after Lavoisier's work led them to describe food needs in terms of energy needs (see Kleiber, 1961).

Interestingly, the first published attempt to grapple with scaling of food needs was due to Jonathan Swift, who faced the problem of the Lilliputians' feeding Gulliver in his splendid satire on Gulliver's Travels (Swift, 1726). In Lilliput, according to Swift:

> His Majesty's Ministers, finding that Gulliver's stature exceeded theirs in the proportion of twelve to one, concluded from the similarity of their bodies that his must contain at least 1728 of theirs, and must be rationed accordingly.

Swift's analysis is a remarkable one – for 1728 is the cube of 12. Thus he is actually arguing from dimensional analysis that Volume (and hence

Weight) varies with the cube of the linear dimension. But, though he has grasped dimensional analysis, he has failed to apply Galileo's Principle of Similitude, and assumed that food needs were simply proportional to mass[1]. The principle of similitude would make it clear that the Lilliputians could not simply be scaled down versions of Gulliver, for, if they were, they were rather 'overdesigned' having, for example, legs that were 12 times too thick. More importantly for Swift's purposes is the nutritional aspect of this. Since heat loss must vary essentially with surface area, that is with the square of the linear dimensions, Lilliputian nutrition had to be incorrect, unless either Lilliputians had much lower body temperatures than Gulliver or were possessed of a different method of losing heat.

In fact, food needs do not increase in proportion to weight, but, as Sarrus & Rameaux (1839) first proposed, increase with a power function of weight. Sarrus & Rameaux argued that heat production, and hence energy intake, had to vary with surface area, because heat is lost through the surface, and heat production must match heat loss. Sarrus & Rameaux argued from first principles that the relationship must be $W^{0.666}$, since surface increases as the square of linear dimensions and mass as their cube. This approximation makes two assumptions, that density does not change with size (so that weight was predicted by volume) and that shape did not alter, so that animals formed a homologous series of geometric shapes. Neither of these assumptions is quite correct. Shape is clearly variable, both for the allometric reasons already noted and for reasons quite apart from the allometric constraints upon it. Variations in body fat (which increases as $0.075W^{1.19}$ in terrestrial mammals, Calder, 1984) and skeletal mass (which has an allometry of $0.061W^{1.08}$ in mammals (Prange, Anderson & Rahn, 1979)) would lead to variations in density. Since fat is less dense than lean tissue, and bone much more dense, animals should become less dense with increasing size, but the allometry would be so close to 1.0 that the approximate equation of volume and weight applies over a wide weight range.

To study empirically the law proposed by Sarrus & Rameaux, it is necessary to define the conditions under which the study is being made, and by the mid-nineteenth century investigations had become concentrated

[1] Interspecific scaling would suggest Gulliver only needed 267 times as much food. But according to Kleiber (1967), Swift may not have been wrong since he seems to have hedged his bets by claiming elsewhere in his book that Gulliver was 26 times taller than the Lilliputians. On this basis Kleiber notes, if Gulliver's food needs were scaled up according to the usual interspecific rules, the ration he was offered was only 13% above predicted so that Swift anticipated Kleiber & Brody by 233 years (Kleiber, 1967).

[2] BMR is more usually called standard metabolic rate by biologists, a procedure which seems to us to substitute one opaque term for another. We will use BMR, which has at least a claim to priority.

upon Basal Metabolic Rate (BMR)[2], the metabolic rate in the post-absorptive state of an inactive, non-stressed animal in a thermally neutral environment. Various nineteenth and early twentieth century physiologists investigated the applicability of the surface rule under these conditions and provided a wealth of data which led to its acceptance in physiology and nutrition books. The first of these was Max Rubner, who derived his Surface Law in a study on dogs ranging in bodyweight from 3 to 31 kg (Rubner, 1883; 1902). The interspecific confirmation of the Surface Law, was made by Voit (1901) and later extended by Benedict (1938).

Nutritional allometry

Despite this distinguished beginning, the science of nutrition does not pay much attention to the need for allometric studies. A major reason for this is that nutritionists tend to study a single species and so do not, in general, face the Lilliputians' problem, since in most animal species the bodysize variation of non-obese adults is very limited.

For men or women, for example, the variation in bodyweight of non-obese adults is two- to threefold, as it is for cats or most domestic animals. For such species the exact allometric considerations used in adjusting nutritional requirements for size are not very important and somewhat difficult to determine empirically. Comparative interspecific studies where bodyweight variation is much wider do require an awareness of the allometric approach (which is probably why most papers on the subject emanate from biology departments). But, for two species, the dog and the horse, the allometric approach is vital: for each of these there is, in contrast with other mammals, a wide range of adult bodyweights. The dog is unique in the extent of this variation, for the range is at least fifty-fold.

The lack of importance of the allometric assumptions in scaling for most species is illustrated in Figure 2 where the values generated by different allometric assumptions are compared. To simplify matters, all values are expressed relative to that generated by an exponent of 0.75. For animals like the cat, which have a normal bodyweight range of 50–150% of the mean, the exact allometric assumption does not matter much. For dogs, on the other hand, with a weight range from at least 5–200% of the mean, the choice of allometric exponent is crucial. The impact of the allometry on the requirement does not need underlining, and, of course, quite where the maximal errors occur depends upon where the reference animal, at which all lines intersect, is positioned: if the reference dog were assumed to be small the maximal errors would be observed in the larger animals, and vice versa.

The extent of this body size variation and its evident impact upon nutrition was presumably one factor which made the dog an attractive test animal, used, for example, by Rubner, Voit & Lusk in their nutritional work, since 100 years ago it was the study of quantitative variation in nutritional

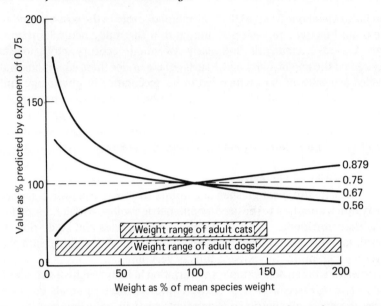

Fig. 2. The greater importance of precise allometric scaling in dogs as opposed to cats. The deviations of using allometric exponents of 0.56, 0.67 and 0.879, from prediction obtained using 0.75 are shown, assuming as a reference value the point where the lines intersect. The relative unimportance of the exact scaling scheme for an animal with a threefold bodyweight range (a generous estimate for the cat) compares with that for a fiftyfold range (a conservative estimate for the dog).

needs that dominated nutrition. Only with the advent of the vitamin theory, and the consequent concentration on qualitative variation, did the laboratory rat become the standard animal.

It is, therefore, particularly ironical that research in pet animal nutrition has largely concentrated upon the qualitative phase. This is perhaps understandable as a result of the many qualitative differences discovered in the nutrition of the cat.

Empirical allometry and allometric theory

Despite the voluminous literature on the surface rule, the norm for twentieth-century nutrition has been to move away from this rule derived from theory, towards a purely empirical allometry, an approach advocated simultaneously by Kleiber (1932) and by Brody & Procter (1932a, b). Both sets of authors determined by regression analysis the best fit solution to Equation 2, and hence, they argued, arriving at an allometry free of potentially misleading *a priori* theoretical considerations.

Both Brody & Kleiber did this by re-analysing BMR data for adult animals of different species, and using this to derive an interspecific allometry of BMR in adults. As is evident from their papers, Kleiber & Brody had a long running disagreement about the precise value for the second decimal place of the exponent. The relationship that has generally become accepted is:

$$BMR = 70W^{0.75} \qquad \text{[Equation 3]}$$

Where BMR is Basal Metabolic Rate in kcal/day, and W is bodyweight in kg. The relationship is illustrated by Figure 3, taken from Kleiber (1961).

This allometry is significantly different from that of 0.67 predicted by the surface rule, (Kleiber, 1961), and the exponent has been empirically confirmed by numerous studies (see, for example, Hemmingsen, 1950, 1960; Henry, 1981, and reviews by Schmidt-Nielsen 1984; Calder 1984, 1987). Interestingly, the exponent has been found to describe the allometry of various other physiological parameters (Table 1), at least some of which might be expected to vary in this way because their magnitude is determined by metabolic rate.

Yet unlike the surface law which it displaces, the allometry of 0.75 has not yet been provided with any agreed theoretical basis, though attempts to provide such have been ingenious. Hemmingsen (1950) who claimed to demonstrate an allometry of metabolic rate related to 0.75 over all the kingdoms of biological creation, concluded that it is a compromise which

Fig. 3. The interspecific allometry of BMR with mature bodyweight as originally described by Kleiber (1932). The straight line relationship between the logarithms indicates an allometric relationship, and the slope of the line, 0.75, is the allometric exponent. The intercept is the logarithm of the allometric co-efficient. *Note:* The diameter of the circles shows a + or − 10% deviation in BMR. (From Kleiber, 1962.)

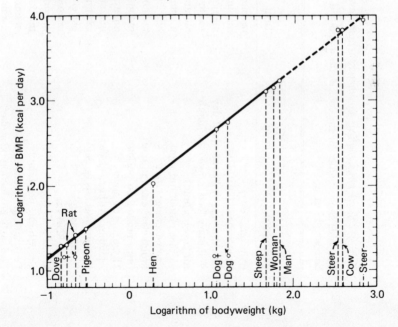

Table 1. *Some physiological parameters which have an interspecific allometry of 0.75*

The table shows values for the coefficients *a* and *b* in the Equation $Y = aW^b$ where Y is the dependent variable and W is bodyweight in kg.

Parameter	Taxa	Allometry observed		Reference
		a	*b*	
BMR (kcal/day)	mammals	70	0.75	1,2
Maintenance (kcal/day)	mammals	107–140	0.75	3
Maximal metabolic rate (kcal/day)	mammals	480	0.73	4
Voluntary food intake (kcal/day)	zoo mammals	146	0.75	5
Air flow (ml/min)	mammals	330	0.74	6
Cardiac output (ml/min)	mammals	205	0.74	7
Cardiac power (watts)	homeotherms	0.038	0.77	7
Inulin clearance (ml/min)	mammals	5.3	0.72	8
Urine production (ml/day)	mammals	60	0.75	8
Thyroxine turnover	mammals	—	0.75	9
Albumin synthesis (g/day)	mammals	0.397	0.76	9
Protein synthesis (g/day)	mammals	18.0	0.75	10
Brain mass (g)	placental mammals	11	0.76	11
Liver RNA (g)	mammals	.256	0.755	9
Wt. of litter (g)	mammals	112	0.767	12
Growth rate (kcal/day)	mammals	9.4	0.72	13
Peak milk production (g/day)	mammals	120	0.73	14
Peak milk production (kcal/day)	mammals	172	0.74	14

Threonine requirement (mg/day)	mammals	19	0.76	9
Methionine requirement (mg/day)	mammals	39	0.78	9
TONL (mg/day)	mammals	250	0.75	15
Nitrogen intake (g/day)	zoo mammals	1.60	0.75	5
Nitrogen excretion (mg/day)	mammals	295	0.74	16
OUNL (mg/day)	mammals	146	0.72	17
TONL (mg/day)	mammals	272	0.75	18
FUNL (mg/day)	mammals	418	0.77	18
Endogenous S losses	mammals	0.0067	0.74	17

Sources:

References:

1: Kleiber, 1932
2: Kleiber, 1961
3: See text for details.
4: Hart, 1971
5: Evans & Miller, 1968
6: Adolph, 1949
7: Gunther, 1975
8: Edward, 1975
9: Munro, 1969
10: Calculated from Waterlow, Garlick & Millward, 1978

11: Martin, 1981
12: Millar, 1981
13: Case, 1978
14: Linzell, 1972
15: Miller & Payne, 1964
16: Stahl, 1962
17: Brody, 1945
18: Henry, 1984

results from the fact that metabolism per unit weight must fall due to the demands of the surface law. McMahon (1973) has developed an ingenious theoretical model based on the principle of elastic similarity, while Smith (1976, 1978) has proposed that the explanation is the energy cost of resisting gravity.

This lack of theory has given rise to the suggestion that no theoretical basis need be sought to what is, after all, merely a prediction equation (Heusner, 1987). We do not accept that empirical predictive allometry is adequate, agreeing with Sir Peter Medawar that 'a science without a theoretical basis is nothing more than a kitchen art' (Medawar, 1967), and with Kleiber (1950) that physiological regression equations must be physiologically meaningful.

A failure to consider theory eventually leads to results like those used by the recent WHO/FAO/UNU consultation on human energy and protein requirements (WHO/FAO/UNU, 1985) which based its requirements on BMR prediction equation presented in a literature review by Schofield, Schofield & James (1985). These authors reviewed all the human BMR data and showed that it could be described at any age by a linear equation of the form

$$\text{BMR} = C + mW \qquad\qquad \text{[Equation 4]}$$

where m and C are constants.

Such an equation is non-allometric because it contains a constant term, C, unrelated to weight. The problem with such equations, as Kleiber (1950) argued, is that they are meaningless: since they suggest that part of the BMR is unrelated to mass, if taken literally they predict a real BMR for a weightless, and therefore non-existent subject. This means that they are limited in their use to the weight range over which they were derived, and are of no use to the comparative nutritionist. Moreover, as McNeill *et al.* (1988) have argued, if they are based on data for the middle of the weight range they will overestimate the BMR at either extreme (Fig. 4).

Some authors, most recently Heusner (1982, 1985, 1987) have dismissed the interspecific allometry as an artefact, arguing that it is a 'grade effect' arising from the fact that, among closely related taxonomic groups, both the coefficent and exponent of the allometric relationship differ from the interspecific line. This phenomenon is illustrated in Figure 5. Grade effects are swamped by the process of regression analysis which ignores the possibility of changes in the coefficient a with bodysize. The most challenging re-analysis along these lines is due to Heusner (1982, 1984, 1985, 1987), who has argued from dimensional analysis that the coefficient a can only be mass independent if the exponent is 0.67, and that the coefficient of this mass-independent metabolism is the variable that must be studied (Figure 5).

Given this, Heusner argues the allometry of BMR should be described by the variation in the coefficient a between different species, while he shows by

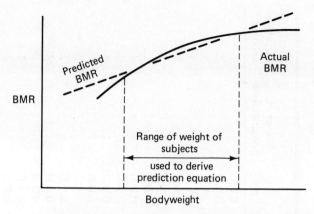

Fig. 4. Use of a linear prediction equation (shown by a discontinuous line) to relate body weight to BMR will result in a tendency for BMR values of both very small and very large animals to be underestimated if the true relationship is allometric with an exponent less than 1.0 (as shown by continuous line). (From McNeill *et al.*, 1987.)

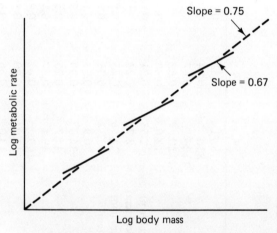

Fig. 5. Grade effects and the potential for errors in allometric interpretation. An interspecific allometric regression could result from the unwarranted pooling of taxa which differ in allometric coefficient and/or exponent. Illustrated here is Heusner's contention that the interspecific exponent for metabolic rate of 0.75 is an artefact resulting from the fact that individual taxa each have an exponent of 0.67 but different allometric coefficients. (From Schmidt-Neilsen, 1984.)

a reanalysis of the published data that this is indeed the allometry of BMR within two species, the dog and rat (Heusner, 1982).

Allometry in the dog

Whilst energy expenditure lends itself to dimensional analysis, most other aspects of anatomy and physiology do not and allometric relationships, must be derived empirically. Relative to other mammal

Table 2. *Some allometric relationships in the dog*

Variation in some physiological and anatomical parameters in the dog. The table shows values for the coefficients a and b in the equation. $Y = aW^b$ where Y is the dependent variable and W is the bodyweight in kg.

Dependent variables	a	b	r	n	p<	Source of original data	Source of calculated allometry
Length (cm)	5.71	0.37	0.97	6	0.01	1	1
Body surface area (cm²)	830	0.74	1.0	8	0.001	1	2
	1120	0.67				3	3
	1120	0.67		6	NS	4	2
Height at shoulder (cm)	12.0	0.50	0.91	69	0.001	5	1
Internal pelvic diameter:lateral (cm)	2.53	0.21	0.92	11	0.001	6	3
Internal pelvic diameter:vertical (cm)	2.75	0.23	0.94	11	0.001	6	3
Distance between ischial tubers (cm)	4.36	0.25	0.91	11	0.001	6	3
Distance between iliac crests (cm)	3.12	0.35	0.93	11	0.001	6	3
Blood weight (g)	70.9	0.93	0.98	9	0.001	1	2
Brain weight (g)	36.0	0.30	0.99	6	0.001	7	3
Brain weight (g)	41.7	0.25	0.57	10		8	2
Brain weight (g)	39.4	0.27	0.95	1037	0.001	9	2
Gut weight (g)	109	0.67	0.95	17	0.001	1	2
Heart weight (g)	11.0	0.90	0.96	36	0.001	1	2
Kidney weight (g)	13.0	0.66	0.88	39	0.001	1	2
Liver weight (g)	68.0	0.69	0.88	39	0.001	1	2
Lung weight (g)	15.9	0.77	0.92	34	0.001	1	2

Spleen weight (g)	2.07	1.06	0.91	38	0.001	1	2
Duration of oestrus cycle (days)	37.4	−0.25	−0.90	7	0.01	10	3
Gestation length (days)	61.3	0.005	0.46	14	NS	11	3
Litter number	2.33	0.32	0.87	51	0.001	12	3
Neonate weight (g)	69.0	0.56	0.97	8	0.001	13,14,15	3
Width of neonate head (cm)	4.05	0.16	0.53	17	0.05	6	3
Width of neonate shoulders (cm)	6.58	0.41	0.78	17	0.001	6	3
$t25$ to $t75$ (days)	93	−0.07	0.34	16	NS	16	3
$t90$ (days)	160	0.18	0.80	16	0.001	16	3
Heart rate (beats/min)	179	−0.19	−0.53	50	0.001	3	3
Maximal lifespan (yr.)	24	0.15	0.92	6	0.01	6	3
Relative incidence of osteosarcoma		2.46	0.84	13	0.001	17	3
Age at onset of osteosarcoma (yr.)	17	0.24	0.94	4	0.01	17	3

Sources:

References:

1: Stewart, 1921a, 1921b
2: Present authors
3: Kirkwood, 1985
4: Lusk, 1928
5: Evans, Curtis & Wilkins, 1982
6: Wright, 1934
7: Comfort, 1960
8: Sacher, 1959
9: Bronson, 1979
10: Sokolowksi & Vanravenswaay, 1983
11: Kryzanowksi, Malinowski & Studnicki, 1975
12: Robinson, 1973
13: Altman & Dittmer, 1982
14: Chakraborty, Stewart & Seager, 1983
15: Harmer, 1971
16: Kirk, 1968
17: Brodey, Sawer & Medway, 1963

Table 3. *The allometry of rates of nutrient in the dog (all metabolic rates and energy intakes are as kcal/day)*

Parameter	a	b	r	n	Data source	Allometry calculated by
BMR	88.7	0.69	0.89	37	1	15
BMR	97.1	0.65		80/1070	2	16
BMR	88.8	0.51		6	3	20
BMR	142	0.61		7	4	17
BMR	111	0.57		26	1,5	17
BMR	82.2	0.69		7	1	20
BMR	82.0	0.55		14	5	20
BMR	86.7	0.67		27	2	20
Total metabolic rate	105.9	0.67		9	2	20
Total metabolic rate	132	0.67			7	18
Total metabolic rate	112	0.65		9	6	20
Voluntary food intake	167	0.73			8	16
VFI, non-protein diet	225	0.64			9	16
Maintenance Energy Intake	99.6	0.879			10	19
Maintenance Energy Intake	121.9	0.83	0.98	15	8,11,12	13
EUNL (mg/day)	178.1	0.81		25	9	16
MFN	61.6	0.75		25	9	16
TEN	240.2	0.80		25	9	16
Creatinine mg/day	6.773	1.11	0.88	11	13	16
Neutral sulphur	5.45	−0.63	0.72	8	13	16

Sources:

1: Kunde & Steinhaus, 1928
2: Not known
3: Steinhaus cited in ref. 1
4: Rubner, 1908
5: Lusk & Dubois, 1927
6: Beer & Hort, 1938
7: Udall *et al.*, 1953
8: Cowgill, 1928
9: Kendall *et al.*, 1982
10: WCPN data

11: Blaza, 1981
12: Wolf *et al.*, 1970
13: Brody, 1945
15: Kirkwood, 1981
16: Abrams, 1977
17: Brody, 1945
18: Present authors
19: NRC, 1985
20: Heusner, 1982

species, there are numerous allometric analyses of anatomy and physiology of the dog, precisely because its fifty-fold bodyweight range makes it a very attractive animal for such studies. In his excellent review, Kirkwood (1985) collated 62 allometric relationships on a range of topics from kidney weight through respiration frequency to gestation time. In Tables 2 and 3, we have drawn heavily on his collation with the addition of some parameters.

The allometries shown are of importance in two ways. First, as Kirkwood has pointed out, various aspects of the allometry of the dog cast light on matters of veterinary importance. For example, Kirkwood draws attention to the allometry of osteosarcoma incidence in the dog, which may reflect the greater forces acting on the bones of large breeds. Again, Kirkwood has pointed out that the allometries of the width of the neonatal head and of the bitch's pelvic canal are such that the two are equal when bitch's weight is about 1.5 kg. Therefore the high incidence of dystocia in toy breeds is the simple outcome of this allometric constraint, which, if it is unavoidable, places a lower size limit on the size of the dog (Kirkwood, 1985).

The second aspect of the allometry of the dog is its basis, for the problem that the dog presents is that, for a variety of organs and physiological parameters, it has, as Kirkwood again pointed out, an allometry which differs significantly from that observed interspecifically (Table 4), although it is not of course known if it is different from other intraspecies allometries. In commenting upon these data it is as well to bear in mind that empirical allometries are only as good as the data on which they are based and it is clear that much of the literature data on dog allometry is poor, often being based upon a very small number of animals of limited size range. The anatomical allometries in particular are dominated by data collected by compilation due to Stewart (1921), excluding the animals which he noted as diseased or immature. Since the size variation that exists among dogs is so great, allometric considerations will dominate our understanding of their biology, and it seems a pity that more data are not collected from post-mortem or clinical material.

This alone, though necessary, would not be sufficient, and it would be preferable if some coherence could be given to the data, lifting it above the category of rules of thumb. Since such a process could have valuable spin-offs for the subject in general, we have attempted a preliminary analysis with the extant data.

This highlights some interesting possibilities. It is clear from Table 4 that, while the allometries of most organ sizes differ in the dog from the interspecific pattern, certain parameters share the interspecific allometry. The most obvious of these are the overall geometry of the dog (height, length and surface area) which is obvious from looking at dogs. A second group are the parameters which relate to the cardiovascular system: blood volume and spleen weight, which are essentially directly proportional to bodyweight in the dog as they are interspecifically. It seems likely that heart rate and heart size also share the interspecific allometry, though it is difficult to be certain: both of them have lower allometric exponents in the dog than interspecifically, and the difference, though non-significant, is on the borderline of statistical significance.

For the cardiovascular system, therefore, there appears to be some strong biological constraint relating to organ size, which does not allow any difference in the dog from the interspecific pattern.

Table 4. *A comparison of some allometries in the dog compared with those observed interspecifically*

Organ/function	Interspecific		Dog		Ratio		p	Reference
	a	b	a	b	a	b		
Length (cm)	7.14	0.31	5.71	0.37	1.25	−0.06	NS	1
Body Surface Area (cm²)	1090	0.67	830	0.76	1.31	−0.07	NS	2
Blood weight (g)		0.99	71	0.93	—	0.06	NS	2
Brain weight (g)	11.6	0.76	41.7	0.25	0.28	0.50	<0.001	2
Gut weight (g)	74.6	0.941	109	0.67	0.68	0.23	<0.001	3
Heart weight (g)	5.88	0.983	11.0	0.90	0.53	0.08	NS	3,4
Kidney weight (g)	9.6	0.84	13.0	0.66	0.74	0.21	<0.001	5
Liver weight (g)	33	0.867	68.0	0.69	0.49	0.18	<0.001	3
Lung weight (g)	11.3	0.986	15.9	0.77	0.71	0.22	<0.001	3
Spleen weight (g)	2.5	1.02	2.1	1.06	1.28	0.04	NS	4
BMR	70	0.75	95	0.655	0.74	0.1	0.01	6
Maintenance	140	0.75	100	0.879	1.4	−0.23		7
Duration of oestrus cycle (days)		0.13	37.4	−0.25	—	0.38		8
Gestation length (days)	63	0.238	61.3	0.005	1.03	0.19		9
Litter number	2.71	0.0	2.33	0.32	1.20	0.32		10
Neonate weight (g)	40	0.94	69.0	0.56	0.66	0.33		11
t90 (days)	440	0.26	160	0.18	2.75	0.08		3

Heart rate (beats/min)	240	−0.25	179	−0.19	1.34	−0.06	3
Maximal lifespan (years)	11.6	0.20	24	0.15	0.48	0.05	12

Notes:

Values of a and b in the equation $Y = aW^b$.

p value indicates the statistical significance of the difference between the allometric exponents, b

Dog allometries from Table 2

Sources of interspecific allometries shown by reference number.

Sources:

References:

1: Economos, 1981
2: Martin, 1981
3: Brody, 1945
4: Stahl, 1985
5: Fujita *et al.*, 1966
6: Kleiber, 1961
7: Kleiber, 1932
8: Kirkwood, 1985
9: Millar, 1981
10: Blueweiss, 1978
11: Cabana *et al.*, 1982
12: Sacher, 1959.

By contrast, a few parameters which show a marked interspecific allometry are essentially size independent in the dog, notably gestation length. Interestingly, Martin & McLarnon (1985) have pointed out that there is, in general, no significant allometry of gestation lengths within closely related taxa, including *Felis* spp. The time taken to grow from 25% to 75% of adult size (t_{25} to t_{75}), is also size independent unlike the interspecific trend, although there is a statistically significant allometry to t_{90}, the age by which 90% of adult size is reached. Lifespan also appears to be size independent in dogs: Bronson (1982) failed to find any relationship and though Comfort (1960) did it was with a very small sample ($n = 6$).

These three parameters are all rates, and in the mammals, in general, rates vary with $W^{-0.25}$, something which has given rise to the notion of physiological time (Schmidt-Neilsen, 1984; Calder, 1985, 1987). It is tempting to explain the lack of allometry in these rates by regarding 'physiological time' as fixed in the dog – perhaps to an ancestral size – and subsequent selection has dissociated bodyweight from this variable.

However, not all rates are constant in the dog, for example, turnover rates of metabolites as measured by energy metabolism or nitrogen metabolism have a positive allometry, though the exponents differ from one another. The variability in allometries for rates makes it clear that no general concept of physiological time can be sustained for a single species.

Superficially, the other anatomical parameters show a diverse pattern of differences from the interspecific pattern, but detailed analysis shows strong similarities can be detected. These are shown in Table 5 where the allometric cancellation technique (ACT) (Stahl, 1965) has been used to reveal inter-organ allometries. The use made of the allometric cancellation technique in Tables 4 and 5 can be illustrated by the following equations:

$$\text{If } A = bW^c$$
$$\text{and } R = sW^d$$
$$\text{then } A/R = b/sW^{(c-d)} \qquad \text{[Equation 5]}$$
$$\text{and since } W = \frac{A^{(1-c)}}{b^{(1-c)}}$$

$$\text{then } A = kR^{(c-d)}$$
$$\text{where } k = s^{(1-d)} \cdot b^{(c-1)} \qquad \text{[Equation 6]}$$

By use of this approach in Table 5, it is clear that, while the allometry of the abdominal organs, liver, kidney, and gut is different from the interspecific allometry for these organs, their *relative* allometries are such that interspecific allometries in the relationships between these organs is preserved in the dog. It is clearly plausible that these relative allometries represent optimal values for these functionally closely interrelated organs.

Finally, there is a residuum of organ allometries which differ from the interspecific pattern with respect to bodyweight and to most other organs.

Table 5. *A comparison of organ size variations obtained by the allometric cancellation technique*

	Brain	Gut	Heart	Kidney	Liver	Lung	Skin	Spleen
Blood	0.43	0.25	0.02	0.09	0.12	0.16	−0.13	−0.10
Brain		−0.18	−0.41	−0.34	−0.31	−0.27	−0.57	−0.53
Gut			−0.19	0.10	−0.09	−0.05	−0.20	−0.31
Heart				0.07	0.08	0.12	0.15	−0.12
Kidney					0.02	0.06	−0.22	0.19
Liver						0.04	−0.24	−0.21
Lung							−0.28	−0.26
Skin								0.03

Notes:

Values given are exponents calculated by the Allometric Cancellation Technique (ACT) for the relative allometries of various organs in the dog. When the exponent is zero the relative sizes of the two organs at any bodysize are the same in the dog as they are interspecifically.

Thus, for example, the first value in the table of 0.43 indicates that, with brain size, the blood volume of the dog rises faster than it does interspecifically, and the disproportion is related to brain weight to the power 0.43.

Notable amongst this group is the brain, which has an allometry in the dog of 0.25, in striking contrast to the interspecific allometry of 0.75.

One reason for this odd allometry of brain size in the dog was proposed by Kirkwood (1985) who noted that post-natal brain growth in the beagle has an essentially similar allometry to the interbreed allometry in dogs, as Figure 6, which is taken from Kirkwood's paper, illustrates. Indeed, when the allometries of post-natal brain growth for any mammal species are compared (Table 5) they cluster close to 0.25, so that inter-breed adult allometries in the dog are those which would result from an extended, or reduced pattern of growth.

The phenomenon is not confined to the brain. Reports from a variety of species for a variety of organs show that the allometry of organ size in post-natal, usually post-weaning, growth in mammals is very similar to that which is observed between breeds of dogs (Table 5). It is worth noting that for many mammal species the allometry of organ size during growth tends to be 0.2 units below the interspecific mean, as is the allometry of organ size in dogs and that no reduction in allometric exponent during growth is observed for those organs for which no reduction in interbreed allometry is observed in the dog. In other words, the allometry of organ size in the dog is explicable if adults from different breeds are treated as growing animals of different sizes. In this context it is worth noting that Brody & Procter (1932) found they could satisfactorily describe the difference in BMR between adult dogs of different size by a non-allometric equation of the same type that in

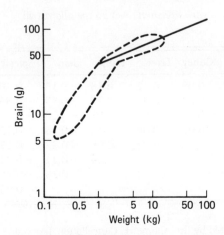

Fig. 6. The interbreed allometry of allometry of brain size in dogs, shown by the solid line, fits well with the postnatal range of brain sizes in growing Beagle pups. (From Kirkwood, 1985.)

other animal species gave good predictions of the changes of BMR during growth.

Besides their intrinsic interest, certain of the unusual allometries found in the dog require comment in a specifically nutritional context. These are brain size and a group of parameters relating to reproduction (litter size, litter number and gestation length), metabolic rate and nitrogen requirements.

Nutritional aspects of the allometry of brain size in the dog

Both the absolute size and the allometry of the brain of the dog are worth noting. Bronson (1979) has studied the brain weight allometry of dogs and cats and pointed out the remarkable fact that, relative to cats, dogs have rather large brains, something which runs contrary to the popular view of the 'intelligence' of the two species. Table 7 shows that, in breeds of comparable size, the dog has a brain that is twice as large as the cat. This is because the size of the cat brain is approximately that which interspecific allometry would suggest, while the allometry of brain size in the dog crosses the interspecific line so that big dogs (> 11.5 kg) have much smaller brains than would be expected while smaller breeds have much bigger ones than expected from interspecific allometries. The fact that dogs have larger brains than cats is thus a consequence of the small size of the cat. Bigger felids have bigger brains than dogs.

The unique allometry of brain size in the dog is of interest in the context of Martin's (1981) suggestion of a nutritional basis for the interspecific allometry of brain size. Martin has found that, for eutherian mammals, non-primates and primates, there is an allometry of brain size of 0.75, though the

Table 6. A comparison of the allometric exponents describing breed variation in the dog with those observed in growing animals

| | Mature | | Immature | | | | | |
| | | | Exponent observed | | | | | |
Organ	Mammals	Dogs	Cats	Humans	Rats	Horses	Cattle	Chickens
Adrenals	1.78	0.78	0.64		0.83	0.71		0.84
Blood	0.99	0.93			0.98		0.93	0.99
Brain	0.76	0.27		0.43	0.24	0.30	0.39	
Gut	0.94	0.67			1.26			
Heart	0.98	0.90	0.87	0.93	0.91	0.93		
Kidney	0.85	0.66	0.63	0.79	0.66	0.51		
Liver	0.87	0.69		0.82	0.61	0.70	0.67	
Lungs	0.99	0.77			0.75	0.58		0.87
BMR	0.75	0.66		0.56	0.56			

Sources:
Mature mammals and mature dogs: Table 4.
Growing animals: Brody, 1945 and authors' own data.

allometric coefficient differs between these taxa. Martin argues that the 0.75 relationship reflects the fact that brain weight varies directly and causally with the mother's BMR, because the metabolic capacity of the mother determines the rate at which the brain can grow in utero and during lactation. By contrast, for egg laying taxa, the birds, reptiles and monotremes, Martin has shown that the allometry of brain size has an exponent of 0.56, which he suggests is related to the fact that the BMR of the mother determines the egg size and this in turn determines the brain size: for 0.56 is 0.75 squared.

Since the brain is the major obligatory net consumer of glucose, it is possible that the change in relative brain size with body size has implications for glucose economy, and hence amino acid catabolism in the fasting dog. In a short-term fast, once glycogen stores are depleted, glucose is provided for the brain by gluconeogenisis mainly from amino acids liberated by protein catabolism. Thus, it might be expected that, with decreasing size, the demand of the brain for glucose would make protein loss in fasting greater. But we have found no information on this point in the literature.

Nutritional aspects of the allometry of reproduction in the dog

We now turn to an apparently completely different area of allometry, the allometry of reproduction in the dog as compared with other mammals. There have been various studies scaling aspects of reproductive function in mammals and fewer, though quite reliable studies have examined the same measures in dogs. The differences in allometry were pointed out recently by Kirkwood (1985) in his excellent review, to which reference has already been made. Table 4 shows that, in most mammals, the weight of the newborn animal varies closely with the weight of the adult, W, with an allometry of 0.9, while the gestation time, like many other physiological times, varies with $W^{0.25}$. The number of pups in a litter is more or less independent of size, that is to say, it is a species-specific characteristic.

Completely different allometries apply in dogs. Litter size is a function of bodyweight in the dog, varying as $W^{0.32}$. Gestation time as noted above, is constant over the whole weight range, while the weight of the newborn pup varies as adult body weight, $W^{0.56}$. At first, these differences, though quite spectacular demonstrations of the distorting effects of artificial selection in the dog, appear to vary in a meaningless manner. However, in our view the very variability is indicating something very important about the way in which nutritional factors constrain reproductive activities.

Martin (1985) has argued that, in closely related taxonomic groups of mammals, where gestation time is usually constant, there are good theoretical grounds for believing that the weight of the newborn animal should vary with $W_m^{0.50}$, with which the observed allometry of 0.56 is in good agreement. A similar allometry is found in other taxa.

The nutritional load imposed by a pregnancy is obviously related to the

Table 7. *Relative brain sizes of dogs and cats (data of Bronson 1979)*

	Bodyweight (kg)		Brain weight, (g)		Number of animals
	Mean	s.d.	Mean	s.d.	
Domestic cats	3.7	1.3	29.6	5.5	571
Siamese cats	3.2	0.9	26.1	3.1	126
Chihuahua	2.6	1.2	53.4	8.1	19
Toy poodle	3.2	1.8	59.1	12.1	13
Toy fox terrier	3.4	0.9	52.3	3.6	6
Pekingese	4.9	1.3	53.4	4.9	8
All dogs	19.9	5.7	85.2	11.0	1,037

weight of the products of conception. If it is assumed, as seems to be the case, that the foetal support system increases in mass in proportion to the size of the foetuses in comparative studies, then the weight of the products of conception will be proportional to the number of foetuses born multiplied by their weight. (Payne & Wheeler, 1967a, 1967b, 1968).

Payne & Wheeler argue that the incremental nutritional load imposed by the pregnancy will be proportional to the production divided by the gestation time. So, from the earlier estimates of allometric exponents for each of the separate aspects of pregnancy, we can estimate the allometry of the total metabolic impact of the pregnancy. As Table 8 shows, interspecifically this is very close to $W^{0.75}$. In other words, the metabolic load imposed by pregnancy represents a constant fraction of the normal voluntary food intake of the mother. Thus pregnancy imposes no greater load for large than small animals.

Despite the different allometries of each component in the dog, the same is true of the incremental load imposed, which varies with $W^{0.875}$, almost exactly as voluntary food intake in the maintaining animal varies with bodyweight (see below).

Thus, it is probable that, given that gestation time is a species characteristic, litter size and number have to vary in order that the load imposed does not exceed some nutritional limit. Although, whether this is energy intake or some more innate metabolic parameter, like protein turnover, we cannot say. Nevertheless we suggest that the allometric generalisation provides an *a priori* case for more research.

The allometry of energy requirements of the dog

Attempts to define the allometry of energy requirements of the dog are still largely empirical. The new NRC report (NRC, 1985) presents data on the observed intakes of maintaining adult dogs (Fig. 7) and fits to it three distinct scaling schemes for maintenance requirements of adult dogs.

Table 8. *Comparative impact of pregnancy: dogs and interspecifically*

$$\text{Metabolic load of pregnancy} = \frac{\text{litter no.} \times \text{litter wt.}}{\text{gestation time}}$$

$$\text{Relative impact of pregnancy} = \frac{\text{metabolic load}}{\text{maintenance}}$$

$$\text{Interspecifically MLP} = \frac{2.71 \times 40 \times W^{0.94}}{6.3 \times W^{0.24}} = 1.72\ W^{0.70}$$

$$\text{RIP} = \frac{1.72 \times W^{0.70}}{140 \times W^{0.75}} = 0.012\ W^{-0.05}$$

$$\text{For dogs MLP} = \frac{2.33 \times W^{0.32} \times 69 \times W^{0.56}}{61.3 \times W^{0.005}} = 2.62\ W^{0.875}$$

$$\text{RIP} = \frac{2.62 \times W^{0.875}}{99.5 \times W^{0.879}} = 0.026\ W^{-0.004}$$

Maintenance (kcal/day) = $132W^{0.75}$	[Equation 7]
Maintenance (kcal/day) = $99.56W^{0.879}$	[Equation 8]
Maintenance (kcal/day) = $144 + 62W$	[Equation 9]

Equation 7 is the allometry used in the previous report (NRC, 1974) which the 1985 NRC committee suggested should be abandoned in favour of either Equation 8 or 9 for the eminently practical reason that the new NRC prediction equations describe more precisely the data used to derive them, and the difference is significant. These data were the intakes of 55 animals of 7 breeds with a bodyweight range from 4 – 36 kg. But they are purely empirical, having no theoretical underpinning in nutritional theory, however rudimentary.

When they are considered as descriptive rather than purely predictive equations neither is satisfactory. The linear Equation 9 can be rejected on the grounds already discussed above, that it is non-physiological, while Equation 8 not only suggests an allometry for maintenance that is not only unique to the dog but also a strange relationship between maintenance and BMR.

Maintenance and BMR
Maintenance is a rather poorly defined state (Abrams, 1977; Milligan & Summers, 1986; Coyer, Rivers & Millward 1987). Maintenance intakes are those intakes required for maintaining a constant body state, with body nitrogen, body energy or bodyweight all being used as criteria for this constancy. For adults, although not for young animals (Coyer, Rivers & Millward 1987), it is probable that all three balances are achieved by the same level of intake, and that any criterion can be used as a marker of maintenance.

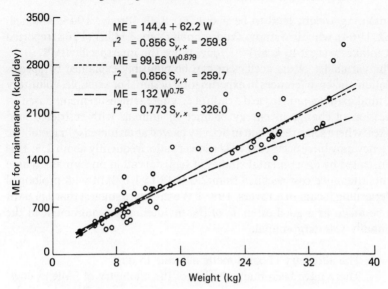

Fig. 7. The voluntary energy intake of adult dogs, used as a measure of the maintenance energy intake by NRC (1985).

Maintenance requirements are derived in two ways, either experimentally by regressing one of the above markers of body state (e.g. energy balance) on to intake and determining, by interpolation or extrapolation, the intake for zero change, (see, for example, Coyer, Rivers & Millward 1987) or observationally by noting the Voluntary Food Intake (VFI) of animals that are maintaining body state (Mitchell, 1964). The maintenance data analysed by the NRC are of the latter kind.

If we consider the intake of the maintaining animal factorially, it is clear that energy is required to match basal expenditure to convert, digest and process food, to meet thermogenesis occasioned by food consumption and to meet expenditure occasioned by activity. There is no guarantee that these avenues of expenditure will be the same under the two types of maintenance conditions, and indeed there is every reason to think that some will not.

There is abundant evidence to suggest that maintenance requirements however defined are closely related to BMR. Brody (1945) and Kleiber (1961) both argued that maintenance varied allometrically as BMR, but neither produced a quantitative relationship. Miller & Payne (1961) and Waterlow & Payne (1975) have suggested that the maintenance intakes vary allometrically as BMR, that is with $W^{0.75}$ and that a mean value is 1.5 BMR, i.e. 105 kcal/$W^{0.75}$/day. Blaxter (1972) has noted that maintenance and BMR have the same allometric exponent of 0.75 with a ratio of coefficients of 1.3–1.6. In laboratory and domestic animals VFI of adults maintaining weight tend to be about 2.0 BMR (Brody, 1945, Mitchell, 1962, 1964), while in a study of zoo animals Evans & Miller (1968) reported that intakes were 140 kcal/$W^{0.75}$/day, twice the interspecific BMR.

The variability of the coefficient can be plausibly explained by species variation and by differences in experimental design. For example, voluntary food intake studies tend to yield a ratio of 2; whereas measurements based on regression of change in energy balance in animals with restricted food intakes (where some reduction in activity as well as an increase in metabolic efficiency may have occurred) yield a lower ratio, frequently about 1.5. This is illustrated by experimental studies of semistarvation on humans where the maintenance cost declined from 2.0 BMR to 1.6 BMR with prolonged underfeeding (Seaman & Rivers, 1988). We therefore suggest that 1.5 BMR can be taken as a good estimate of the maintenance requirement in the essentially *sedentary* animal.

The allometry of basal metabolic rate in dogs

There have been many studies of the allometry of BMR in dogs, classic data which go back into the nineteenth century. There have also been three important collations of various items of this data, by Brody, Abrams & Heusner.

Brody, in 1932, performed an analysis on the classical data of Rubner, Krause's compilation of Fasting Metabolic Rates (not resting), and on Kunde & Steinhaus's studies of BMR in the dog. As was his wont he all but buried the data in the Bulletin of the Minnesota Agricultural Station (Brody & Proctor, 1932*b*), presenting his analysis as a graph that is almost indecipherable. Figure 8 is a redrawn version of his diagram showing the two curves that Brody fitted to the data – one describing Rubner's data which were probably not strictly basal, and one the other measurements of Kunde, Kunde & Steinhaus (1924) and Lusk & Du Bois (1924). Idiosyncratically, Brody chose non-allometric formulae to fit these data, but in his later magnum opus (Brody, 1945) presents the allometric formulae which were, for the basal animals (of weight range 9 to 30 kg):

$$\text{BMR (kcal/day)} = 111 \ W^{0.567} \qquad \text{[Equation 10]}$$

and for the Rubner's data of animals in which basal conditions were not enforced:

$$\text{Heat Production (kcal/day)} = 142 \ W^{0.612} \qquad \text{[Equation 11]}$$

Abrams (1977) reported the analysis of 1070 measurements relating to a total of 80 dogs, of weight range 3.4 to 31 kg, and obtained the allometry:

$$\text{BMR (kcal/day)} = 97 \ W^{0.655} \qquad \text{[Equation 12]}$$

He also reported a slight sex difference in the coefficient with bitches 2.6% lower than male dogs. Though Abrams is very critical of other workers, it is

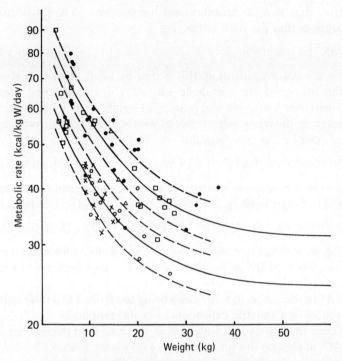

Fig. 8. Brody's analysis of data on the metabolic rate of mature dogs of different weights. The upper curve is $BMR/W = 68e^{-0.092W} + 31$, and is fitted to the data of Rubner, 1883; Kraus 1928 and Kunde & Steinhaus, 1926. According to Brody in these observations the animals may not always have been strictly at rest. The lower curve ($BMR/W = 47e^{-0.099W} + 24.5$) is fitted to the data of Boothby and Lusk and Du Bois whose animals were basal. An allometric fit to these data gives $BMR = 95.7W^{0.63}$. The dashed lines show a $\pm 10\%$ range from the regressions. Data sources:

□ Rubner, (1883)   ○ Kunde & Steinhouse (1926)
△ Boothby (from Krans, 1928)   ● Krauss, (1928).
× Dubois & Lusk (1924)

unfortunate that he neither makes clear what were his criteria for including animals, nor the extent to which his result replicates others, since he uses the data collected by other authors.

Heusner (1982) has also undertaken a recalculation of the allometry of BMR in dogs using published values. Heusner obtains essentially similar allometries for the data of Kunde, Kunde & Steinhaus (1926) and Lusk, as well as presenting a best fit equation for the measurement of Beer & Hort (1938), for which the best fit line is:

$$BMR \text{ (kcal/day)} = 88.8^{0.51} \qquad \text{[Equation 13]}$$

Heusner argues on the basis of theoretical arguments that the allometry of BMR should be $W^{0.67}$, and fits his data to this exponent, obtaining a coefficient (a) that is 86.7 kcal/day and independent of weight. Thus Heusner suggests that the BMR of the dog is:

$$\text{BMR (kcal/day)} = 86.7 \; W^{0.67} \qquad \text{[Equation 14]}$$

One other study deserves mention, this is that by Udall, Rankin & Moss (1953) who measured the metabolic rate of 160 anaethetised post-absorptive dogs over 1 year-old and ranging in weight from 2.5 to 30 kgs. The authors claim the relationship between weight and metabolic rate was linear, they describe it by the formula:

$$\text{Metabolic Rate (kcal/day)} = 15 \; W - 1.70 \qquad \text{[Equation 15]}$$

If the solution for this equation is used to predict the requirement every 5 kg, over the weight range 5–30 kg, the result gives an allometric relationship

$$\text{Metabolic Rate (kcal/day)} = 132 \; W^{0.678}, \; r = 0.99 \qquad \text{[Equation 16]}$$

In Table 3, as well as these allometries, we present those from as many individual studies reported in the literature as we have been able to re-examine. As individual studies they are each limited by the number of animals and the size range, in some cases being too limited to enable quite large differences in allometric exponents to be differentiated.

But it is clear that the overall pattern from these, as from the reviews, is that the BMR of the dog has an allometry of no higher than 0.63 – 0.67 and perhaps lower. For the rest of this paper we will use Abrams' values of 0.655 for the exponent and 97 for the coefficient. It is clear that, though the precise allometry of BMR in the dog has yet to be established, nevertheless the allometry has a significantly lower exponent, and a significantly higher coefficient, than the interspecies value.

It is interesting that a possible metabolic basis for this lower allometric exponent may be found in the allometry of the brain and abdominal organs, which are metabolically the most active ones. In Table 9, we have used the data of Krebs (1950), Field *et al.* (1939) and Martin & Fuhrmann (1955) to calculate the metabolic rate for abdominal organs in the dog, and hence the overall allometry of oxygen consumption due to these organs. While these *in vitro* figures are potentially subject to considerable error, it is nevertheless encouraging that the overall allometric coefficient is 0.69, so that, with increasing bodysize, there is no significant shift in the fraction of energy expenditure due to abdominal organs from a value of 32%, which is comparable to the value for the rat shown in the table although much lower than is observed for man.

In the context of Heusner's theory, the close agreement of the BMR of the dog to $W^{0.67}$ suggests that the species as a whole has a constant mass-independent metabolism, and, in this regard, the dog is like the cat and the rat (Heusner, 1982) and the cow (Agricultural Research Council, 1980).

In practical terms, the differences between it and the interspecific

Table 9. The relationship between Organ Metabolic Rate (OMR) and Basal Metabolic Rate (BMR) in dogs of different sizes, rats and humans.

Species Bodyweight kg Organ	Dog 15.6 OMR per g	Dog 15.6 Organ wt	Dog 15.6 Total OMR	Dog 5.0	Dog 15.6	Dog 50.0	Dog 100	Rat 0.15	Human 70
				OMR:BMR					
Brain	2.12	83.3	177	5.4	3.4	1.9	1.6	3.8	19
Gut	0.68	675.1	452	8.5	8.9	9.8	10.0	7.3	15
Heart	1.06	130.1	140	2.1	2.7	3.8	4.1	1.2	7
Kidney	2.70	76.6	207	4.1	4.1	3.9	3.8	5.2	10
Liver	1.17	446.7	523	10.0	10.3	10.4	10.4	14.9	27
Lung	0.49	135.4	66	1.2	1.3	1.5	1.5	1.0	4
Spleen	0.66	39.6	26	0.3	0.5	1.0	1.1	0.5	
OMR % BMR				31.6	31.2	32.3	32.5	33.9	82
BMR			5091	2416	5091	10,918	17,192	110.1	15,625

Notes:
Organ Metabolic Rates (OMR) and Basal Metabolic Rates (BMR) are as ml oxygen per hour.
Sources: see text

regression is that the two predictive equations reverse in direction at 30 kg – in small dogs the interspecific equation underestimates BMR by as much as 20% at 3 kg, while it progressively overestimates above 30 kg, though not by much: by 60 kg, for example, Equation 3 overestimates the BMR of dogs by only 8% as Figure 9 shows. If the true allometry of BMR in dogs is to be clarified, it obviously calls for studies in miniature breeds, not large ones.

BMR and the allometry of maintenance in the dog

The fact that the bodyweight exponent for the variation of BMR is about 0.67 is what generates the problem in the interpretation of the maintenance needs of dogs. For, if the allometry of maintenance is correct, then there is a weight-dependent constant of proportionality between Maintenance and BMR in the dog. Comparison of Equations 8 and 12 suggests there should be an allometry of the maintenance : BMR ratio

$$\text{Maintenance} : \text{BMR} = 1.0W^{0.224} \qquad \text{[Equation 17]}$$

Figure 9 shows the BMR allometry for the dog using Abrams' value, which derived from animals of the same bodyweight range as the Maintenance intakes. In small animals, the Maintenance : BMR ratio approaches unity. At 4 kg it is only about 1.1 : 1, while at 40 kg it is 2.3 : 1, and at 60 kg it is 2.6 : 1. There is no compelling physiological reason why this should be so; indeed, were the explanation to be purely physiological, the variability of the ratio would indicate that the net efficiency of utilisation of energy for maintenance approached 100% in small animals and declined with size, to 40% at 60 kg.

Fig. 9. A comparison of the allometry of BMR in the dog with the allometry of voluntary food intake at maintenance. BMR is predicted from Equation 12, the best fit lines used to describe VFI and the data points are from NRC (1985).

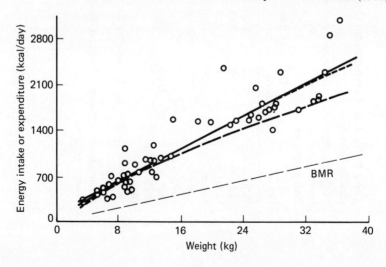

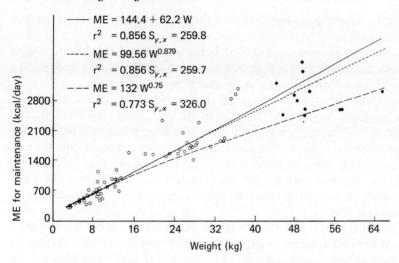

Fig. 10. The Voluntary Food Intake of maintaining dogs. The NRC (1985) data and regression lines (as in Fig. 7) with the addition of WCPN data for large breeds. The regression lines are those fitted by NRC, and do not accurately describe the new data.

There is a further problem evident from the data points plotted by the committee, the data are extremely variable with confidence limits for Equation 8 being 500 kcal per day. What this means, as Figure 9 also shows, is that the range of intakes at which Maintenance can be achieved is very variable at any weight. The range of intakes among the ten animals weighing approximately 10 kg, for example, is nearly threefold from 450 to 1120 kcal per day, among the six largest animals there is still a 1.5-fold variation.

To put it another way the Coefficient of Variation (CV) of the maintenance requirement per unit of $W^{0.879}$ is about 15%, while the CV of the BMR data (as estimated from the residual error of Abrams' allometric analysis) is only about 8%. Maintenance then is twice as variable as BMR. Again there is no compelling reason why this should be so.

The NRC Committee (NRC, 1985) noted that their data only extend over a limited weight range and it may be thought that adding data for heavy dogs might clarify the allometry. It does not, however, do so. We have added some such data on large animals from the Waltham Centre for Pet Nutrition (WCPN) collection to the NRC graph, and, as Figure 10 suggests, they are far from simply providing a resolution to the problem. The allometry is not now significantly reduced from 0.879. But the data do not suggest that a straight line relationship necessarily exists: it is possible, as we show, to draw a fairly constant plateau, parallel with the X-axis through these data for heavy dogs, indicating, in other words, no detectable bodyweight trend of maintenance over the weight range 40–65 kg. And there is no indication of any reduction in variability.

Clearly more research is needed, but it is important that it is not simple allometric number collecting, but that the questions to be asked reflect a sound theoretical approach to the problem.

We believe that part of the problem of variability lies with the definition of maintenance noted above. The data in the NRC report are VFIs of animals that are maintaining weight, and it would seem reasonable to expect a dog on a restricted intake to have a lower maintenance as the interspecific data suggest. This seems to occur: Cowgill (1928) reported that, by restricted feeding, he was able to get maintenance 20–25% below VFI for a maintaining animal, though he does not publish his data except as an illegible graph.

As noted above, when food intake is restricted to achieve maintenance, there are two avenues of expenditure that may be reduced: facultative dietary induced thermogenesis and activity. We are not in a position to comment on the quantitative importance of facultative dietary induced thermogenesis in the dog on a normal diet, beyond noting that a recent study in Beagles failed to demonstrate it although cold-induced thermogenesis was detected (Crist & Romsos, 1987).

It does seem to us that the far less controversial subject of activity could provide valuable insights into the apparent anomalies of maintenance in the dog. The possible importance of this in the maintenance costs of the dog was noted first by Benedict (1938), who felt that the spontaneous activity of the dog, particularly after a good meal, made it an unsuitable animal for metabolic experiments. If activity accounts for the higher maintenance costs on the dog consuming at VFI, it might be expected that maintenance costs on VFI might be more variable, which Cowgill's (1928) results certainly support.

There is some evidence to suggest that in the absence of activity the allometric exponent for maintenance is the same as that for BMR. In Cowgill's (1928) study, he measured VFI in 15 dogs which were not active, and had a weight range of 3.4 to 15.4 kg. The allometry of VFI was:

$$VFI = 149.5 \ W^{0.69} \qquad \text{[Equation 18]}$$
$$(r = 0.92, \ n = 15)$$

In a non-protein feeding experiment where dogs were in metabolism cages so that activity was severely restricted, Kendall *et al.* (1982) had eight animals that maintained weight to within 0.5 kg. For these animals the allometry of digestible energy (DE) intake paralleled Cowgill's

$$\text{Intake (kcal/day)} = 225 \ W^{0.64} \qquad \text{[Equation 19]}$$
$$(r = 0.808)$$

The exponent is not significantly different to Cowgill's, nor from the weight exponent of BMR. The higher coefficient may have been a result of the non-protein diet on which metabolic rate is increased in many animals (Coyer, Rivers & Milward, 1987).

The maintenance requirement of a non-active animal may therefore appear to be a multiple of BMR, as is the interspecific rule and the different allometry for VFI would therefore result from the importance of activity in the dog and its distinct allometry. We have attempted to investigate this possibility by estimating the incremental costs of activity. This, under the conditions we are concerned with, consists of standing, horizontal motion – walking and running, the cost of which has been extensively analysed, and jumping, the energetics of which have been less extensively studied.

The energy cost of standing

Standing involves supporting the body against gravity so that theoretical calculations would suggest that it should vary directly with bodyweight (Schmidt-Neilsen, 1984; Calder, 1984). The energy cost of standing has been studied in farm animals, and humans. In cattle and sheep the best estimate of the incremental cost (i.e. the increase of metabolic rate over the value for lying down) appears to be 0.1 kcal/W/h (Agricultural Research Council, 1980). Reported values for humans are somewhat higher, for example, the FAO/WHO (1973) requirement estimates suggest a value of up to 0.62 kcal/kg/h, but this excessive value may be due to bipedalism. We have assumed the value of 0.1 kcal/W/h for dogs.

The amount of time spent supporting the bodyweight varies between species. Sheep in calorimeters tend to spend 4 hours per day reclining, cattle 8 hours (Blaxter, 1967). Observation suggests that it is unlikely that the time spent standing by the dog exceeds 8 hours. On this basis, we estimate the Incremental Cost of Standing (ICS) for the dog as

$$\text{ICS (kcal/day)} = 0.8W \qquad \text{[Equation 20]}$$

Equation 20 suggests that standing is a very minor avenue of energy expenditure in the dog; for a 10 kg dog it is 8 kcal per day, 1% of VFI at maintenance, for a 70 kg dog it would still be under 2% of maintenance cost. The cost of standing is thus sufficiently low that any allometry or breed variation in the propensity to stand is negligible.

The energy cost of horizontal motion

For a given animal, the extra energy expended in horizontal motion is clearly the product of the level of activity (the distance covered) and the energy cost of covering unit distance.

Although the incremental energy expenditure (over the cost of standing) per unit time increases with speed, it does so in a way that the incremental energy cost per unit distance is essentially independent of speed, so that for a given animal we can think solely in terms of distance covered in calculating energy costs (Schmidt-Nielsen, 1984). Figure 11a illustrates this principle from comparative data and shows that the dog fits well on the interspecific line.

The cost of covering unit distance depends upon two components: first, the load carried, which is bodyweight, and second, the cost of covering unit distance, which, it can be shown on theoretical grounds, increases with the number of steps taken, i.e. varies inversely with the length of the stride (Schmidt-Nielsen, 1984).

Now, in equally shaped animals, stride length will be in proportion to linear dimensions, and, since weight will rise as the cube of the linear dimensions, then stride length will rise with $W^{0.33}$. Thus the cost of covering unit distance should vary as $W^{1.0} \times W^{-0.33}$, i.e. $W^{0.67}$.

The line in Figure 11b fits well with this prediction having a slope of 0.4, so the cost per unit distance would be related to $W^{0.60}$; the coefficent (as kcals estimated from an energy equivalence of oxygen of 4.8) was 1.8. Later and more extensive interspecific studies yield exponents closer to 0.33, so that the cost per km is related to $W^{0.67}$ (see Schmidt-Nielsen, 1984 for review).

An unpublished re-analysis by Blaza & Kendall at WCPN of Slowtzoff's (1903) data on the incremental costs of running (ICR) of seven dogs (weight range 5 – 37 kg) gave:

$$ICR \; (kcal/horizontal \; km) = 4.5 \; W^{0.612} \qquad [Equation \; 21]$$

Thus this allometric exponent, is virtually the same as that obtained by Taylor interspecifically, and not significantly different to 0.67, although the reason for the higher coefficient is unclear. Nevertheless, Blaza & Kendall's reanalysis of Slowtzoff's results yields an answer that is also theoretically reasonable.

Potentially, the impact of horizontal motion is quite great. For example, using Blaza & Kendall's estimates, every km covered would increase daily energy expenditure by 4–5% of BMR. Its impact on maintenance costs would be weight related but considerable. For each km, a 10 kg dog would expend 18 kcal, 2.5% of VFI at maintenance intake, a 4 kg dog would expend 3.1%, and a 40 kg dog would expend 1.7%, of maintenance per km covered.

The amount of exercise undertaken is more difficult to scale. Common experience suggests that it would be subject to wide individual and possibly breed variation, but that big animals walk further than small ones. As a first approximation we have accepted Brody's (1945) suggestion that the amount of voluntary activity varies allometrically with BMR as do many other frequencies.

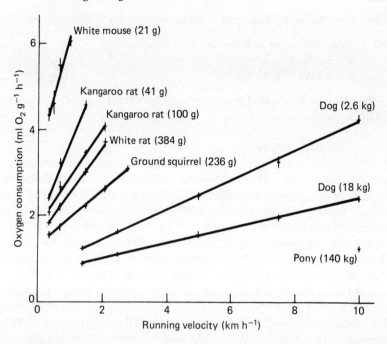

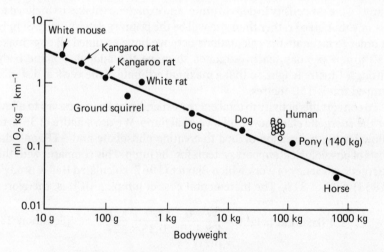

Fig. 11(a). The energy cost of movement at different speeds. The metabolic rate of various animals rises linearly with speed, thus the incremental cost per unit distance travelled remains constant. This generality fits dogs of the two sizes that have been studies. (b). The incremental cost of running. The cost of running expressed as the amount of oxygen needed to transport 1 kg over 1 km declines regularly with increasing weight. The line, which is fitted to the data for quadrupeds, has a slope of -0.4, and the results for dogs fit on the regression line. (From Schmidt-Nielsen, 1980.)

Thus, the total incremental cost of horizontal motion (ICHM) is given by:

ICHM = distance covered × cost per unit distance

$\qquad$ = 4.5 × $W^{0.613}$ × k × $W^{0.655}$

$\qquad$ = 4.5 $kW^{1.268}$ kcal/day $\hspace{3cm}$ [Equation 22]

Where k is the constant of proportionality between voluntary activity and BMR in the dog.

Vertical activity

Some breeds of dogs jump a lot. Apart from breed variation it is obvious that there will be allometric variations of costs of doing this. Modelling this, though superficially daunting, is rather simple. The jumps we are dealing with are standing jumps and, remarkably, the height of a standing jump is almost independent of size. As Professor Schmidt-Nielsen (1984) has pointed out, a man can jump to a height of about 1.6 m, and a flea to a height of 60 cm. Even a galago monkey, which is well adapted to vertical jumps, can only reach a height of about 2 m.

Most of the cost of the jump is the cost of lifting up the mass of the animal: for a creature the size of a dog, air resistance is negligible. The work done is the product of the weight lifted with the height of the jump (divided by 427 to convert it to kcal from kg). We have located no information on the allometry, if any, of the propensity in dogs to jump, and experience makes us inclined to assume that breed rather than size will be the primary determinant. Simply in order to facilitate the calculations here we have assumed all dogs make 150 jumps per day, each estimated to move the midpoint of the body through 1 metre height, so that a maximal estimate of the vertical distance jumped totals 150 metres.

To convert this activity to total energy expenditure it is necessary to allow for the energetic efficiency, E, of a vertical jump. We use a figure of 35%, to allow for the conversion of food to creatine phosphate and ATP, and the costs of physiological support systems for the jump. This compares with the energetic efficiency of work which Blaxter (1967) calculated Hall & Brody's (1933) data of 31%. The incremental cost of jumping (ICJ) is, therefore:

$$\text{ICJ (kcal/km height)} = \frac{W \times H}{E \times 427} = \frac{W \times 1000}{0.35 \times 427} \hspace{1.5cm} \text{[Equation 23]}$$
$$= 6.7W$$

This can be compared with the value for the cost of vertical movement (which should be the same as ICJ) obtained from Slowtzoff's figures, from which it may be calculated that the energy cost of vertical movement is 6.8 kcal/km/kg.W, a figure derived from the somewhat less energetic pastime of running up a slope (Blaza & Kendall, unpublished), to the value given by Blaxter (1967) from an unknown source of 7.26 for dogs, or the interspecific mean of 6.5 (Schmidt-Nielsen, 1984).

Table 10. *Estimates of the energy cost of maintenance by factorial calculations*

Weight	Predicted by		
	NRC [Equation (8)]	Equation 25	Equation 25 as % NRC
kg	kcal/day		%
5	410	491	120
10	754	818	108
30	1,979	1,916	97
50	3,101	2,910	94
70	4,168	3,868	93
100	5,703	5,277	93

It is worth noting that the amount of energy expended in jumping is quite small. Using the jump parameters we have assigned, a 10 kg dog would expend less than 10 kcal on jumping, about 1.5% of the energy expenditure, so that quite large errors on the estimated values assumed would have minimal effects on the calculation of total expenditure.

Thus the incremental cost of voluntary activity (ICVA) in the maintaining adult dog should be the sum of Equations 20, 22 and 23:

$$ICVA \ (kcal/day) = 4.5k \cdot W^{1.268} + 7.5 \ W \qquad \text{[Equation 24]}$$

Assuming a minimal energy cost of maintenance of 1.5BMR, and that BMR varies according to Equation 12, the VFI of a maintaining dog will be:

$$VFI = 146 \ W^{0.655} + 4.5k \cdot W^{1.268} + 7.5W \qquad \text{[Equation 25]}$$

It still has one unknown, that is k, the constant of proportionality of horizontal movement with BMR in the dog. We can estimate this, however, by using the fact that, at least over the weight range 4 – 36 kg, the best fit allometric equation for describing maintenance intakes of dogs is Equation 8, $99.56W^{0.879}$. Equating this with the expression derived factorially, the value of k can be found. It is clear that k cannot be quite weight independent, not surprisingly, given the somewhat arbitrary nature of the model, but, if the value $k = 1.0$, which fits the 30 kg animal, is used, estimates of maintenance intakes that do not differ greatly from the NRC prediction are obtained, differing as Table 10 shows, from 108% NRC at 10 kg bodyweight to a fairly stable 93% of NRC in dogs above 50 kg. The value of $k = 1.0$ predicts a voluntary activity for the 30 kg dog of 9 km/day, a modest 6.3 metres/minutes, or if activity is confined to 12 hours per day, 0.75 km/hour.

It is obviously possible to improve the fit at any range by slightly altering the assumptions made about constants of proportionality. For all that the model is not to be dismissed as a set of fantasies. It now avoids Medawar's condemnation of empiricism and, though we have inserted arbitrary best guesses of various constants to test the model, we have a theory of dog requirements that can be tested. Moreover, the model not only resolves the apparent paradox of differences in allometries of BMR and maintenance in the dog, but allows a simple explanation of the different variabilities of these two measures and of observed interindividual and possible interbreed variations in requirements, which can be seen as a result of different activity patterns.

The allometry of protein requirements in the maintaining dog

Protein requirements in the maintaining animal are required for the replacement of loss of body protein in the obligatory nitrogen losses and the efficiency with which dietary protein is incorporated into body protein, that is the Net Protein Utilisation (NPU) of the protein. NPU is in part dependent upon the digestibility and amino acid content of the protein, and the protein energy ratio of the diet, and in part upon the physiological status of the organism. There is no suggestion either interspecifically or from the data gathered on different breeds of dogs, that NPU has any allometry, and we will therefore neglect this aspect of protein requirements in the rest of this paper.

In order to eliminate the effects of variation in quality in the consideration of protein requirements, it is therefore usual to express requirements as net nutrients, that is as the amounts that would be needed if efficiency of utilisation were perfect. This is the concept of net dietary protein as used by Payne (1965) in his assessment of the protein requirements of the dog.

The Obligatory rate of Nitrogen Loss has been determined interspecifically to vary with $W^{0.75}$ (Brody, 1945; Miller & Payne, 1961; Henry, Rivers & Payne, 1985) but, although many studies on protein metabolism have made use of the dog, there is little information on the allometry of Obligatory Nitrogen Loss (ONL) in the dog, because most authors have used growing animals or too narrow a weight range to permit allometric analysis.

The exception is a paper by Kendall *et al.* (1982) on the allometry of total nitrogen loss in Beagles on a protein-free diet, which yields an allometry:

$$\text{ONL (mg N/day)} = 240.2 \text{ N}/W^{0.80} \qquad \text{[Equation 26]}$$

This exponent is not significantly different to the interspecific 0.75, so that the authors express their data with this allometry in preference which gives

$$\text{ONL (mg N/day)} = 250 \text{ mg N } W^{0.75} \qquad \text{[Equation 27]}$$

Dividing their data by sex shows the allometric exponent is 0.84 for males and 0.76 for females. All these exponents are significantly greater than 0.655, so that these data suggest that ONL does not vary with BMR in the dog. Neither Lusk's (1928) nor Ward's (1975) data allow allometric analyses, but they also suggest an allometric exponent of nitrogen loss of at least 0.75.

It is clear then that the generalisation due to Terroine & Sorg Mater (1927), and Terroine, (1933) that ENL is related to BMR, does not appear to apply to the dog, but that rather the ENL:BMR ratio has an allometry of at least 0.1–0.2 depending on the exact values for the allometries of BMR and ENL in the dog. Thus from Equations 12 and 26 the ONL in mg per basal kcal would rise from 2.8 for a dog of 3 kg to 3.9 at 30 kg and 4.5 at 80 kg.

The dietary implications of this depend upon the allometry of energy intake. The interspecific parallelism in the allometries of ONL, BMR and hence VFI has lead to the suggestion that protein requirements expressed as a protein:energy ratio are constant independent of size (Miller & Payne 1961; Mitchell, 1962; Payne, 1965). It is more equivocal as to whether this is true in the dog.

From the argument advanced above the allometry of energy intake of idle dogs should vary with BMR, an allometric exponent of 0.655. In this case there ought to be a need for significantly higher protein diets in maintenance in larger animals. Using an assumed energy cost of maintenance of 1.5BMR, the ND-p:E ratio for maintenance would rise from 7.0 at 3 kg, to 9.8 at 30 kg and to 11.2 at 80 kg. On the other hand, the ratio of obligatory nitrogen loss to VFI in maintenance is much more constant with an allometry of 0.079. It may well be therefore that no allometry in the protein:energy ratio required in the diet actually exists in normally active animals, with a maintenance ND-p:E of 6.0, close to the interspecific value and that suggested by Payne (1965).

These surmises, which have implications for husbandry of the dog, could be easily tested in a feeding study on a group of dogs.

The nutritional allometry of growth in the dog

There is little published information on the allometry of nutritional requirements in the dog, and the example that is frequently cited, due to Payne (1965) is misleading.

The energy and protein requirements of the growing dog will depend not only upon the relative allometries of obligatory losses but on the allometry of the rate of gain in growth. We have no independent data on any of these variables, where the independent variable is the weight of the growing animal. (In Table 2, the allometry of the time taken to grow to 90% of adult size has an allometry of 0.18 in the dog where the independent variable is, of course, the adult weight.)

Payne's factorial analysis assumed that the obligatory losses of the growing dog had the same allometry as the adult obligatory losses (which he took to be the interspecific allometric exponent of 0.75), and he argued that, though the rate of growth declined with age, at comparable stages of growth, weight gain of different animals had an allometric exponent of 0.75.

These assumptions were almost certainly an over-simplification. Preliminary information on the allometry of bodyweight gain suggests that the interbred allometry is highly variable, and within animals of a single breed no consistent allometry can be observed at different stages of growth. If the growing animal shows the same disparity between obligatory losses of energy and protein as the adult, considerable variation in the optimal protein:energy ratio for growth of different breeds may occur. Thus we feel it is not possible to discuss the allometry of the overall requirement of the growing animal, but the singular lack of this information makes the urgency of the need for future studies apparent.

Conclusion

In his essay 'On Being the Right Size' the eminent biologist J.B.S. Haldane remarked that, given that the most important difference between animals was that of size, it was surprising that biologists paid so little attention to it. It is only fair to remark that, in the 60 years since he wrote, biologists have done a great deal to make amends in this regard, but, as we have tried to show in this essay, not only could Haldane's adage be applied to the study of the nutrition of the dog, but rectifying the situation is clearly a matter of practical importance.

REFERENCES

Abrams, J.T. (1977). The Nutrition of the Dog. In *CRC handbook series in nutrition and Food. Section G: Diets, Culture Media and Food Supplements*. (M. Rechcigl, ed.) pp. 1–27. Florida: CRC Press.

Adolph, E.F. (1949). Quantitative relations in the physiological constitutions of animals. *Science*, **109**, 579–85.

Agricultural Research Council (1980). *The Nutrient Requirements of Ruminant Livestock*. Slough: Commonwealth Agricultural Bureaux.

Altman, P.L. & Dittmer, D.S. (1962). *Growth*. Washington, D.C.: Fed. Amer. Soc. Exptl. Biol.

Beer, E.J. de, & Hjort, A.M. (1938). An analysis of the basal metabolism, body temperature, pulse rate and respiratory rate of a group of pure bred dogs. *American Journal of Physiology*, **124**, 517–23.

Benedict, F.G. (1938). *Vital Energetics: a Study in Comparative Basal Metabolism*. Carnegie Inst. of Washington Publ. No 503 Washington: Carnegie Institute.

Blaxter, K.L. (1967). *The Energy Metabolism of Ruminants (Revised Edition)*. London: Hutchinson.

Blaxter, K.L. (1972). Fasting metabolism and the energy required by animals for maintenance. In *Festskrift til Knut Breirem*. Giovik: Mariensdal Boktrykkerie.

Blaza, S. (1981). The nutrition of giant breeds of dog. *Pedigree Digest*, **8**, 8–9.

Brodey, R.S., Saver, R.M. & Medway, W. (1963) Canine Bone Neoplasms. *Journal of the American Veterinary Medical Association*, **143**, 471–81.

Brody, S. (1945). *Bioenergetics and Growth*. (facsimile reprinted 1968). New York: Hafner & Co.

Brody, S. & Procter, R.C. (1932a). Growth and Development with special reference to domestic animals. Relation between basal metabolism and body weight in laboratory Animals. *University of Minnesota Agricultural Research Station Bulletin*, **166**, 83–8.

Brody, S. & Procter, R.C. (1932b). Growth and Development with special reference to domestic animals. Relation between basal metabolism and mature body weight in different species of animals and birds. *University of Minnesota Agricultural Research Station Bulletin*, **166**, 89–101.

Bronson, R.T. (1979). Brain weight-body weight scaling in breeds of dogs and cats. *Brain Behaviour Evolution*, **16**, 227–36.

Bronson, R.T., (1982). Variation in Ages at Death of dogs of different sexes and breeds. *American Journal of Veterinary Research*, **43**, 1891–901.

Cabana, G., Frewin, A., Peters, R.H. & Randall, L. (1982). The effect of sexual size dimorphism on variations in reproductive effort of birds and mammals. *American Naturalist*, **120**, 17–25.

Calder, W.A. (1984). *Size, Function and Life History*. Cambridge, Mass: Harvard University Press.

Calder, W.A. (1987). Scaling energetics of homeothermic vertebrates: an operational allometry. *Annual Review of Physiology*, **49**, 107–120.

Case, T.J. (1978). On the evolution and adaptive significance of postnatal growth rates in terrestrial vertebrates. *Quarterly Reviews of Biology*, 53, 243–282.

Chakraborty, P.K. Steward, A.P. & Seager, S.W.J. (1983). Relationship between growth and serum growth hormone concentration in the prepubertal labrador pup. *Laboratory Animal Science*, **43**, 51–5.

Christiansen, I.J. (1984). *Reproduction in the Cat & Dog*. London: Bailliere, Tindall.

Comfort, A. (1960). Longevity and mortality in dogs of four breeds *Journal of Gerontology*, **15**, 126–9.

Cowgill, G.R. (1928). The energy factor in relation to food intake experiments on the dog. *American Journal of Physiology*, **85**, 45–64.

Cowgill, G.R. & Drabkin D.L. (1927). The energy factor in relation to food intake experiments on the dog. *American Journal of Physiology*, **81**, 36–61.

Coyer, P.A., Rivers, J.P.W., & Millward, D.J. (1987). The effect of dietary protein and energy restriction on heat production and growth costs in the young rat. *British Journal of Nutrition*, **58**, 73–85.

Crist, K.A. & Romsos, D.A. (1987). Evidence for cold-induced but not diet induced thermogenesis in dogs. *Journal of Nutrition*, **117**, 1280–6.

Du Bois, E.F. & Lusk, G. (1924). Constancy of basal metabolism. *Journal of Physiology*, **59**, 213–16.

Economos, A.C. (1981). The largest land mammal. *Journal of Theoretical Biology*, **89**, 211–15.

Edwards, N.A. (1975). Scaling of renal function in mammals *Comparative Biochemistry and Biophysics A.*, **52**, 63–6.

Evans, E. & Miller, D.S. (1968). Comparitive nutrition, growth and longevity. *Proceedings of the Nutrition Society*, **27**, 121–9.

Evans, J.M. Curtis, R. & Wilkins, I.M. *The Henston Veterinary Vade Mecum.* London: Henston Ltd.

FAO/WHO (1973). *Energy and Protein Requirements.* WHO Tech. Rep. Series No. 522. Geneva: WHO.

FAO/WHO/UNU (1985). *Energy and Protein Requirements.* WHO Tech. Rep. Series No. 724. Geneva: WHO.

Field, J. Belding, H.S. & Martin, A.W. (1939). An analysis of the relation between basal metabolism and summated tissue metabolism in the rat. I The post pubertal albino rat. *Journal of Cellular and Comparative Physiology*, **14**, 143–57.

Fujita, M., Iwamoto, J. & Kondo, M. (1966). Comparative metabolism of caesium and potassium in mammals. Interspecies correlation between body weight and equilibrium level. *Health Physics.*, **12**, 1237–7.

Galilei, Galileo (1638). *Dialogues Concerning Two New Sciences.* (translated by H. Crew & A. DeSalatio, 1933). London: MacMillan.

Galvao, P.E. (1947). Heat production in relation to bodyweight and body surface. Inapplicability of the surface law on dogs of the tropical zone. *American Journal of Physiology*, **148**, 478–9.

Gould, S.J. (1978). Size and Shape. In *Ever Since Darwin. Reflections on Natural History.* (by S.J. Gould) pp 171–8 London: Pelican Books.

Gunther, B. (1975). Dimensional analysis and the theory of biological similarity. *Physiological Reviews*, **55**, 659–99.

Haldane, J.B.S. (1928). On being the right size. In *On Being the Right Size.* (J. Maynard Smith ed.) pp. 1–8. Oxford: Oxford University Press.

Hall, W.C. & Brody, S. (1933). Growth and Development with special reference to domestic animals. XXVI. The energy increment of standing over lying and the energy cost of getting up and lying down in growing ruminants (cattle and sheep); comparison of pulse rate, respiration rate, tidal air and minute volume of pulmonary ventilation during lying and standing. *University of Minnesota Agricultural Research Station Bulletin*, **180**, 1–33.

Harmer, H. (1971). *Chihuahuas.* London: Foyles Ltd.

Hart, J.S. (1971). Rodents. In *Comparative Physiology of Thermoregulation. Vol. II* (editor, G.C. Whittow.) pp. 2–149. New York: Academic Press.

Hemmingsen, A.M. (1950). The relation of standard (basal) energy metabolism to total fresh weight of living organisms. *Reports Steno Memorial Hospital*, **4**, 7–58.

Hemmingsen, A.M. (1960). Energy metabolism as related to body size and respiratory surfaces, and its evolution. *Reports Steno Memorial Hospital,* **9**, part II, 6–110.

Henry, C.J.K. (1984). *Protein-Energy Interrelationships and the Regulation of Body Composition.* PhD Thesis University of London.

Henry, C.J.K. Rivers, J.P.W. & Payne P.R. (1985). The relationship between fasting urinary nitrogen loss and obligatory nitrogen loss in rats. *Nutrition Research,* **5**, 1131–41.

Heusner, A.A. (1982a). Energy metabolism and body size. I. Is the 0.75 mass exponent of Kleiber's equation a statistical artefact? *Respiratory Physiology,* **48**, 1–12.

Heusner, A.A. (1982b). Energy metabolism and body size. II. Dimensional analysis and energetic non-similarity. *Respiratory Physiology,* **48**, 13–25.

Heusner, A.A. (1984). Biological similitude: statistical and functional relationships in comparative physiology. *American Journal of Physiology,* **15**, R839–45.

Heusner, A.A. (1985). Body size and energy metabolism. *Annual Review of Nutrition,* **5**, 267–93.

Heusner, A.A. (1987). What does the power function reveal about structure and function in animals of different size? *Annual Review of Physiology,* **49**, 121–33.

Huxley, J.S. (1932). *Problems of Relative Growth.* London: Methuen & Co.

Huxley, J.S. & Tessier G. (1936). Terminology of relative growth. *Nature,* **137**, 780–1.

Kendall, P.T., Blaza, S.E. & Holme, D.W. (1982). Assessment of endogenous nitrogen output in adult dogs of contrasting size using a protein-free diet. *Journal of Nutrition,* **112**, 1281–6.

Kirk, R.W. (1968). *Current Veterinary Therapy.* Philadelphia: W.B. Saunders.

Kirkwood, J.K. (1985). The influence of size on the biology of the dog. *Journal of Small Animal Practice,* **26**, 97–110.

Kleiber, M. (1932). Body size and metabolism. *Hilgardia.,* **6**, 315–53.

Kleiber, M. (1950). Physiological meaning of regression equations. *Journal of Applied Physiology,* **2**, 417–23.

Kleiber, M. (1961). *The Fire of Life.* New York: John Wiley and Sons.

Kleiber, M. (1967). An old professor of animal husbandry ruminates. *Annual Review of Physiology,* **29**, 1–20.

Kraus, E. (1928). *Lerbuch der Stoffwechsel methodik.* (Vol. I) Leipzig: von S. Hirzel Verlag.

Krebs, H.A. (1950). Body size and tissue respiration. *Biochimica et Biophysica Acta,* **4**, 249–69.

Krzyzanowski, J., Malinowski, E. & Studnicki, W. (1975). Examinations on the period of pregnancy in dogs of some breeds. *Medycyna Weterynarjna,* **31**, 373–80.

Kunde, M.M. Steinhaus, A.H. (1926). Studies of metabolism, IV. The basal metabolic rate of normal dogs. *American Journal of Physiology,* **78**, 127–35.

Linzell, J.L. (1972). Milk yield, energy loss in milk and mammary gland weight in different species. *Dairy Science Abstracts*, **34**, 351–60.

Lusk, G. (1928). *The Elements of The Science of Nutrition*. (3rd edition), Philadelphia: W.B. Saunders Company.

Martin, A.W. & Fuhrman, F.A. (1955). The relationship between summated tissue respiration and metabolic rate in the mouse and dog. *Physiological Zoology*, **28**, 18–34.

Martin, R.D. (1981). Relative brain size and basal metabolic rate in terrestrial vertebrates. *Nature, London*, **293**, 57–60.

Martin, R.D. & MacLarnon, A.M. (1985). Gestation period, neonatal size and maternal investment in placental mammals. *Nature, London*, **313**, 220–3.

McMahon, T.A. (1973). Size and shape in biology. *Science*, **179**, 1201–4.

McNeill, G., Rivers, J.P.W., Payne, P.R., de Britto, J.J. & Abel, R. (1987). Basal metabolic rate of Indian men: no evidence of metabolic adaptation to a low plane of nutrition. *Human Nutrition: Clinical Nutrition*, **41C**, 473–83.

Medawar, P.B. (1967). *The Art of the Soluble*. Methuen & Co. Ltd.

Millar, J.S. (1981). Post partum reproductive characteristics of eutherian mammals. *Evolution*, **35**, 1149–63.

Miller, D.S. & Payne, P.R. (1963). A theory of protein metabolism *Journal of Theoretical Biology*, **5**, 1398–412.

Milligan, L.P. and Summers, M. (1986). The biological basis of maintenance and its relevance to assessing responses to nutrients. *Proceedings of the Nutrition Society*, **45**, 185–93.

Mitchell, H.H. (1962). *Comparative Nutrition of Man and Domestic Animals. Vol I*. New York: Academic Press.

Mitchell, H.H. (1964). *Comparative Nutrition of Man and Domestic Animals. Vol II*. New York: Academic Press.

Munro, H.N. (1969). The evolution of protein metabolism in mammals. In *Mammalian Protein Metabolism. Vol. III*. (H.N. Munro, ed). pp. 132–82. Academic Press, New York.

National Research Council (1974). *Nutrient Requirements of Dogs*. Washington, D.C.: National Academy of Sciences.

National Research Council (1985). *Nutrient Requirements of Dogs*. Washington, D.C.: National Academy of Sciences.

Payne, P.R. (1965). Assessment of the protein values of diets in relation to the requirements of the growing dog. In *Canine and Feline Nutritional Requirements* (O. Graham-Jones, Editor) pp. 19–31. Oxford: Pergamon Press.

Payne, P.R. & Waterlow, J.C. (1975). Relative energy requirements for maintenance, growth and physical activity. *Lancet*, **ii**, 210–11.

Payne, P.R. & Wheeler, E.F. (1967a). Growth of the foetus. *Nature, London*, **215**, 849–50.

Payne, P.R. & Wheeler, E.F. (1967b). Comparative nutrition in pregnancy. *Nature, London*, **215**, 1134–6.

Payne, P.R. & Wheeler, E.F. (1968). Comparative nutrition in pregnancy and lactation. *Proceedings of the Nutrition Society, 27,* 129–38.

Peters, R.H. (1983). *The Ecological implications of Body Size.* Cambridge: Cambridge University Press.

Prange, H.D., Anderson, J.F. and Rahn, H. (1979). Scaling of skeletal mass to body mass in birds and mammals. *American Naturalist.* **113,** 103–22.

Robinson, R. (1973). Relationship between litter size and weight of dam in the dog. *Veterinary Record,* **92,** 221–6.

Rubner, M. (1883). Uber den Einfluss der Korpergrosse auf Stoffund Kraft-weschel. *Zeitschrift für Biologie,* **19,** 535–62.

Rubner, M. (1902). *Die Gesetze des Energieverbrauchs bei der Ernahrung.* Leipzig und Wein: Franz Deuticke.

Sacher, G.A. (1959). Relation of lifespan to brain weight and body weight in mammals. In *The Lifespan of animals.* (G.A. Wolstenholme & M. O'Connor, eds). CIBA Colloquia on Ageing, No. 5. pp. 115–41. Churchill Livingstone, London.

Sarrus, F. and Rameaux, J.F. (1839). Memoire adresse a l'Academie Royale. *Bulletin de l'Academie Royale de Medicine,* **3,** 1094–100.

Schmidt-Nielsen, K. (1984). *Scaling. Why is animal size so important?* Cambridge: Cambridge University Press.

Schofield, W.N., Schofield, C. & James, W.P.T. (1985). Basal metabolic rate – review and prediction, together with an annotated review of source material. *Human Nutrition: Clinical Nutrition,* **39C,** Suppl. 1, 1–96.

Seaman, J.S. & Rivers, J.P.W. (1988). Strategies for the distribution of relief food. *Journal of the Royal Statistical Society,* **188,** 100–14.

Slowtzoff, B. (1903). *Archives Gesamte Physiologie,* **95,** 158–91.

Sokolowski, J.N. & Vanravenswaay, F. (1983) cited from Kirkwood, 1985.

Stahl, W.R. (1962). Similarity and dimensional methods in biology *Science,* **137,** 205–12.

Stahl, W.R. (1963). Similarity analysis of physiological systems. *Perspectives in Biology and Medicine,* **6,** 291–321.

Stahl, W.R. (1965). Organ weights in primates and other mammals. *Science,* **150,** 1039–42.

Stewart, G.N. (1921a). The pulmonary circulation time, the quantity of blood in the lungs and the output of the heart. *American Journal of Physiology,* **18,** 20–30.

Stewart, G.N. (1921b). Possible relations of the weight of the lungs and other organs to body-weight and surface area (in dogs). *American Journal of Physiology,* **18,** 45–52.

Swift, J. (1726). *Travels into several remote nations of the world. In four parts by Lemuel Gulliver, first a surgeon and then a captain of several ships.* 2 vols, London: Richard Sympson. (Everyman edition, 1966).

Terroine, E.F., (1933). *Le metabolisme de l'azote.* Paris: Les Presses Universitaires de France.

Terroine, E.F. & Sorg-Matter, R.H. (1927). Loi quantitative de la depense azotee minime des homeothermes interspecifique. *Archives of Internal Physiology*, **15**, 121–32.

Udall, R.H., Rankin, A.D. & Moss, L.C. (1953). Energy requirements of dogs. *Veterinary Medicine*, **48**, 111–14.

Voit, E. (1901). Uber Die Grosse Des Energiebedarfs Der Tiere in Hungerzustande. *Zeitschrift für Biologie*, **41**, 113–54.

Ward, J. (1975). Investigation of the Amino Acid Requirements of the Adult Dog. PhD Thesis, University of Cambridge.

Waterlow, J.C., Garlick, P.J. & Millward, D.J. (1978). *Protein Turnover in Mammalian Tissues and the Whole Body*. Netherlands: Elsevier, North Holland.

Wolf, H.G., Della Rosa, R.J. & Corbin, J.E. (1970). Nutrition. In *The Beagle as an Experimental Dog*. (A.C. Anderson, ed.) pp. 22–32. Iowa: Iowa State University Press.

Wright, J.G. (1934). Some aspects of canine obstretrics. Diagnosis of pregnancy and Dystokia – its causes. *Veterinary Record*, **14**, 563–8.

7

Bodyweight changes and energy intakes of cats during pregnancy and lactation

G.G. LOVERIDGE AND J.P.W. RIVERS

Introduction

Though nutritionists tend to concentrate upon the nutritional demands of growth, the impact of reproduction can be equally large for the mother as for the offspring. For mammals, pregnancy is a period of increased nutritional demand when both mother and foetus are in positive energy and nitrogen balance, while, during lactation, the fact that the nutritional needs of the growing offspring must be met from the mother's milk, can place great nutritional demands on the dam, which is of course generally in negative balance at this time (see, for example, Mepham, 1987).

Although the most basic information on pregnancy has been gathered for many mammal species, very little is known of the nutritionally related changes that occur in cats. As a first step in providing this information data have been collected on a colony of respiratory virus-free cats, established at the Waltham Centre for Pet Nutrition (WCPN) in 1979. Full details of these colonies have been given elsewhere (Loveridge, 1985) however, some is clearly pertinent to the present discussion.

The colony, which is barrier maintained, is the primary breeding group from which healthy kittens are produced for feeding and growth studies at WCPN. Offspring of these animals have been used to establish a second colony which, while it is not barrier maintained, is disease free, all cats being vaccinated at 12 weeks against feline viral rhinotracheitis, feline calicivirus and feline infectious enteritis. Also, entry is restricted to authorised staff who have to change outer clothing and wear protective overalls to enter. The management and general husbandry of both colonies has been described elsewhere (Loveridge, 1984, 1986).

Pregnant queens are generally fed in groups during the first seven weeks of pregnancy, thereafter being individually fed with intakes recorded until the end of lactation. Food is freely available for 18 hours each day, intakes being calculated from the difference between food served and food weighed back. Tap water is available at all times. The diet fed is a standard commercial catfood, each batch of which is analysed for moisture, protein,

113

fat, ash, essential fatty acids, minerals and vitamins. Adult cats are weighed immediately after mating, weekly during pregnancy, within 24 hours of giving birth to give a parturition weight, and thereafter at weekly intervals. The data presented in this paper are for a group of 75 cats, drawn equally from both colonies, and all of which successfully reared kittens. Results were randomly selected from records to provide data on gestation and lactation for 15 queens producing each litter size of 1 to 5 kittens. To provide food intakes during pregnancy, a sub-group of ten animals were fed singly *ad libitum* throughout pregnancy and lactation.

Bodyweight changes in pregnancy

Figure 1 shows the mean bodyweight changes of the group during gestation and lactation, regardless of litter number, and Figure 2 shows the same data divided by litter number. The mean weight gain during pregnancy was 1208.7 gms (SD = 309.0), which was 39% of the mean mating weight of 3,082 grams. Both weight gain during pregnancy and litter number were unrelated to the weight of the queen at mating. Although mean weight gain rose with the number in the litter, even animals producing a single kitten increased their bodyweight by 32% of premating weight during pregnancy, a much larger value than the 25% gain reported by Kronfeld (1982).

The total weight gain during pregnancy was a linear function of litter size, the regression equation being:

Fig. 1. Mean weight changes during gestation and lactation in 160 successful pregnancies in the colony. (From Loveridge, 1986.)

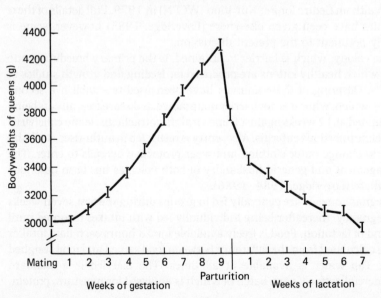

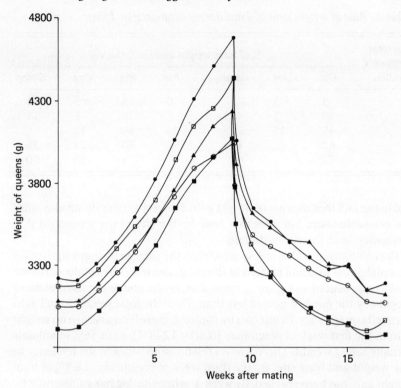

Fig. 2. Weight during gestation and lactation in queens with different sizes of litter. Data show the means for 15 queens in each group. ■ 1 kitten; ○ 2 kittens; △ 3 kittens; □ 4 kittens; ● 5 kittens.

$$\text{Weight Gain (g)} = 888.9 + 106.5N \qquad \text{[Equation 1]}$$
$$n = 5 \qquad r = 0.954$$
(where N = number in litter)

The slope of this equation, the increase in weight gain per extra kitten (106.5 g) is almost exactly equal to the birth weight of kittens in our colony which is 106.2 g.

The pattern of weight gain during pregnancy, illustrated in the curves in Figures 1 and 2, is unusual in several respects. Weight gain in pregnant mammals is notoriously variable and subject to nutritional influences. However, as Table 1 shows, there appear to be two broad patterns of weight gain during pregnancy. In humans and dogs as well as in many farm mammals most weight is gained by the dam late in pregnancy, so that the rate of weight gain increases during pregnancy. However, the cat, like the pig, shows a different pattern where weight gain is more linear. Though this generalisation is by no means comprehensive it does suggest that the cat may be unusual in the way in which weight is gained early in pregnancy,

Table 1. *Rate of weight gain of dams during pregnancy in different species*

% of total pregnancy duration	% of total weight increase achieved						
	Cat	Dog	Human	Rat	Pig	Cow	Sheep
20	8	< 5	2	10	15	< 5	< 2
35	20	5	5		30	5	15
50	35	15	25	35	50	12	
65	62		45		60	25	36
80	77		65	70	75	50	50

and in the fact that the rate of weight gain does not accelerate steadily until just pre-parturition, but rises to a peak by the end of the second third of pregnancy, and thereafter declines.

The comparison with the dog, which has the same gestation length of 63 days makes the unusual pattern in the cat quite clear. The results of Holme (1982) reproduced as Figure 3, show that by the end of the third week of pregnancy the dog has gained less than 5% of the total weight it will gain during the pregnancy. In our cats by contrast, there is on average no weight gain in the first week of pregnancy [0,SD = 124.07], with 56% of animals actually losing weight. There was no relationship between the tendency to lose weight and litter size. By the third week of pregnancy, 18.8% of total weight gain has been reached. By week 4, when the dog has gained only 6–10% of final weight gain, the cat has already reached 28%. This unusual pattern of weight gain does not appear to be unique to the WCPN colony. The unrelated colony kept by Rivers & Frankel at the Zoological Society of

Fig. 3. Weight changes in gestation in the bitch. (From Holme, 1982.)

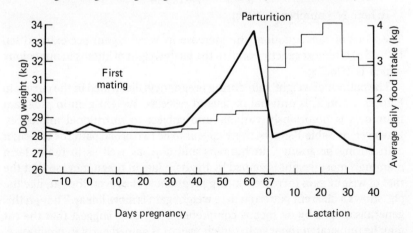

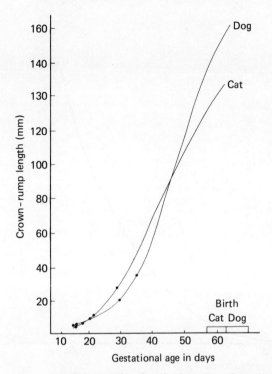

Fig. 4(a). Growth in length of the cat and dog foetus. Crown-rump lengths at various stages of gestation. (Redrawn from the diagrams in Christiansen, 1984.)

London, had a similar pattern of weight gain with their normally fed control animals gaining an average of 20% of final weight by the end of the first third of gestation (Frankel, 1979; Rivers & Frankel, unpublished data).

This more rapid weight gain in the cat does not appear to be associated with more rapid foetal growth. Implantation of the cat foetus does not occur until day 15 of gestation, and by day 17 the foetal length is only 4% of its final length and hardly any increase in uterine horn size has occurred (Christiansen, 1984). In fact, as Figure 4a shows, the growth curves of the cat and dog foetuses in early pregnancy broadly parallel one another. The published data are for length, but if as Rivers & Burger point out in Chapter 6, weight is regarded as the cube of linear dimension, then weight curves for cat foetuses can be calculated as in Figure 4b. They make it clear that, as several other animals, including humans, the cat foetus gains most weight in the last third of gestation.

In this context it is interesting to note that, when mean weight gained by the queen by successive weeks of pregnancy was analysed by the number in the litter, there was no relationship until the last third of the gestation (Table

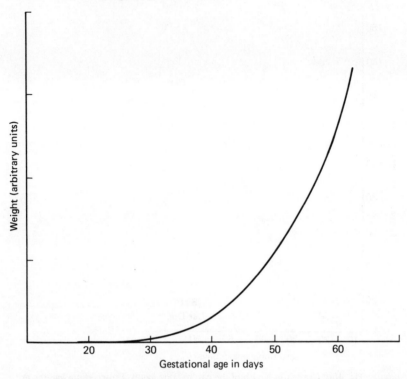

Fig. 4(b). Calculated growth in weight of the cat foetus assuming that weight is predicted by the cube of crown–rump length.

2). Even when weight gained by each week was related to the total weight gain, the coefficient of correlation in early pregnancy, though statistically significant, was very weak. Only by week 4 did r^2 approach 0.5 (Fig. 5).

For each litter size, total weight gain of the queen was related to gain in weight by week 4 of pregnancy [Table 3]. Including litter number as an independent variable did not improve the coefficient of correlation.

The nature of this early weight gain can only be surmised. Part of it may be due to fluid retention associated with the increases in blood volume and tissue hydration that occur in pregnancy in mammals (Schmidt–Nielsen, 1985). But we find it inconceivable that so large a weight gain could be accounted for simply by positive water balance, and feel that it must represent tissue deposition in the cat. Since the increase in weight of the uterine contents are so small at this time we feel that the tissue deposition must be extra-uterine.

Such extra-uterine deposition of fat and of protein has been suggested to be a frequent concomitant of pregnancy in other mammal species, and in particular Naismith (1969) has suggested from his studies on laboratory rodents and humans, that specific extra uterine retention of protein occurs

Table 2. *Cumulative weight gains of queens during gestation and the subsequent litter size. (Weight gains are as a percentage of weight at mating.)*

Week of gestation	Number of kittens in litter				
	1	2	3	4	5
1	−0.3	−0.4	0.2	−0.2	−0.5
2	3.1	2.7	3.4	2.8	2.3
3	6.7	7.4	7.9	7.1	6.7
4	11.1	10.4	12.0	11.2	11.4
5	17.2	15.7	17.0	16.4	18.5
6	21.9	21.7	25.1	24.2	26.0
7	27.4	28.6	30.8	30.6	33.9
8	29.7	34.3	35.9	34.1	40.3
9	32.4	37.5	40.2	39.4	45.7

Fig. 5. The correlation between weight gain at any week of gestation and weight gain by week 9 of gestation. Open circles show determined values of the coefficient of linear correlation (r), the lower line shows the calculated values of r^2, the coefficient of determination. The value of the latter indicates the fraction of the total weight gain by week 9, which can be predicted from the gain at any earlier week.

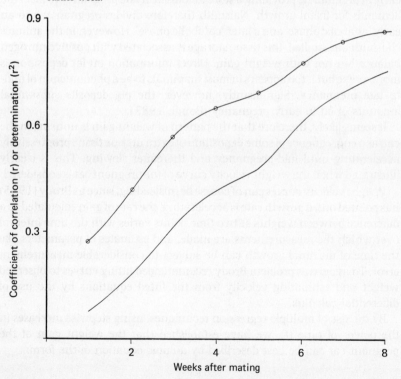

Table 3. *The relationship between weight gain by week 4 of pregnancy and total weight gain in pregnancy*

Values for intercept and slope in the linear equation predicting total weight gain from week 4 weight gain (Values in grams).

Number in litter	Intercept	Slope	Coefficient of Correlation (r)	n
1	715.6	0.724	0.61	15
2	725.2	1.342	0.92	15
3	739.2	1.233	0.93	15
4	971.5	0.791	0.57	15
5	892.5	1.581	0.91	15
All cats	835.1	1.054	0.65	75
	(173.1)*	(0.289)*		

Notes:
All values of r were significant at $p < 0.001$
*95% confidence limits

early in pregnancy, providing a store mobilised in late pregnancy to meet the demands for foetal growth. Naismith therefore divides pregnancy into an early 'anabolic phase' and a later 'catabolic phase'. However, in the animals Naismith has studied, this tissue storage is associated with positive nitrogen balance but not with weight gain. Direct information on fat deposition is more sparse but it too appears in most mammals to be a phenomenon of mid- to late pregnancy. Significantly, however, the pig deposits substantial amounts of fat in early pregnancy (Bondi, 1987).

It seems likely, therefore that the pattern of weight gain is unusual in the cat, as a consequence of some deposition of extra uterine tissue progressively accelerating until mid-pregnancy and thereafter slowing. This is clearly illustrated when the weight velocity curve of the pregnant cat is considered.

Weight velocity curves can of course be misleading, since as Brody (1945) has pointed out, if growth rate is accelerating, the rate of gain calculated as a difference between weights at two time points varies with the time interval over which the measurements are made, and estimates of parameters like the time of maximal growth can be subject to considerable measurement error. To avoid this problem Brody recommended fitting curves to observed weight and estimating velocity from the fitted equations by the use of differential calculus.

By the use of multiple regression techniques, using stepwise increases in the power of time (t), we have established that the weight gain of the pregnant cat can be best described by a cubic equation of the form:

Table 4. *Polynomial equations describing weights of cats during gestation of litters of different sizes.*

(Equations are polynomials fitted by least squares regression to mean weights of 15 cats at each week and litter size. Including extra powers of t beyond t^2 did not improve the value of R.) Maternal wt at time $t = At^3 + Bt^2 + Ct + D$

	Number of kittens in litter				
	1	2	3	4	5
Coefficient of t^3 (A)	−2.67	−1.95	−2.06	−2.68	−2.77
Coefficient of t^2 (B)	38.92	34.46	34.91	44.92	49.95
Coefficient of t (C)	−23.53	−21.28	−16.74	−46.13	−57.38
Intercept (D)	3046.4	3083.0	2910.6	3182.7	3225.8
D as % weight at mating	100.2	100.1	100.0	99.8	100.0
Predicted time of maximal weight gain (weeks)	4.9	5.9	5.6	5.6	6.0
Predicted maximal rate of weight gain (g)	166.6	181.7	180.5	209.7	281.1

$$Wt = At^3 + Bt^2 + Ct + D \qquad \text{[Equation 2]}$$

Where Wt is weight at time t during gestation.

Such equations were fitted by multiple regression to the means of the weekly weights of animals for each size of litter. Cubic equations of the form in Equation 2 gave significantly better fit to the data than a quadratic, but the use of a quartic did not significantly improve the coefficient of correlation. The resultant best fit cubics are shown in Table 4. It is clear that these equations are of essentially similar form. They give good prediction of weight over the whole range of gestation time and are within 2% of mean weight at both weeks 0 and 9.

We do not regard the description of the growth curve by a cubic as fortuitous. Various authors (Blaxter, Wheeler & Payne, Calder, 1982, 1984) have pointed out that the growth curve of the foetus follows a cubic rule and Payne & Wheeler in particular have pointed out that the weight gain of the foetus in a variety of mammal species varies according to a cubic law of the form:

$$Wt = k(t - a)^3 \qquad \text{[Equation 3]}$$

Where a is the time after which rapid growth begins. Payne & Wheeler have suggested that this is implantation time (15 days in the cat). Calder's (1982,

1984) allometric relationship for *a* in Equation 3 would suggest an essentially similar value (17.5 days for neonatal weight of 106 g.) Payne and Wheeler (1968) have argued that it is reasonable to expect the weight of the other uterine products of conception to vary in the same way as the foetus. Thus if the weight gain of the pregnant animal consisted of nothing but the products of conception, we could expect maternal weight gain to be described by an equation of the form of Equation 3. Given that early weight gain was unrelated to litter size, a modified cubic therefore seems reasonable.

Equation 3 may be expanded, of course, to yield:

$$Wt = kt^3 + 3akt^2 - 3a^2kt - ka^2 \qquad \text{[Equation 4]}$$

and in this form the structural similarity to Equation 2 and therefore the equations in Table 4, is obvious.

The first-order differential (dw/dt) of the weight gain equations in Table 4 is the prediction equation for the rate of weight gain at any time t, and the maximal value of this can be estimated from the time at which the second-order differential (d^2w/dt^2) is zero. Peak growth velocities and their timings calculated in this way are shown in Table 4.

It can be seen that maximal rates of weight gain estimated for each litter size in this way occurred in weeks 4 to 5 of gestation. The apparent tendency for the maximum to be reached earlier the fewer in the litter, was not statistically significant. There was a marginally significant positive correlation between the time of maximal weight gain and the rate of maximal weight gain.

Using a value of $a = 15$ days (2.14 weeks), and regarding the weight loss of the queen at parturition as a measure of the weight of the foetus and uterine contents, it is possible to derive approximate values for Equation 3 at each litter size. These equations of presumptive foetal amniotic fluid and placental growth are shown in Table 5, and as with the maternal growth equations in Table 4, the broad similarity of these equations for different litter sizes does not need underlining.

By subtracting the appropriate equations in Table 5 from those in Table 4, a prediction of the growth rate of uterine wall and extra-uterine tissues can be obtained and these equations are given in Table 6. These too are a closely similar set of cubic equations that form a regular 'homologous' series with the constants changing in a remarkably systematic fashion with progression in litter size. Their second-order differentials predict accelerating deposition until the fifth week (4.2 to 4.6 weeks, unrelated to litter size) of pregnancy while the rate of deposition is negative during the first week (until 0.4 to 0.7 weeks of gestation) and after 7.5 to 8.5 weeks. In other words, growth of extra uterine tissue rises to a maximum when rate of weight gain of the queen is maximal, and there is mobilisation of extra

Table 5. *Prediction equations for the growth of the foetus and uterine contents*

Equation are solutions to Equation 3: Wt of uterine contents $= k(t-a)^3$ where t is time in weeks. Solutions are obtained from weight loss at birth and assuming $a = 2.14$ weeks (15 days). Thus the equation can be expanded into Wt of uterine contents $= kt - 3kat^2 + 3ka^2t - ka^3$

	Number of kittens in litter				
	1	2	3	4	5
k	0.787	1,056	1.642	2.026	1.976
a	2.14	2.14	2.14	2.14	2.14
Coefficient of t^3	0.787	1.056	1.642	2.026	1.976
Coefficient of t^2	-5.050	-6.870	-10.542	-13.007	-12.686
Coefficient of t	10.820	14.519	22.576	22.855	22.168
Intercept	-7.735	-10.377	-16.135	-19.908	-19.417

Table 6. *Estimating equations for weight of extra-uterine tissue in pregnant cats*

The equations are the difference between the polynomials in Table 4, which predict weight gain of queen and those in Table 5 which predict weight gain of foetus and uterine contents. Values shown are for coefficients in the equation: Wt gain of extra uterine contents $= Et^3 + Dt^2 + Ft + G$ (where t is time in weeks.)

	Number of kittens in litter				
	1	2	3	4	5
Coefficient of t^3(J)	-3.5	-3.0	-3.7	-4.7	-4.8
Coefficient of t^2(K)	44	41	45	58	63
Coefficient of (L)	-34	36	39	74	85
Intercept	2	6	14	26	-20

Notes:
Omitting the intercept, the equations can be rearranged to:
Litter number
1. $-3.5t\,(t+13.5)\,(t-0.74)$
2. $-3.0t\,(t+14.5)\,(t-0.83)$
3. $-3.1t\,(t+13.1)\,(t-0.81)$
4. $-4.7t\,(t+13.5)\,(t-1.17)$
5. $-4.75t\,(t+14.4)\,(t-1.24)$

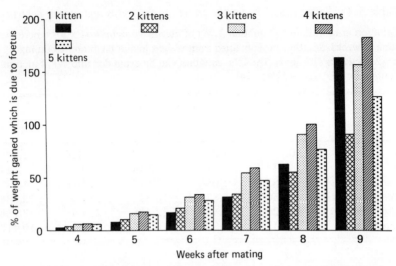

Fig. 6. Predicted percentage of total weight gain at various stages of gestation that is attributable to the weight of the foetus and uterine contents. The weight of the foetus and uterine contents at each stage of lactation was predicted from the equations in Table 5.

uterine tissues during the first week (which is obvious from the fact that the cat loses weight) and during the last week of gestation. This is compatible with the view that tissue deposition in early gestation is providing a nutrient store for late pregnancy.

Figure 6 shows the calculated fraction of weight gain which is due to the uterine products of conception, and it can be seen that this dominates weight gain from the 7–8th week onwards. The fact that more than 100% of the gains in the final week of gestation are attributed to foetal gain is a reflection of the fact that extra-uterine tissue is mobilised then to provide for foetal growth.

Food intake in pregnancy

A sub-group of ten queens were fed individually from the beginning of pregnancy, so that changes in food intake and weight gain could be compared. Overall, the metabolisable energy (ME) intake of the cats was 90 kcal per kg weight per day (125 kcal/$w^{0.75}$/day) throughout gestation. Scott (1960) suggested that cats require 100 kcal/per kg wt per day during pregnancy, and since recent work by Kendall, Burger & Smith (1985) has shown that true ME intakes of cats are only 88% of those obtained using conventional prediction factors, Scott's estimate is equal to 88 kcal true ME, which is good agreement with the value observed by us. It can be seen from the results, which are shown in Figure 7, that food intake increased from the very beginning of pregnancy in cats, and rose broadly in parallel with increasing rates of weight gain.

Prior to mating these animals had an estimated mean daily energy intake of 246 kcal per day. As Figure 7 shows, this rose progressively during pregnancy to a peak value of 415 kcal per day, 70% above the non-pregnant figure, in weeks 6 and 7 of gestation. This is somewhat higher than the dog in which Holme (1982) considered an increase of 50% by week 8 was satisfactory. Another difference from reported dog data is that according to the results reported by Anderson (1983). Romsos *et al.* (1981), intakes of pregnant bitches do not rise above maintenance levels until the fifth week of pregnancy, whereas the cat data show a continuous rise in intake from mating onwards. From their peak value the intakes of the queens declined somewhat, to a value of 371 kcal per day in the week prior to parturition, a decline which explains the slowing in growth rate discussed above. The extra energy consumption throughout the pregnancy was equal to 10,300 kcal; in the ten animals studied, the mean weight gain during pregnancy was 1.5 kg, an apparent cost of weight gain of 6.9 kcal per gram gained.

To estimate the incremental energy costs of weight gain, allowance must be made for the fact that maintenance requirements will be increased during pregnancy as a result of increased weight of the queen and possibly also as a result of increased metabolic rate per unit size. To allow for the weight changes weight gain per unit weight was regressed onto intake per unit weight, as is shown in Figure 8. The mean cost of growth calculated in this

Fig. 7. Variation in food intake of cats during gestation and lactation. Data are for 10 cats fed singly. Their patterns of weight change parallel those observed in the main group of animals. Food intakes after the fourth week of lactation probably include significant amounts of food eaten by kittens. (From Loveridge, 1986.)

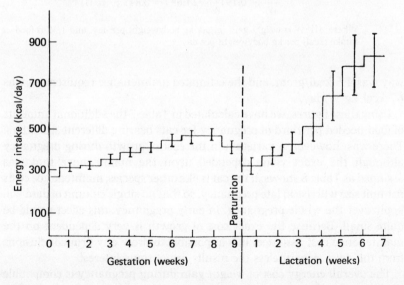

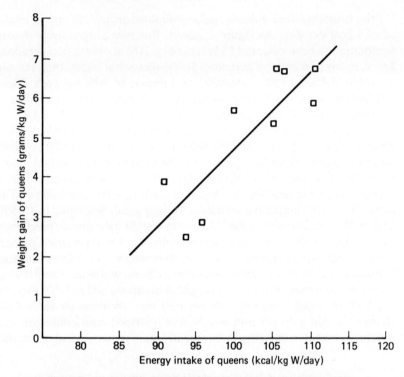

Fig. 8. The energy cost of weight gain in the pregnant cat. Data points are means for each week of gestation for a group of 10 queens fed singly. The equation of the line of best fit is:

$$\frac{\Delta W}{W} = 0.194\frac{I}{W} - 14.68 \ (r = 0.845 \ p < 0.01)$$

Where $\Delta W/W$ is weight gain (g) per kg bodyweight per day. and $1/W$ is food intake (kcal) per kg bodyweight per day.

way was 4.5 kcal/gram, and the estimated maintenance requirement was 67 kcal/kg W/day.

Using these figures, we have calculated in Table 7 the additional amounts of food needed per third of pregnancy for cats bearing different size litters. There was, however, a variation in the cost of growth during pregnancy although the exact value depended upon the maintenance that was assumed as Table 8 shows. If the cat is like other species, maintenance costs per unit size will rise in late pregnancy, so that no single column of data will apply over the whole pregnancy. In early pregnancy, this effect would be quite small. Because the exact cost of growth is very dependent on the maintenance cost assumed, it is impossible to draw any firm conclusions from the data, nevertheless the results are of some interest.

The overall energy cost of weight gain during pregnancy is compatible

Table 7. *Estimates of the energy cost of weight gain by week of pregnancy assuming different maintenance costs*

Week after mating	Maintenance cost assumed (kcal/kg)		
	70	75.7	82.7
		Energy cost of weight gain (kcals/g)	
1	9.5	7.2	4.46
2	9.0	7.1	4.62
3	5.3	4.3	3.04
4	6.6	5.5	4.21
5	5.5	4.6	3.58
6	6.0	5.2	4.17
7	6.9	5.9	4.73
8	5.2	4.4	3.38
9	5.4	3.9	2.13

with that reported for other species (Bondi, 1987). If the estimate of Burger *et al.* (1984) is used and maintenance of the adult is close to 70 kcal/W/day, then the energy cost of weight gain in early pregnancy is above this mean value, suggesting that it may well be due to adipose tissue rather than lean tissue deposition. It may be that, as Naismith (1969) has suggested occurs in other species, there is also some retention of nitrogen in early pregnancy not associated with weight gain in the cat. There is clearly a need for simultaneous energy and nitrogen balance studies in pregnant cats to clarify exactly what occurs.

Weight loss at birth

In these studies, measures of the weight loss at birth have a small error since they are the difference between the weight on day 63 of gestation and the post-partum weight. Errors introduced by fluctuation in gestation length and the fact that post-partum weighing could occur up to 24 hours after parturition may well have accounted for the fact, that although a correlation was detectable between weight loss at parturition and litter size, there was, as Table 9 shows, considerable variation in the weight loss per extra kitten born. Interestingly, though, a linear regression of weight loss at parturition upon litter size gave a best estimate of

$$\text{Weight loss at Parturition} = 159.1 + 108.1N \qquad \text{[Equation 5]}$$
$$(n = 5 \qquad r = 0.955)$$

This equation, like equation 1, fits remarkably well with the observed mean weight of newborn kittens in our colony of 106.2 g.

By comparing Equations 1 and 5, it is clear that there is a retention after parturition of 729.8 g (888.9 − 159.1) of the weight gained, regardless of

Table 8. *Estimated additional energy intake required by pregnant cat*

Litter number	First	Second	Third
		Third of pregnancy	
		Additional kcal/day required	
1	50	120	75
2	50	125	120
3	50	125	115
4	50	140	125
5	50	160	160

Values approximated to ± 5 kcal, calculated from wt gains and assume a cost of gain of 5.2 kcal/g.

Table 9. *Weight loss at parturition*

	1	2	3	4	5	all
	Number of kittens in litter					
Wt loss at parturition	254	341	530	654	638	483
Increase in wt loss per extra kitten	(254)	87	189	124	−16	
Wt loss as % of wt gain in pregnancy	26	30	45	52	43	41

litter size. Although the fact that heavier cats had larger litters meant that the post-partum weight as a percentage of the pre-mating weight declined slightly with increasing litter size, the range was small, animals being 119% to 126% of pre-mating weight. This is another contrast with the dog in which, according to Holme (1982), all the weight gained in pregnancy was lost at parturition. In these cats, however, the weight loss which occurred within 24 hours of parturition was much less, being a mean value of 512 grams, only 40% of the weight gained in pregnancy.

All this excess weight was lost during lactation, so that by the end of lactation the queen was within 2% of the pre-mating weight, regardless of litter size. Both relative and absolute rates of weight loss during lactation were faster in queens with big litters, though the relationship between total weight loss during lactation and litter size was not strong.

$$\text{Weight loss in lactation} = 669.5 + 33.1N \qquad \text{[Equation 6]}$$
$$(n = 5 \qquad r = 0.70)$$

Quantitatively, this fraction of weight gained retained after parturition was approximately equal to that which had been gained by week 5 to 6 of pregnancy, that is, by the point at which maximal growth was occurring and weight gain was beginning to be dominated by the growth of the litter

Table 10. *Weekly food intakes of lactating queens by litter size*

Week of lactation	Number of kittens in litter				
	1	2	3	4	5
	Intake (kcal/kg W/day)				
1	58	78	87	99	118
2	69	92	123	124	146
3	79	110	141	151	183
4	80	125	165	187	223
5	92	140	191	228	295
6	97	157	230	297	354

and uterine contents. This makes it plausible, but by no means certain, that the early weight gain provided an endogenous nutrient supplement for lactation. How far the weight retained at parturition was physically the same tissues as that gained in early pregnancy is a matter of conjecture; it is nevertheless clear that it would have provided some sort of supplement to the nutritional demands that lactation made upon the queen.

Food intakes of 15 individual queens with each litter number were measured during lactation, and the data are shown in Table 10 separated by litter size. It can be seen that intakes rise progressively during lactation, and, throughout, both intake and rate of increase of intake are clearly related to litter size. For a queen rearing a single kitten, the apparent increase was from 51 kcal/W/day in week 1, to 85 kcals/W/day in week 7, an increase of 67%. For a queen rearing five kittens, the intake at week 1 was twice as high, and the apparent increase was three times as great to an intake of 312 kcals/W/day in week 8 of lactation, an apparent increase in intake of 200% during lactation. However, in our colony, the energy-dense and highly palatable food made available encourages kittens to begin to eat from the third to fourth weeks of age, thus reducing the demands made on the dam and making weaning less stressful. Thus from week 3 onwards the intakes are to an increasing degree that of queen and kittens together. Subsequent calculations are therefore confined to the first three weeks of lactation.

During these three weeks the weight loss was much more closely related to litter size:

$$\text{Weight loss, weeks 0--3 of lactation} = 339.2 + 58.8 \text{ N} \qquad \text{[Equation 7]}$$
$$(n = 5 \qquad r = 0.84)$$

Comparison of Equations 6 and 7 makes clear certain facts about weight loss. It is clear that, during the first three weeks, more than half the weight loss occurs and that the fraction of the loss rises with litter size so that for a queen suckling five young the ratio rises to 76%. A comparison of slopes of

Equations 6 and 7 suggests that weight loss is much more affected by litter size in the first three weeks. This is not paradoxical, it is merely a reflection of the fact that animals suckling large litters lose weight more rapidly in early lactation and then their weight 'plateaus'. The existence of the intercepts in Equations 6 and 7 suggests that the lactating queen has a relative anorexia, that, even if no demands were being made for milk production for the young, i.e. $N = 0$, then food intake is inadequate for maintenance (i.e. 51% of weight would be lost).

It is clear, however, that the weight lost by the queen must provide somewhere between 9 kcal/g if it is fat and 1.0 kcal/g if it is completely lean tissue (and assuming a protein:water ratio in that tissue of 1:4). If lean tissue is being lost therefore, then the contribution of weight loss to energy expenditure is less than 30 kcal/day and in excess of 90% of the energy expenditure during the first three weeks of lactation comes from food intake. However, in this case, the lean tissue could provide a significant contribution to the protein demands of lactation. If, on the other hand, the energy is coming from adipose tissue, it could provide 50–75% of the energy costs of lactation during the first three weeks of lactation, but the diet would be an important protein source.

The energy expenditure by the queen during the first three weeks of lactation have been calculated using energy intakes and four different assumptions about the energy value of weight loss. These assumptions are (i) that the loss represented lean tissue with an energy value of 1.2 kcal/g, (ii) that it was a mixture of approximately 15:85 fat to lean with an energy value of 2.5 kcal/g (often taken as the mean value for body tissues), (iii) that it had an energy value of 6 kcal/g (approximately 40:60 lean to fat) and finally, (iv) that it was fat with an energy value of 9 kcal/g. The impact of these assumptions on energy expenditure is shown in the equations below. In each case the values have been regressed on to litter size, generating the following equations:

$$EE(lt) = 193.2 + 71.4N \quad (r = 0.982) \; [1.2 \text{ kcal/g}] \qquad \text{[Equation 8]}$$
$$EE(lt) = 217.6 + 75.6N \quad (r = 0.987) \; [2.5 \text{ kcal/g}] \qquad \text{[Equation 9]}$$
$$EE(lt) = 274.7 + 85.3N \quad (r = 0.991) \; [6 \text{ kcal/g}] \qquad \text{[Equation 10]}$$
$$EE(lt) = 323.6 + 93.6N \quad (r = 0.989) \; [9 \text{ kcal/g}] \qquad \text{[Equation 11]}$$

where EE(lt) is the energy yield from tissue catabolism.

The correlation coefficients of these equations are essentially similar and provide no grounds for discriminating between them, however, the intercepts are markedly different. The intercept is the energy expenditure if $N = 0$, i.e. if no milk is produced. Thus the intercept is a maintenance requirement for the lactating animals. In dairy animals, lactating animals have a maintenance requirement that is 10–20% higher than that of the dry animal (Bondi, 1987). This would suggest a maintenance of 290–300 kcal, on which basis Equations 10 and 11 are the most likely, and the tissue lost is,

Table 11. *Energy expenditure of queens during lactation (based on measured energy intake and weight loss; (as kcal/day)*

		Number of kittens in litter				
		1	2	3	4	5
Assuming 6 kcal/g of weight loss	Week 1	130	145	155	180	210
	2	85	110	135	165	170
	3	95	130	175	190	215
	4	105	130	155	180	195
Assuming 9 kcal/g of weight loss	Week 1	165	175	190	225	255
	2	95	120	140	190	215
	3	100	145	190	205	230
	4	120	140	175	205	225

consequently, largely or totally adipose tissue and the energy yield per gram of tissue lost is 6–9 kcal per gram. On this basis the energy expenditure during the first three weeks of lactation has been calculated and the results are shown in Table 11. The demands of lactation during the first three weeks per extra kitten would be 85 to 94 kcal. This is presumably the energy available for the production of milk for the young kitten. Literature evidence suggests that a requirement for the kitten at this time of about 60 kcal/day would be appropriate (NRC, 1986), so that an efficiency of milk production of about 70% would be indicated. This is broadly comparable with the values reported for other non-ruminant species (Blaxter, 1968).

Conclusion

The pattern of weight gain in the pregnant cat is unusual, though perhaps not completely unique for a mammal. It is, however, explicable as a reflection of a tendency to deposit tissue in early pregnancy which is mobilised to a degree in late pregnancy and in lactation. This would result in the fact that food intake in early pregnancy would be higher than a simple factorial analysis of foetal growth would suggest, while in late pregnancy and lactation it would be insufficient to meet the needs for energy expenditure and production.

REFERENCES

Anderson, R.S. (1983). Nutrition and reproduction in the bitch. *Pedigree Digest*, **10**, No. 3, 8–10.
Brody (1945). *Bioenergetics and Growth*. New York: Haffner & Co.
Blaxter, K.L. (1968). The effect of dietary energy supply on growth. In *Growth and Development of Mammals*, ed. G.A. Lodge & G.E. Lamming, pp. 329–44. London: Butterworths.
Bondi, A.A. (1987). *Animal Nutrition*. New York: John Wiley & Sons.

Calder, W.A. (1982). The pace of growth: an allometric approach to comparative embryonic and post-embryonic growth. *Journal of Zoology London*, **198**, 215–25.

Calder, W.A. (1984). *Size, Function and Life History*. Cambridge, Mass.: Harvard University Press.

Christiansen, I.J. (1984). *Reproduction in the Dog and Cat*. London: Bailliere, Tindall.

Frankel, T.L. (1979). Essential fatty acids in Cats. PhD Thesis University of Cambridge.

Holme, D.W. (1982). The practical use of prepared foods for dogs and cats. In *Dog and Cat Nutrition*, ed. A.T.B. Edney, pp. 47–59. Oxford: Pergamon Press.

Kendall, P.T., Burger, I.H. & Smith P.M. (1985). Methods of estimation of the metabolizable energy content of cat foods. *Feline Practice*, **15**, no. 2, 38–44.

Kronfeld, D.W. (1982). Optimal regimes based on recipes for cooking in home or hospital or on proprietary petfoods. In *Proceedings of the First Nordic Symposium on Small Animal Veterinary Medicine*, ed. R.S. Anderson, Oxford: Pergamon Press.

Loveridge, G.G. (1986). Bodyweight changes and energy intake of cats during gestation and lactation. *Animal Technology*, **37**, 7–15.

Loveridge, G.G. (1984). The establishment of a barriered respiratory disease-free cat breeding colony. *Animal Technology*, **35**, 7–15.

Mepham, T.L., (1987). *Physiology of Lactation*. Open University Press.

Naismith, D.J. (1969). The Foetus as Parasite. *Proceedings of the Nutrition Society*, **28**, 25–31.

NRC (1986). *Nutrient Requirements of Cats*. National Research Council. Washington: National Academy of Sciences Press.

Payne, P.R. & Wheeler, E.F. (1967a). Growth of the foetus. *Nature, London*, **215**, 849–50.

Payne, P.R. & Wheeler, E.F. (1967b). Comparative Nutrition in Pregnancy. *Nature, London*, **215**, 1134–6.

Payne, P.R. & Wheeler, E.F. (1968). Comparative nutrition in pregnancy and lactation. *Proceedings of the Nutrition Society*, **27**, 129–38.

Romsos, D.R. Palmer, H.J., Muiruri, K.L. & Bennink, M.R. (1981). Influence of a low carbohydrate diet on performance of pregnant and lactating dogs. *Journal of Nutrition*, **111**, 678–85.

Schmidt-Nielsen, K. (1985). *Animal Physiology.: Adaptation and Environment*, 3rd edn, Cambridge: Cambridge University Press.

Scott, P.P. (1966). Nutrition. In *Disease of the Cat*, ed. G.T. Wilkinson, pp. 1–31. Oxford: Pergamon Press.

8

Nutrition, anaerobic and aerobic exercise, and stress

D.S. KRONFELD, T.O. ADKINS AND R.L. DOWNEY

Introduction

Muscle contraction involves the conversion into mechanical energy of chemical energy released during the splitting of the high-energy phosphate ($\sim$P) bond in adenosine triphosphate (ATP). ATP is present in minute amounts in muscle cells, probably sufficient only for one or a very few brief contractions. It diminishes only during maximal efforts, however, because it is replenished continuously from phosphocreatine (PCr), from the breakdown of glycogen and glucose to lactic acid (anaerobic metabolism), or from oxidation of lactic acid and fatty acids (aerobic metabolism). This essay focusses on the ways that nutrition can influence these three systems for generating ATP.

These three sources of $\sim$P are distributed characteristically between the three types of muscle fibre. Phosphagens and glycogenolysis predominate in the white, fast-twitch fibres. Both glycogenolysis and the oxidative system are highly active in the red, fast-twitch fibres. In the red slow-twitch fibres, phosphagens, glycolysis and myosin-ATPase are at a low ebb, while oxidative enzymes are most active.

Whole muscles vary in the relative abundance of types of muscle fibres, and intensity of contraction affects the recruitment of fibre types. These influences combine with the well-known temporal sequence of use of energy sources to determine the mixture utilised at any moment (McGilvery, 1973).

The three metabolic systems are amenable to training, and, since this interacts with nutrition, this discussion will have to extend somewhat into training.

Exercise involves primarily the muscles, which are supported by the cardiovascular, respiratory, metabolic and endocrine systems. Training involves all of these systems acutely and affects tendons, ligaments, bones and blood composition more chronically. In discussing exercise, it is necessary to orientate muscular activity in relation to its supporting systems, and it becomes tempting to expand the subject. Such digressions will be resisted while the three energy systems are being addressed, then will be indulged in discussion of stress.

133

The phosphagens

The phosphagen system is most important for single convulsive (supramaximal) efforts, such as a jump or weight-lifting. We have found little convincing evidence for muscle phosphagen concentration being increased by training or diet.

Intense training, such as progressively fewer repetitions or isometrics of supramaximal effort, will increase muscle bulk mainly in the form of enlarged white, fast-twitch fibres that contain abundant phosphagen. This response is retarded by dietary protein insufficiency and is enhanced by testosterone. It may be promoted by anabolic steroids, and excess dietary protein has favoured the response, though only in about half of the studies published (Ryan, 1981).

Interval-training consists of alternating periods at near-maximal effort and effort just below the anaerobic threshold. It has been suggested as a way to increase the 'phosphagen capacity' in human athletes (Fox, 1973). The estimate is based on net oxygen consumption adjusted by assumptions for depletion of oxygen stores and efficiency of repayment of oxygen debt. It was increased by sprint repetitions (19×30 s) but not endurance intervals (7×2 min). These results make good sense but remain inconclusive because the method is so indirect.

Direct measurements of ATP, PCr and inorganic phosphate (Pi) are now possible in intact muscles by means of nuclear magnetic resonance (NMR). The shift of the Pi peaks towards the PCr peak allows calculation of changes in pH. The ratio, Pi/PCr, may be used to estimate ADP and the 'energy cost index' of metabolic activity in mitochondria. During moderate and near-maximal arm exercise PCr decreases sharply and Pi increases while ATP remains unchanged (Chance *et al.*, 1983). Pi/PCr increases with increasing work rate; the slope of Pi/PCr plotted against work rate is a measure of energetic efficiency. It is linear for sub-maximal work, but decreases in a non-linear manner when work approaches maximal. The linear component would be expected to become more steep with training. It has been found to be more steep in an Olympic athlete than in an untrained individual (Chance *et al.*, 1983). Trials comparing responses to training and diet are now possible.

The exercise protocol in NMR studies may be designed to focus attention on each or all of the energy sources (Chance, *et al.*, 1983). Maximal or supramaximal efforts lower ATP as well as reducing PCr well below the Pi level. The recovery rate of the Pi/PCr ratio following a maximal effort may be a guide to the capacity of the muscle to generate phosphagens. Difficulties of calibration and determination of absolute values are avoided in the Pi/PCr ratio. This is about 0.05 or less at rest. Work that increases it to about 1.0 can be maintained for a long period but higher work rates rapidly lead to fatigue. This value, Pi/PCr = 1.0, may reflect a threshold beyond which emphasis shifts from aerobic to anaerobic metabolism.

The anaerobic system

The anaerobic system produces only one-eighteenth the ATP that the aerobic system generates from glucose (McGilvery, 1973). Its use brings on fatigue, a cumulative phenomenon that includes metabolic components, such as muscle glycogen depletion and lactic acid accumulation in muscle and blood. [Other components, such as proprioceptive and psycho-social inputs, are just as important to a competitor but less relevant to the present discussion.]

The anaerobic system is boosted by *carbohydrate-loading*, a regimen that combines training and diet. The objective is to raise muscle glycogen concentration before the bout of exercise. Massive ingestion of carbohydrate releases insulin which makes more glucose available for glycogen synthesis and boosts intracellular glucose-6-phosphate, the allosteric activator of glycogen synthase (Ivy *et al.*, 1983).

One attention-gaining study was conducted on cross-country skiers who averaged 30 km in 135 min when carbohydrate-loaded but required 143 min when consuming an ordinary mixed diet (Karisson & Saltin, 1971). Such studies are confounded by psycho-social influences, but nevertheless have engendered widespread acceptance among endurance athletes.

One classical laboratory study involved an abrupt change from a mixed diet to a high-carbohydrate diet, which delayed exhaustion, or to a high-fat diet, which precipitated exhaustion [Bergstrom *et al.*, 1967). Although the investigators attributed the adverse effect of the high-fat diet to low muscle glycogen, we would have been concerned more with ketosis due to the abrupt change of diet.

Carbohydrate-loading probably caused exertional rhabdomyolysis in racing sled dogs (Kronfeld, 1973). The regimen mimics the sequence of events in the classical 'Monday morning disease' of draft horses. Muscle glycogen accumulation (on Sunday) is followed by rapid breakdown (on Monday morning) to lactic acid which accumulates and damages muscle membranes (Carlstrom, 1931; Lindholm *et al.*, 1974).

Lactic acid accumulation has numerous adverse effects (Sahlin, 1983), including the following:

o Inhibition of glycogenolysis and glycolysis.
o Inhibition of lipolysis.
o Oedema in muscles.
o Diminished lactic acid efflux from muscle cells.

Lactic acid accumulation begins sharply in an individual as workload increases progressively. This point of inflection is called the *lactate threshold*. Similarly, ventilation increases to a more rapid rate at about the same workload or rate of oxygen consumption, and this point of inflection is called the *ventilatory threshold* (Brooks, 1985). These two points presumably reflect the same metabolic phenomena, and so both have been taken to represent an *anaerobic threshold*.

A proponent of the anaerobic threshold concept has defined it more narrowly as the highest oxygen uptake during exercise before blood lactate rises (Davis, 1985). Despite a vigorous debate about the terminology as it may or may not relate to metabolic events (cp. Brooks, 1985, and Davis, 1985), the anaerobic threshold has proved to be a useful concept in research and fitness index in athletics. We predict that kinetic analysis of the plot of Pi/PCr against work rate will provide more direct information on the same metabolic phenomena.

Anaerobic threshold is about 50 to 60% of maximal oxygen consumption rate in untrained subjects, and it improves to about 70 to 75% with training. It has been improved by a high fat meal, from about 54% to about 60% (Ivy *et al.*, 1981). It predicts the racing pace of long-distance runners: for example, in one study of 13 marathon runners, the correlation coefficient was 0.98 ($P < 0.001$) (Farrell *et al.*, 1979).

Postulated mechanisms for the anaerobic threshold include the following:

o Inadequate oxygen delivery to working muscles.
o Inadequate oxidative capacity of working muscles.
o Recruitment of fast-twitch muscle fibres.
o Substrate preference – glucose over fatty acids.
o Reduced lactate removal by muscle or liver.

Participants in the anaerobic threshold controversy usually discuss each of the above five mechanisms as if they are mutually exclusive, which is not necessary or even likely. One of our studies on 18 racing sled dogs suggested a variation of the fourth mechanism (Hammel *et al.*, 1976): elevation of blood glucose concentration was associated ($r = 0.70$, $P < 0.002$) with elevation of blood lactic acid concentration. Both events appeared to occur in the most extreme form in the most excited dogs. Release of adrenalin was presumed to cause the hyperglycaemia, hence to promote glucose utilisation and lactic acid production.

Anaerobic threshold may also be influenced by extra- and intra-cellular buffer systems. Ingestion of about 20 g sodium bicarbonate 4 to 5 hours before running improved the mean time of six skilled athletes by 2.9 s over 800 meters (Wilkes *et al.*, 1983). This benefit of 'soda-loading' was attributed to facilitated efflux of lactic acid from muscle cells. We have shown that 2% sodium bicarbonate in the diet of a filly prevented exertional rhabdomyolysis which otherwise occurred consistently in response to mild exercise (Robb & Kronfeld, 1985). We are currently testing the hypothesis that dietary sodium bicarbonate may raise the anaerobic threshold of horses.

The aerobic system

An alternative to carbohydrate-loading as a nutritional strategy for stamina is fat-adaptation (Kronfeld & Downey, 1981). Studies of substrate preferences of heart and skeletal muscle indicate that increased availability of free fatty acids suppresses utilisation of glycogen and glucose (Randle *et al.*, 1963; Newsholme, 1983). With this in mind, we started in 1973 to test the hypothesis that dogs trained aerobically on a high-fat diet would develop more stamina. Supporting evidence was obtained in the field with racing sled dogs (Hammel *et al.*, 1976) and in the laboratory with Beagles running on a treadmill (Downey, Banta & Kronfeld, 1980).

Similar results were obtained with trail-ride horses (Slade *et al.*, 1976; Hintz *et al.*, 1978). But all studies on humans, until recently, appeared to be negative, probably, we suggest because the switch to a high-fat diet was abrupt and caused ketosis, natriuresis, sodium depletion and dehydration. During the last few years, reports from Boston have indicated that adaptation to a 'eucaloric ketogenic diet' (EKD) has the potential for prompting aerobic metabolism (Phinney *et al.*, 1983; Fisher *et al.*, 1983). Glucose oxidation and hexokinase activity decrease, while fatty acid oxidation and carnitine palmitoyl transferase increase.

Preliminary reports of improved performances of endurance athletes consuming the EKD have appeared in magazines (e.g. *American Health*, page 30, March 1984). Competitive road-racing cyclists have reported informally that their feeling of power declines for about two weeks after the introduction of a high-fat, zero-carbohydrate diet, then returns. From then on, their ability to sustain aerobic work appears to be improved, for they never 'hit the wall', but they have difficulty in the anaerobic efforts of sprinting or hill-climbing. These subjective observations are consistent with enhanced fatty acid oxidation associated with diminished carbohydrate oxidation and capability for anaerobic work (S.D. Phinney & W.J. Evans, personal communication).

Stress

Strenuous exercise, like all other extra demands on the body, causes *stress*. Selye's central concept of stress has had remarkable implications in health and disease (Selye, 1976). Some contemporary workers would like to expand the concept of stress to include more than the central component (Moberg, 1985; Levine, 1985). Every stressor elicits more than the central response (Fig. 1). The central pituitary–adrenal response most characteristically follows emotional perturbations (such as uncertainty or social dislocation), and it may be diminished or eliminated by reduction of the emotional component of any stressful situation. Moreover, the relative importance of the central component diminishes 'after repeated experience' (Levine, 1985), i.e. as a consequence of training.

Our studies of stress in racing sled dogs have involved 'performance ratings' intended to reflect a dog's progressive inability to cope with further stress. The scheme embodies Selye's concepts of eustress, distress and cumulative stress. A litte stress is not harmful, indeed it may tone up mechanisms of adaptation, hence be beneficial – eustress. More stress may

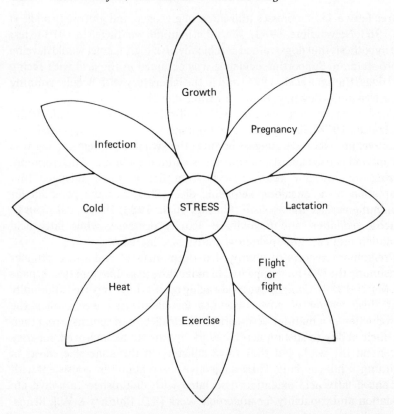

Fig. 1. The classical concept of *stress*. This is confined to a set of homeostatic adaptations that occurs in response to all extra demands on the body and emotions (Selye, 1976). It is represented by the centre of this 'daisy'. More recently, concepts of stress are becoming more expansive, and the non-central responses in each situation might be represented by a petal of the daisy. The central pituitary – adrenocortical part is associated closely with emotional perturbation, and its relative importance appears to become attenuated by accustomisation or training for any given extra demand.

become detrimental – distress. It depletes body reserves of protein, ascorbic acid and probably other nutrients, and diminishes immune competence.

Other general concepts of stress depend on its time-course, an initial phase of depletion followed by a phase of over-compensation (Fig. 2). Repeated stresses during overcompensatory phases will form an upward staircase, a phenomenon known as training. This is a series of graded and tolerable exposures to a specified type of demand for the purpose of improving ability to cope with this type of demand.

In contrast, repeated exposures during the depletion phase form a downward staircase, the phenomenon known as cumulative stress or in the sporting world as over-training.

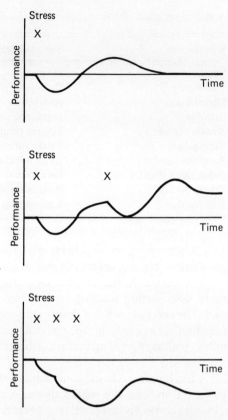

Fig. 2. Performance in relation to stress. Each bout of stress has a hypothetical time course (upper figure). The frequency and intensity of a series of stresses may improve (middle figure) or diminish (lower figure) the capability of responding to another stress. Improvement represents training, while diminishing capability represents over-training or cumulative stress.

Sound training of a team involves decisions about the frequency and intensity of exposures to stress. What is best for most of the team may be too much for an individual. Such individuals should be recognised as soon as possible. In sled dog racing, however, many inexperienced and a few seasoned 'hard' drivers overtrain the whole team before becoming aware of the problem.

We have observed that eustressed and distressed dogs show characteristic physical and behavioural signs (Table 1). Similar schemes for three syndromes or levels of stress have been proposed for farm animals (Ewbank, 1985). The groupings are typical, but not all of the listed signs are exhibited in every case.

A dog exhibiting three or more signs of severe distress (Table 1) is assigned a performance rating of 1.0 (and a rest). One or two signs of severe distress

Table 1. *Typical physical and behavioural signs of stress observed in the dog*

Eustress	Mild distress	Severe distress
Barks	Snarls	Silent
Alert	Inattentive	Unresponsive
Behaves well	Misbehaves	Apathetic
Exhilarated	Irritable	Depressed
Seeks company	Avoids company	Ignores company
Runs well	Runs poorly	Shirks running
Recovers quickly	Recovers slowly	[Does not run]
Hydrates well	Dehydrated slightly	Dehydrated
Drinks well	Drinks poorly	Refuses drink
Eats well	Eats poorly	Refuses food

will warrant a rating of 1.0 or 1.5, depending on its general demeanor. A dog that exhibits signs of mild distress (but not severe distress) is given a rating of 2.0 to 3.0, and its work is relaxed for two or three days. Signs of eustress are exhibited by many dogs during training and racing, and warrant a rating from 3.5 to 4.5. The accolade of 5.0 is reserved for the few dogs that run powerfully and continually in a calm and purposeful manner, and recover rapidly and smoothly, with no sign of agitation during or after the effort.

Dogs assigned a rating of 2.0 or less have been found consistently to have significantly depressed red blood cell indices. We regard this as a stress anaemia, for several reasons. Plasma cortisol is increased two- to three-fold in these dogs following a strenuous run (Hammel *et al.*, 1976). Cortisol would be expected to diminish formation of red blood cells. The increase in red blood cells during training of these dogs involves enhanced formation, as indicated by the increase in mean cell volume and reticulocytes (Kronfeld *et al.*, 1976). Repeated excessive cortisol release would retard this adaptation to hard work. Moreover, other stressful conditions, such as diarrhoea and kennel cough, contribute to the degrees of distress and anaemia (Adkins & Kronfeld, 1982; Kronfeld, 1984).

This anaemia in racing sled dogs tends to develop in the whole team if the diet contains less than 32% of available energy in the form of good-quality protein (Kronfeld *et al.*, 1976; Kronfeld & Downey, 1981; Adkins & Kronfeld, 1982).

Another study of dogs during the Iditarod Trail race from Anchorage to Nome, about 2,000 km, indicates a linear response of haematocrit to dietary protein that predicts prevention of anaemia by a diet containing 31% protein (Fig. 3). If these data are pooled with those for two diets containing 32 and 28% protein in a previous study (Kronfeld *et al.*, 1976), the regression of haematocrit (H, % of initial value) on dietary protein (P, % metabolisable energy) is significant ($P < 0.01$):

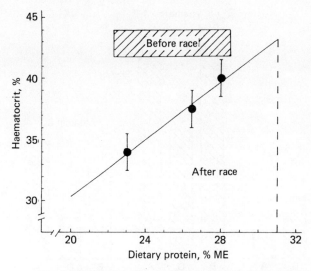

Fig. 3. Haematocrit relates to dietary protein. Haematocrit declined more in racing sled dogs fed lower levels of protein during the Iditarod Trail race from Anchorage to Nome (Adkins & Kronfeld, 1982, and unpublished data.)

$$H = 45.6 + 1.67P \pm 5.50, \; r^2 = 0.96$$

Proponents of high-fat, low-protein diets for highly stressed dogs, e.g. the military stress diet known subsequently as maximal stress diet, have suggested that anaemia was prevented in our first study by a high content of fat rather than protein. This suggestion makes no sense mechanistically, and a regression like that above of haematocrit on dietary fat has an r^2 of 0.10 which is not significant.

Others argue that our studies did not differentiate between dietary protein and iron, and that dietary iron may have contributed to the response. Iron was supplemented to reach 149 to 166 mg/kg dry matter in these diets, about five times the usually recommended level for diets made from whole chicken, pork lungs and a little rice. Moreover, in the classical studies of stress anaemia conducted by Whipple & associates, protein was more effective than iron in the repletion of red blood cells in dogs (Miller & Whipple, 1940; Yuile *et al.*, 1953).

The performance ratings have correlated in a single trial with plasma ascorbic acid concentration (Kronfeld, 1983). Plasma ascorbate declines during the cumulative stress of the racing season, along with vitamin E (Fig. 4). Stress depletes ascorbic acid in many species, including the horse (Jaesche & Kellner, 1978). Supplementation of vitamin C and, perhaps, vitamin E deserves further trials in highly stressed animals even though a review of the trials to date, especially in human athletes, was not encouraging (Dwyer & Brotherhood, 1981).

In practice, we supplement all micronutrients well above the usual

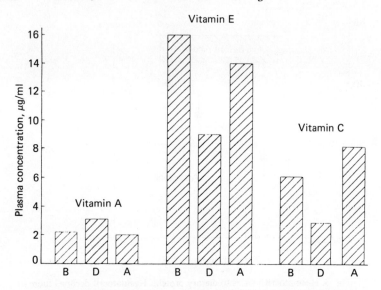

Fig. 4. Vitamin concentrations in blood: before (B), during (D) and after (A) a period of stress. Significant depressions of serum concentrations of alpha-tocopherol and ascorbic acid (but not retinol) occurred during the racing season in a champion team of racing sled dogs (Donoghue & Kronfeld, unpublished data.)

standards for growth when formulating diets for stressed animals. Under various conditions, stress has been shown to decrease blood concentrations or urine outputs of at least zinc, copper and iron, vitamins A and E, ascorbic acid and riboflavin. Given the enormous number of interactions among micronutrients, we are wary of increasing six or seven instead of the lot. So we increase all recommended micronutrients by factors of 1.3 to 2.0 and add 'unrequired' ascorbic acid, about 1 mg/kcal ME.

Conclusion

Diets for athletes are variations on four themes:

o Maintenance only – more of the same.
o Nutrient dense – less bulk.
o Enriched for work only.
o Enriched for work and stress.

Our studies on racing sled dogs suggest that the diet that promotes maximal endurance performance should have minimal bulk and maximal fat content for stamina. It also needs at least 32% ME in the form of good-quality protein and supplementary ascorbic acid, about 1 mg/kcal, to help the dog cope with stress.

REFERENCES

Adkins, T.O. & Kronfeld, D.S. (1982). Diet of racing sled dogs affects erythrocyte depression by stress. *Canadian Veterinary Journal,* **23**, 260–3.

Bergstrom, J., Hermansen, L., Hultman, E. & Saltin, B. (1967). Diet, muscle glycogen, and physical performance. *Acta Physiologica Scandanavia,* **71**, 140–50.

Brooks, G.A. (1985). Anerobic threshold: review of the concept and directions for future research. *Medicine and Science in Sports and Exercise,* **17**, 22–31.

Carlstrom, B.S. (1931). The etiology and pathogenesis in horses with haemoglobinaemia paralytica. *Skandanavian Archiv Physiologica,* **62**, 1–69.

Chance, B., Sapega, A., Sokolow, D., Eleff, S., Leigh, J.S., Graham, T., Armstrong, J. & Warnell, R. (1983). Fatigue in retrospect and prospect: 31P NMR studies of exercise performance. In *Biochemistry of Exercise,* ed. H.G. Knuttgen, J.A. Vogel & J. Poortmans, pp. 895–908. Champaign IL: Human Kinetic Publishers.

Davis, J.A. (1985). Anerobic threshold: review of the concept and directions for the future. *Medicine and Science in Sports and Medicine,* **17**, 6–18.

Downey, R.L., Kronfeld, D.S., & Banta, C.A. (1980). Diet of Beagles affects stamina. *Journal of the American Animal Hospital Association,* **16**, 273–7.

Dwyer, T., & Brotherhood, J. (1981). Long-term dietary considerations in physical training. *Proceedings Nutrition Society of Australia,* **6**, 31–40.

Ewbank, R. (1985). Behavioural responses to stress in farm animals. In *Animal Stress,* ed. G.P. Moberg, pp. 71–9. Bethesda MD: American Physiological Society.

Farrell, P.A., Wilmore, J.H., Coyle, E.F., Billings, J.E. & Costill, D.L. (1979). Plasma lactate accumulation and distance running performance. *Medicine and Science in Sports and Exercise,* **11**, 338–44.

Fisher, E.C., Evans, W.J., Phinney, S.D., Blackburn, G.L., Bistrian, B.R. & Young, V.R. (1983). Changes in skeletal muscle metabolism induced by a eucaloric ketogenic diet. In *Biochemistry of Exercise,* ed. H.G. Knuttgen, J.A. Vogel & J. Poortmans, pp. 497–501. Champaign IL: Human Kinetic Publishers.

Fox, E.L. (1973). Difference in metabolic alterations with sprint versus endurance interval training programs. In *Metabolic Adaptation to Prolonged Physical Exercise* ed. H. Howald & J.R. Poortmans, pp. 119–26. Magglingen.

Hammel, E.P., Kronfeld, D.S., Ganjam, V.K. & Dunlap, H.L. (1976). Metabolic responses to exhaustive exercise in racing sled dogs fed diets containing medium, low or zero carbohydrate. *American Journal of Clinical Nutrition,* **30**, 409–18.

Hintz, H.F., Ross, M.W., Lesser, F.R., Lieds, P.F., White, K.K., Lowe, J.E., Short, C.E. & Schryver, H.F. (1978). The value of dietary fat for working horses. 1. Biochemistry and hematological evaluation, *Journal of Equine Medicine and Surgery*, **2**, 483–8.

Ivy, J.L., Costill, D.L., VanHandel, P.J., Essig, D.A. & Lower, R.W. (1981). Alteration in the lactate threshold with changes in substrate availability. *International Journal of Sports Medicine*, **2**, 139–42.

Ivy, J.L., Sherman, W.M., Miller, W., Farrell, S. & Fishberg, B. (1983). Glycogen synthesis: effect of diet and training. In *Biochemistry of Exercise*, ed. H.G. Knuttgen, J.A. Vogel & J. Poortmans, pp. 291–6. Champaign IL: Human Kinetic Publishers.

Jaesche, G. & Kellner, H. (1978). Beitrag sum ascorbinsaurestatus des pferdes 2. Mitteilung: Klinsche aspecte und mangelsituationen. *Berlin-Munchen Tierartzlichen Woschrift*, **91**, 375–9.

Karlsson, J. & Saltin, B. (1971). Diet, muscle glycogen, and endurance performance. *Journal of Applied Physiology*, **31**, 203–6.

Kronfeld, D.S. (1973). Diet and the performance of racing sled dogs. *Journal of the American Veterinary Medical Association*, **162**, 470–3.

Kronfeld, D.S. (1983). Stress supplements: Protein and vitamin C. *Pure-Bred Dogs/Kennel Gazette*, **100**. 10, 8–9.

Kronfeld, D.S. (1984). Kennel cough: The 1984 Alaskan epidemic. *Team & Trail*, **21**. 7, 3–9.

Kronfeld, D.S. & Downey, R.L. (1981). Nutritional strategies for stamina in dogs and horses. *Proceedings Nutrition Society of Australia*, **6**, 21–9.

Kronfeld, D.S., Hammel, E.P., Ramberg, C.F. & Dunlap, H.L. (1976). Hematological and metabolic responses to training in racing sled dogs fed diets containing medium, low and zero carbohydrate. *American Journal of Clinical Nutrition*, **30**, 419–30.

Levine, S. (1985). A definition of stress. In *Animal Stress*, ed. G.P. Moberg, pp. 51–69. Bethesda MD: American Physiological Society.

Lindholm, A., Johansson, H.E. & Kjaersgaard, P. (1974). Acute rhabdomyolysis in Standard bred horses. A morphological and biochemical study. *Acta Veterinaria Scandanavia*, **15**, 325–39.

McGilvery, R.W. (1973). The use of fuels for muscular work. In *Metabolic Adaptation to Prolonged Physical Exercise* ed. H. Howald & J.R. Poortmans, pp. 119–26. Magglingen.

Miller, L.L. & Whipple, G.H. (1940). Chloroform injury decreases as protein stores decrease. Studies of nitrogen metabolism in these dogs. *American Journal of Medical Sciences*, **199**, 204–16.

Moberg, G.P. (1985). Biological response to stress: key to assessment of animal well-being? In *Animal Stress*, ed. G.P. Moberg, pp. 27–49. Bethesda MD: American Physiological Society.

Newsholme, E.A. (1983). Control of metabolism and the integration of fuel supply for the marathon runner. In *Biochemistry of Exercise*, ed. H.G. Knuttgen, J.A. Vogel & J. Poortmans, pp. 144–50. Champaign IL: Human Kinetic Publishers.

Phinney, S.D., Bistrian, B.R., Evans, W.T., Gervino, E. & Blackburn, G.L. (1983). The human metabolic response to chronic ketosis without caloric restriction; preservation of submaximal exercise capability with reduced carbohydrate oxidation. *Metabolism*, **32**, 769–75.

Randle, P.J., Garland, P.B., Hales, C.N. & Newsholme, E.A. (1963). The glucose–fatty acid cycle. *Lancet i*, 785–9.

Robb, E.J. & Kronfeld, D.S. (1985). Dietary sodium bicarbonate as a treatment for equine exertional rhabdomyolysis. *Journal of the American Veterinary Medical Association*, **188**, 602–7.

Ryan, A.J. (1981). Anabolic steroids are fool's gold. *Federation Proceedings*, **40**, 2682–8.

Sahlin, K. (1983). Effect of acidosis on energy metabolism and force generation in skeletal muscle. In *Biochemistry of Exercise*, ed. H.G. Knuttgen, J.A. Vogel & J. Poortmans, pp. 151–60. Champaign IL: Human Kinetic Publishers.

Selye, H. (1976). *The Stress of Life*. New York: McGraw-Hill.

Slade, L.M., Lewis, L.D., Quinn, C.R. & Chandler, M.L. (1976). Nutritional adaptation of horses for endurance performances. *Proceedings Equine Nutrition Symposium*, **4**, 114–21.

Wilkes, D., Gledhill, N. & Smyth, R. (1983). Effect of acute induced metabolic alkalosis on 800-m racing time. *Medicine and Science in Sports and Exercise*, **15**, 277–80.

Yuile, C.L., Lucas, F.V., Jones, C.K., Chapin, S.J. & Whipple, G.H. (1953). Inflammation and protein metabolism studies of C-14 labeled proteins in dogs with sterile abscesses. *Journal of Experimental Medicine*, **98**, 173–94.

9

Feeding behaviour of the cat

EDWARD KANE

Introduction

As a result of studies conducted on palatability and feeding patterns of adult cats and kittens, knowledge of the feeding habits of cats has increased in recent years.

Cats are regarded as among the domestic animal species, having evolved distinct nutritional needs and feeding-behaviour patterns, presumably because they are strict carnivores. Cats like to consume freshly killed carcasses rather than carrion (Scott, 1968). As Table 1 shows, different surveys show quite wide variation in the diets of cats but clearly small mammals – rats, mice and rabbits – make up the bulk of their diet. Birds are much less important as are reptiles, frogs, and insects (Dilks, 1979; Fitzgerald & Karl, 1979; Jones & Coman, 1981; Karl & Best, 1982). Large cats (*Panthera* and *Felis* species) prey mainly on herbivores or primates, especially the young of the species (Scott, 1968). Because of the limited availability of prey, large cats do not necessarily eat often. Among the large cats, some species hunt by day, others by night. In some species, the males hunt; in other species, the females do the killing. Large cats in the wild can be intermittent feeders, with a tremendous capacity to store temporary excesses of food as fat, so that adults (not pregnant or lactating) can go without food for long periods of time (Scott, 1968).

A small rodent, as consumed, has a nutrient content of approximately 64–76% water, 14–18% protein, 6–18% fat, 1–5% minerals, the vitamins mostly obtained from the liver, and other internal organs (Brown, Frahm & Johnson, 1977; Aubert, Suquet & Lemonnier, 1979; Bulbulian, Gunewald & Haack, 1985). Cats kept as pets have more varied dietary habits and are liable to be fed what and when their owners see fit; the availability of food is seldom limiting. Many cats have adapted to coerce their owners to feed them according 'to their schedule'. Cats are frequently said to be difficult to feed, finicky and selective. When compared with other species, however, although there are some differences, there are many more similarities. It has been shown that cats are similar to other species in that their energy intake

147

Table 1. *Food intake of feral cats. Approximate percentage or relative importance in diet*[A]

Mammals (primarily rats, mice, rabbits)	55.0	55.0	*	dominant	88.0	dominant (wild rabbits, rodents)	trace
Birds	4.0	8.5	—	secondary	3.5	minor	92
Reptiles, frogs	2.0	10.6	—	—	1.2	—	trace
Insects	12.5	19.0	—	present	0.5	—	trace
Garbage	26.0	—	—	present	—	—	—
Carbohydrate (vegetation)	trace	6.4	—	present	—	—	—
Reference	1	2	3	4	5	6	7

Notes:

*Species listed, no numerical account

[A] – As inferred from relative volume of stomach contents

1 – McMurry & Sperry, 1941
2 – Parmalee, 1953
3 – Toner, 1956
4 – Errington, 1936
5 – Coman & Brunner, 1972
6 – Liberg, 1984
7 – Kirkpatrick & Rauzon, 1986

is controlled and by similar neural pathways. For example, lesions in the VMH (ventromedial hypothalamus) cause hyperphagia, LH (lateral hypothalamus) lesions cause aphagia in cats as in other mammals (Anand & Brobeck, 1951).

Palatability

Palatability, a concept which includes aroma, texture, consistency and taste, comprises several unique features in the cat. They can be extremely sensitive to very minute differences between diets. In our experience at Carnation, where we create a large variety of recipes, meat-based and dry expanded, varying only a small percentage of one ingredient will often have a marked effect on palatability. Palatability testing procedures assume that cats choose the diet most agreeable to their palates. Their decisions are used as a development research tool in formulating diets, but also illustrate the sensitivity of cats to the palatability of diets they will consume. Various studies have been conducted in an attempt to define such factors. Previous eating experience, texture, physical nature, odour and taste are among those currently thought to be most important.

Previous eating experience

Whether early experience in the kitten markedly affects later feeding behaviour is of great interest, although the evidence is sufficiently confused to warrant further investigation. Hart (1974) reported that cats will eat a variety of foods later in life if they were first presented with similar foods early in life – especially directly after weaning. Kuo (1967) fed specific diets to three groups of new-born kittens and later tested their response to foods. One group was fed a diet solely of soybeans; the second group, mackerel and rice; the third group, a variety of foods. All diets were supplemented with vitamins and minerals. After six months each group would only consume diets similar to the diet on which they had been raised. Mugford (1977) could not replicate such enhanced preference, but the cats in his study were introduced to their diets at weaning, not immediately after birth.

According to Adamec (1976), when cats were given a choice, they preferred meat-based cat food to rats, cold rats to warm ones, and salmon to the cat food fed. These results may well depend on prior prey-eating experience and the specific cat food fed.

It has been shown by Mugford & Thorne (1980) that cats will eat a novel diet rather than a familiar diet unless they are put into a new environment or stressful situation, in which case they will choose the familiar diet. It is often assumed that, as hunger increases, poor palatability becomes less important in limiting food intake. This is assumed to be the case in, for example, man in famines, and the dog. In cats, however, in the laboratory (Kane, unpublished observations) and in the home environment, a nutritionally complete but relatively unpalatable diet will be refused for long periods.

Texture and physical nature

Beauchamp, Maller & Rogers (1977) have suggested that cats prefer solutions of greater density for their textural qualities, e.g. cats preferred whole milk to dilute milk, because it is thicker. Hirsch, Dubose & Jacobs (1978) have shown that cats first fed a diet of dry-expanded, commercial cat food, will decrease their food intake when this substance is ground to a powder. One can observe, in the home, a cat's dislike for the smaller particles by noting that fines will sometimes remain in the food bowl even when cats are anxious for food. From observations in the laboratory, texture of the diet is known to be very important.

A purified diet prepared of a dough or paste consistency is cautiously accepted (Kane *et al.*, 1981*a*). As the diet becomes powdery or dry in appearance, due to varying ingredient content or the result of mixing ingredients in different ways, acceptance decreases. Kane *et al.* (1986) found that food intake was significantly greater for a pelleted, purified, 15% fat diet, than for a 10% fat diet that was not pelleted (i.e. powdery). A fat-free diet prepared as a gel containing 50–60% water was readily acceptable to kittens (McDonald *et al.*, 1983). Relatively low-fat diets are, therefore, palatable as long as the physical nature of the diet is not rejected by the kittens. Commercial, expanded dry diets contain approximately 10% fat and can, of course, be highly palatable.

Flavour and nutrient content

Cats seem to prefer a variety of foods in their diet and to discriminate against specific ingredients e.g. saccharin and cyclamate (Bartoshuk *et al.*, 1975), casein (Cook *et al.*, 1985), medium-chain triglycerides and caprylic acid (McDonald, Rogers & Morris, 1983). They also discriminate against certain specific ingredient combinations (Kane, Morris & Rogers, 1981*a*; Kane, unpublished observations). Studies have been conducted concerning the cat's selection of carbohydrates, fats and protein.

Carbohydrates

Early studies indicated that cats did not detect sweet substances (Carpenter, 1956). It appears reasonable to expect this in an animal such as the cat, which is a strict carnivore. However, Bartoshuk, Harned & Parks (1971) disputed this observation. With reference to the cat's negative response to sugar in pure water solution, they suggested that water, via water-sensitive nerve fibres, interferes with the preference for sucrose, through their ability to mask the taste of sucrose in the cat. Preference for sugar in dilute milk (1:4) had previously been shown for both lactose and sucrose (Frings, 1951). Boudreau, *et al.* (1971) observed that common cat foods mixed in distilled water caused discharges in most tongue nerve fibres. He also found that only very high concentrations of solutions of sucrose would elicit a response. Boudreau, Anderson & Oravec (1975), testing tongue ganglion discharge, reported that substances commonly associated with a carnivorous cat diet (i.e. pork, kidney, pork liver, tuna, cod, egg yolk and chicken) elicited the strongest response.

In view of conflicting evidence concerning the cat's response to sweetness, Beauchamp *et al*, (1977) concluded that cats showed no preference for sugars or artificial sweeteners, whether in water or saline. Beauchamp suggested that, although sweet carbohydrates do not stimulate ingestion, several substances associated with the cat's carnivorous diet do (e.g. hydrolysed protein and emulsified fats). Cats preferred hydrolysed soy protein, lactalbumin, casein, L-alanine, L-proline and butterfat in deionised water solution. Further support for the suggestion that cats prefer substances associated with meat-based diets came when Mugford (1977) demonstrated that cats positively respond to odour from meat passed through dry cat food.

Fats

Fat is present in relatively high amounts (35–40% of the dry matter) in small animals such as rodents, which are components of feral cat diets. Extensive use is made of supplemental fat, especially tallow, in commercial dry and canned meat-based diets. Kane *et al*, (1981a) studied acceptability of diets made with various sources and levels of fat. Seven diets contained 25% each of bleached tallow, unbleached tallow, chicken fat, yellow grease, lard, butterfat or hydrogenated vegetable oil. Cats exhibited marked preferences for diets based on certain fats as demonstrated by the greater intake (significant, $P < 0.001$) of diets made from bleached tallow rather than chicken or butterfat. Five of eight cats preferred unbleached tallow to bleached tallow; 5 of 8 cats preferred bleached tallow to lard; 6 of 8 preferred yellow grease to bleached tallow; all these were not significant. Cats preferred diets made with 25% yellow grease over 15% ($P < 0.001$) or 50% ($P < 0.02$). Recent work (Kane *et al.*, 1986) suggests that these preferences were the result of differences in texture of the diets caused by different levels of fat. In another study comparing fats, Beauchamp *et al.* (1977) reported that cats preferred a solution of butter fat to corn oil. Clearly, fat is a substance that elicits a response to a feedstuff, and cats do prefer some fats over others. But fat content *per se*, independent of flavour components and its effect on consistency, has little effect on dietary preference of the cat (Kane *et al.*, 1986). Cats showed no significant preference between a 15% and a 45% fat diet.

Proteins

Cook *et al.* (1985) investigated kittens' preference for semipurified diets made with various levels of two sources of protein. Unlike rats which avoid protein-free diets (Osborne & Mendel, 1918; Sanahuja & Harper, 1963;

Table 2. Daily food and water intake

Diet	Commercial		Purified			
	Dry[1,a]	Canned[1,b]	Casein[1,c]	Amino acid[1,d]	Low fat[2,e]	High fat[2,f]
Daily food intake g gDM	86.3±6.8 / 79.7	256.1±25.1 / 57.4	75.6±12.8 / 72.2	73.7±6.6 / 70.5	75.8±11.2 / 71.8	51.2±7.4 / 49.9
kcal/kgBW	74.9±7.9	73.1±7.2	71.8±5.5	65.8±3.9	63.6±4.7	63.7±6.5
Daily meal frequency	15.7±1.4	16.6±1.8	16.8±2.3	17.4±3.5	11.4±1.2	10.4±1.1
Light/meal frequency	8.6±0.9	10.7±1.2	10.4±1.2	10.6±2.2	6.5±0.8	6.7±0.6
Dark/meal frequency	7.1±0.7	5.9±0.8	6.3±1.3	6.7±1.5	4.8±0.6	3.9±0.6
Daily water intake ml	147.3±15.7	0	79.4±10.6	98.1±25.8	112.8±23.4	99.4±20.8
No. bouts	16±1.3	0	12.5±1.8	12.4±2.1	9.9±1.5	10.8±2.3

Notes:

[1] Kane et al., 1981b

[2] Kane et al., 1986

[a] DM 92.3%, GE 4.5 kcal/g.

[b] DM 23.4%, GE 1.5 kcal/g.

[c] DM 95.5%, GE 5.4 kcal/g.

[d] DM 95.7%, GE 5.0 kcal/g.

[e] DM 94.7%, GE 4.85 kcal/g.

[f] DM 97.4%, GE 6.52 kcal/g.

Rogers, Tannous & Harper, 1967; Leung, Rogers & Harper, 1968) and choose diets of low or moderate protein in preference to high-protein diets, (Harper, 1967; Peng *et al.*, 1975; Leung, Gamble & Rogers, 1981) selection by kittens of diets of varying protein content did not provide for consistent protein intake. When kittens were offered a choice of casein-based diets containing 18%, 36% and 54% protein, they selected the diets of lower casein, i.e. protein, content. Total food intake was reduced when the choice was only between 36% and 54% protein. Diets of soy protein at 16%, 31% and 63% showed no significant preference. When the soy diets were offered with a protein-free diet, kittens selected similar amounts of each, resulting in low protein consumption for the 16% protein-free group.

Although cats exhibited these choices when fed purified diets, when they are fed commercial meat or fish-based cat food diets, or 'natural' diets, they would be consuming a diet of approximately 40–80% protein on a dry matter basis. Perhaps, during the evolution of the cat, there has been no selective pressure on control systems to ensure an adequate protein intake, since presumably, cats have always ingested relatively constant and greater than adequate levels of protein.

Feeding patterns

Cats as strict carnivores are somewhat unique in regard to diurnal feeding patterns and to circadian rhythms such as sleep–wake and activity. It has been suggested that cats are intermittent feeders, eating once in 24 hours (Scott, 1968, 1971); or feed predominantly during the dark period (Kanarek, 1975). However, Mugford (1977) and Mugford & Thorne (1980) showed that cats ate frequently throughout the 24-hour period and there was no direct relationship between the size and timing of meals. In a study of feeding and drinking patterns of cats fed laboratory and commercial diets, Kane *et al.* (1981*b*) also reported that cats ate many small meals throughout the 24-hour period (Table 2). In this study cats were each fed *ad libitum* either a commercial dry, expanded product, a high-moisture, canned gourmet product, a casein-based purified diet, or an amino-acid-based, purified diet. Deionised water was available *ad libitum*. The cats were housed in individual metabolism cages, modified to allow food intake measurement. The data were continuously recorded utilising a computer. Eight cats were fed each of four diets during eight consecutive 7-to-10-day periods, in a randomised design, except that the cats were fed the two commercial diets or two purified diets in sequence to enhance acceptance. The room environment was maintained at 22°C, 50% relative humidity with a 12-hour, light–dark cycle. The cats were fed at 12:00 PM daily as the lights turned on.

The cats ate small meals randomly throughout the 24-hour period, with a mean of 16 meals per day. Mugford (1977) reported 13 meals daily and Kanarek (1975) reported ten meals per day. Significant correlations were found between meal size and pre-meal intervals for the cats when fed either the canned or dry commercial diet. No significant correlations were found between post intermeal intervals and meal size. This agrees with the observations of Kanarek (1975) for the cat, but not with the work of Kaufman *et al.* (1980) using an uncaged cat, or with work conducted with other species (Levitsky, 1974). Mean daily caloric intake was 71 kcal/kg Body Weight (BW). This agrees with other published data obtained on the adult cat (NRC, 1978). There were no differences in daily caloric intake for cats fed diets ranging in caloric value from 4.9 to 6.5 kcal/g Dry Matter (DM). From these results it appears that cats are able to, and do, regulate energy consumption. Others previously had not agreed. For example, Hirsch *et al.* (1978) had suggested that cats do not regulate caloric intake. They fed a group of cats powdered dry commercial cat food diluted with 0–40% kaolin and found no increase in food intake with increase in dilution with kaolin. Kanarek (1975) fed two cats ground dry commercial cat food (3.4 kcal/g) or ground dry commercial cat food diluted with cellulose (2.7 kcal/g) and found no difference in energy intake between the two diets. The poor palatability of the two diluents, kaolin and cellulose, may well have limited food intake in these experiments.

Although the cats did not drink water when fed the canned diet, their total water intake exceeded the amount consumed while eating the dry diets. This is in agreement with the observations of other workers (Seefeldt & Chapman, 1979). Hirsch *et al.* (1978) reported the total water intake of cats fed a dry diet was 2.5 g water per gram dry matter. Kanarek (1975) reported a value of 1.7:1. Thrall & Miller (1976) reported total water-to-dry-matter-intake ratios of 2.3:1 and 3.3:1 for cats fed dry and canned commercial diets. Prentiss *et al.* (1959) reported cats surviving without drinking water when fed diets of cod fillets, ground salmon, or beefsteak. Seefeldt & Chapman (1979) reported that cats fed dry food drank six times as much water as when fed a canned diet. Total water intake, though, was significantly greater while eating canned food. Total water intake to DM ratios were: dry food, 2.27; canned, 3.94.

In an additional study of simultaneous monitoring of food and water intake of adult cats, Kane *et al.* (1986) fed diets containing 15% and 45% fat (Table 2). The diets were fed to each cat in sequence, 30 days – 15% fat, 21 days – 45% fat. During low-fat feeding, food intake (75.8 g) and feeding patterns were consistent in meal size (7.3 g/meal) and frequency (11.4 meals/day). When switched from the low, to the high-fat diet, cats decreased their daily food intake (light–dark cycle total) significantly ($P < 0.05$), primarily by eating smaller meals ($P < 0.01$), since daily meal frequency fluctuated only very slightly. High-fat substitution caused reduction of both the light and dark-cycle food intake, but only the latter was significant ($P < 0.01$). Representative feeding patterns illustrated topical responses and showed that the reduction in food intake was due mainly to a decrease in meal size. Mugford & Thorne (1980) reported that, when cats were fed isocaloric diets, they decreased their intake of low-palatability diets by decreasing both meal size and meal frequency. As in the previous study (Kane *et al.*, 1981*b*), the cats ate meals randomly throughout the 24-hour period, although the mean daily meal frequency (meals/day) was 11.1; mean daily meal size (g/meal) was 7.3. These studies also demonstrated that cats regulate their energy intake via a physiological mechanism. Both gross energy intake per meal (7.3 g × 4.8 kcal = 35.04 kcal, low-fat; 5.2 g × 6.5 kcal = 33.8 kcal, high-fat) and daily energy intake (kcal/kg bodyweight: 63.6 low-fat; and 63.7 high-fat) were similar.

Summary

In summary, the feeding behaviour of the cat is unique and complex. The cat has evolved from an animal that once protected the granaries of Egypt and helped contain the Great Plague of Europe. Its feeding habits, therefore, have changed somewhat from an active prey-catching, strict carnivore to a domesticated pet, fed both meat-based and grain-based diets. Throughout this transition period, though, the cat has maintained both a unique nutrient need and specific feeding behaviour with a keen sense of palatability. It can regulate energy intake via a mechanism which includes many small meals and lacks circadian rhythmicity.

REFERENCES

Adamec, R.E. (1976). The interaction of hunger and preying in domestic cat (*Felis catus*): An adaptive hierarchy? *Behavioural Biology,* **18**, 263–72.

Anand, B.K. & Brobeck, J.R. (1951). Hypothalamic control of food intake in rats and cats. *Yale Journal of Biology and Medicine,* **24**, 123–40.

Aubert, R., Suquet, J. & Lemonnier, D. (1979). Long-term morphological and metabolic effects of early under- and over-nutrition in mice. *Journal of Nutrition,* **110**, 649–61.

Bartoshuk, L.M., Harned, M.A. & Parks, L.H. (1971). Taste of water in the cat: Effects on sucrose preference. *Science,* **171**, 699–701.

Bartoshuk, L.M., Jacobs, H.L., Nichols, T.L., Hoff, L.A. & Ryckman, J.J. (1975). Taste rejection in cats. *Journal of Comparative Physiological Psychology,* **89**, 971–5.

Beauchamp, G.K., Maller, O. & Rogers, J.G. Jr. (1977). Flavor preferences in cats (*Felis catus* and *Panthera* sp). *Journal of Comparative Physiological Psychology,* **91**, 1118–27.

Boudreau, J.C., Bradley, B.E., Bierer, P.R., Kruger, S. & Tsuchitani, C. (1971). Single unit recording from the geniculate ganglion of the facial nerve of the cat. *Experimental Brain Research,* **13**, 461–88.

Boudreau, J.C., Anderson, W. & Oravec, J. (1975). Chemical stimulus determinants of cat geniculate ganglion chemoresponsive group II unit discharge. *Chemical Senses and Flavor,* **1**, 495–517.

Brown, M.A., Frahm, R.R., & Johnson, R.R. (1977). Body composition

156   E. Kane

of mice selected for preweaning and postweaning growth. *Journal of Animal Science,* **45,** 18–23.

Bulbulian, R., Grunewald, K.K. & Haack, R.R. (1985). Effect of exercise duration on feed intake and body composition of Swiss albino mice. *Journal of Applied Physiology,* **58,** 500–6.

Carpenter, J.A. (1956). Species differences in taste preferences. *Journal of Comparative Physiological Psychology,* **49,** 139–44.

Coman, B.J. & Brunner, H. (1972). Food habits of the feral house cat in Victoria. *Journal of Wildlife Management,* **36,** 848–53.

Cook, N.E., Kane, E., Rogers, Q.R., & Morris, J.G. (1985). Self-selection of dietary casein and soy protein by the cat. *Physiology and Behavior,* **34,** 583–94.

Dilks, P.J. (1979). Observations on the food of feral cats on Campbell Island *New Zealand Journal of Ecology,* **2,** 64–6.

Errington, P.O. (1936). Notes on food habits of southern Wisconsin house cats. *Journal of Mammalogy,* **17,** 64–5.

Fitzgerald, B.M. & Karl, B.J. (1979). Foods of feral house cats (*Felis catus* L.) in forest of the Orongorongo Valley, Wellington, *New Zealand Journal of Zoology,* **6,** 107–26.

Frings, H. (1981). Sweet taste in the cat and the taste spectrum. *Experientia,* **7,** 424–6.

Harper, A.E. (1967). Effects of dietary protein content and amino acid pattern on food intake and preference. *Handbook of Physiology,* **1,** 399–410.

Hart, B.L. (1974). Feline behaviour. Feeding behaviour. *Feline Practice,* **4,** 8.

Hirsch, E., Dubose, C. & Jacobs, H.L. (1978). Dietary control of food intake in cats. *Physiology and Behavior,* **20,** 287–95.

Jones, E., & Coman, B.J. (1981). Ecology of the feral cat, Felis catus (L.), in southeastern Australia. I. Diet. *Australian Wildlife Research,* **8,** 537–47.

Karl, B.J., & Best, H.A. (1982). Feral cats on Stewart Island; their foods, and their effects of kakapo. *New Zealand Journal of Zoology,* **9,** 287–94.

Kanarek, R.B. (1975). Availability and caloric density of diet as determinants of meal patterns in cats. *Physiology and Behavior,* **15,** 611–18.

Kane, E., Leung, P.M.B., Rogers, Q.R. & Morris, J.G. (1986). Diurnal feeding and drinking patterns in adult cats as affected by changing the level of fat in the diet. *Appetite,* (in press).

Kane, E., Morris, J.G. & Rogers, Q.R. (1981*a*). Acceptability and digestibility by adult cats of diets made with various sources and levels of fat. *Journal of Animal Science,* **53,** 1516–23.

Kane, E., Rogers, Q.R., Morris, J.G. & Leung, P.M.B. (1981*b*). Feeding behavior of the cat fed laboratory and commercial diets. *Nutrition Research,* **1,** 499–507.

Kaufman, L.W., Collier, G., Hill, W.G. & Collins, K. (1980). Meal cost and meal patterns in an uncaged domestic cat. *Physiology and Behavior*, **25**, 135–7.

Kuo, Z.Y. (1967). *The Dynamics of Behavior Development, an Epigenetic View*. Random House, NY.

Leung, P.M.B., Gamble, M.A. & Rogers, Q.R. (1981). Effect of prior protein ingestions on dietary choice of protein and energy in the rat. *Nutrition Reports International*, **240**, 257–66.

Leung, P.M.B., Rogers, Q.R., & Harper, A.E. (1968). Effect of amino acid imbalance on dietary choice in the rat. *Journal of Nutrition*, **95**, 483–92.

Levitsky, D. (1974). Feeding conditions and intermeal relationships. *Physiology and Behavior*, **12**, 779–87.

Liberg, O. (1984). Food habits and prey impact by feral and house-based domestic cats in a rural area in Southern Sweden. *Journal of Mammology*, **65**, 424–32.

McDonald, M.L., Rogers, Q.R. & Morris, J.G. (1983). Role of linoleate as an essential fatty acid for the cat independent of arachidonate synthesis. *Journal of Nutrition*, **113**, 1422–33.

McMurry, F.B. & Sperry, C.C. (1941). Food of feral house cats in Oklahoma, a progress report. *Journal of Mammology*, **22**, 183–90.

Mugford, R.A. (1977). External influences on the feeding of carnivores. In *The Chemical Senses and Nutrition*, ed. M. Kare & O. Maller, pp. 25–50. New York: Academic Press.

Mugford, R.A., & Thorne, C.J. (1980). Comparative studies of meal patterns in pet and laboratory housed dogs and cats. In *Nutrition of the Dog & Cat*, ed. R.S. Anderson, pp. 3–14. Oxford: Pergamon Press.

NRC (1978). Nutrient requirements of domestic animals, No. 13. *Nutrient Requirements of Cats*. National Academy Sciences, National Research Council, Washington, DC.

Osborne, T.B. & Mendel, L.B. (1918). The choice between adequate and inadequate diets, as made by rats. *Journal of Biological Chemistry*, **35**, 19–27.

Parmalee, P.W. (1953). Food habits of the feral house cat in East–Central Texas. *Journal of Wildlife Management*, **17**, 375–6.

Peng, Y., Meliza, L.L., Vavich, M.G., & Kemmerer, A.R. (1975). Effects of amino acid imbalance and protein content of diets on food intake and preference of young, adult and diabetic rats. *Journal of Nutrition*, **105**, 1395–1404.

Prentiss, P.G., Wolf, A.V. & Eddy, H.A. (1959). Hydropenia in cat and dog. Ability of the cat to meet its water requirements solely from a diet of fish or meat. *American Journal of Physiology*, **196**, 625–32.

Rogers, A.R., Tannous, R.I. & Harper, A.E. (1967). Effects of excess leucine on growth and food selection. *Journal of Nutrition*, **91**, 561–72.

Sanahuja, J.C. & Harper, A.E. (1963). Amino acid balance and imbalance. XII. Effect of amino acid balance on self-selection of diet by the rat. *Journal of Nutrition*, **81**, 363–71.

Scott, P.P. (1968). The special features of nutrition of cats with observations on wild felidae nutrition in the London Zoo. *Symposium of the Zoological Society*, London, 21–36.

Scott, P.P. (1971). Dietary requirements of the cat in relation to practical feeding problems. *Paper on Small Animal Nutrition Workshop*, University of Illinois, College of Veterinary Medicine.

Seefeldt, S.L., & Chapman, T.E. (1979). Body water contents and turnover in cats fed dry and canned rations. *American Journal of Veterinary Research*, **40**, 183–5.

Thrall, B. & Miller, L.G. (1976). Water turnover in cats fed dry rations. *Feline Practice*, **6(1)**, 10–17.

Toner, G.C. (1956). House cat predation on small animals. *Journal of Mammalogy*, **37**, 119.

10

Protein in the nutrition of dogs and cats

MONICA C. SCHAEFFER, QUINTON R. ROGERS AND
JAMES G. MORRIS

Introduction

One hundred and seventy years ago Magendie postulated that dogs
require a source of nitrogen in their diet (Magendie, 1816). Yet the dietary
levels at which protein and its constituent amino acids are required by
canines and felines are still subjects of research and debate. For example, the
1985 NRC Subcommittee on Dog Nutrition has estimated the protein
requirement of growing dogs at 11.4% of dietary energy (NRC, 1985), just
half the value of 22% proposed by the previous committee (NRC, 1974). The
new value was based on the surge of research on amino acid requirements
which occurred during the intervening 11 years as well as an attempt to
establish a minimum requirement rather than a recommended allowance
(see chapter 3). A vivid illustration of the renewed research is provided by
the number of reference citations on amino acid nutrition in the 1985
report: nearly *half* (28 out of 59) of the papers cited had been published in the
11 years since the previous committee's report. Only a quarter of the
citations in the 1974 report had been as relatively recent. An even stronger
surge of research on protein and amino acid nutrition of the cat occurred
during this time.

Some of the earliest work on protein nutrition, and a number of the
milestone experiments, were done using dogs. However, in the early
twentieth century, use of dogs became less common, and rats became the
preferred experimental animal because of expense, convenience, time, and
other practical factors. Use of cats as experimental animals never became
widespread in nutrition, probably because many types of purified diets are
not readily accepted by them. These factors explain why, until recently, little
information existed about the protein and amino acid nutrition of dogs and
particularly cats.

Many factors – environmental, physiological and genetic, as well as
dietary – affect protein requirements, and have made a consensus difficult to
reach. Therefore, we shall begin by putting the general problem of defining
protein requirements in an historical perspective (a more detailed treatment

159

is given by Munro, 1964). After this we outline potential pitfalls in experiments intended to determine protein and amino acid requirements, and conclude with a review of the literature specifically relating to dogs and cats.

Historical perspective

Daniel Rutherford is generally credited with the discovery of gaseous nitrogen, which he called 'aer malignus' in his doctoral dissertation at the University of Edinburgh in 1772 (cited in Munro, 1964). He showed that it was incapable of supporting life or flame, work later extended by the French chemist Lavoisier in 1790. The identification of a biological role for the 'inert' element emerged only after the development of a new approach to the analysis of organic compounds by Gay-Lussac and Thenard in 1810 (cited in Munro, 1964), in which measurable quantities of gaseous nitrogen were liberated from organic compounds. Only a few years after this Magendie (1816) performed his experiments. He fed dogs either sugar or olive oil exclusively, and found these substances incapable of supporting life. Eggs or cheese, on the other hand, would support the dogs apparently indefinitely, though in a weakened state. Magendie concluded that dietary nitrogen was essential, and, moreover, that it was the ultimate source of nitrogen in the body. He thereby made the first real distinction between nitrogen and non-nitrogen containing foods. While some of his results were no doubt complicated by unrecognised multiple vitamin and mineral deficiencies, his insights were germinal to the development of modern nutrition.

Further clarification of the role of nitrogen in nutrition and metabolism came from the German chemist Justus von Liebig, who focused on urea as the end-product of tissue protein degradation. His work with dogs laid the groundwork from which, in 1860, Carl Voit developed the concept of nitrogen balance. Voit improved the methods for measuring nitrogen in the urine and then systematically studied the effect of varying levels of dietary protein on urinary nitrogen output of dogs. He found that healthy adult dogs were in *nitrogen balance*, that is, the amount of nitrogen excreted was equivalent to the amount of dietary nitrogen consumed. If animals were deprived of food, the amount of nitrogen excreted exceeded intake, i.e. there was a net loss of nitrogen from the body, a state of *negative nitrogen balance*. Voit also studied the effects of other dietary constituents on the excretion of nitrogen by dogs.

The technique of nitrogen balance became widely accepted as useful both in the evaluation of nutritional status and in the determination of dietary protein requirements. To this day it is almost unchallenged as the most sensitive approach for assessing dietary nitrogen and amino acid requirements.

Voit's work did not immediately lead to meaningful estimates of protein requirements. From the mid-nineteenth century onwards it was recognised that proteins differed in quality (Munro, 1964), that is, in the amount needed to meet requirement. A full understanding of the concept of requirement had to await the clarification of the chemical composition of proteins. The association between protein quality and amino acid composition was first shown by Willcock & Hopkins (1906) when they postulated the essentiality of tryptophan. The discovery of all the amino acids, however, was a protracted affair, beginning with the isolation of glycine and leucine by Braconnot in 1820 (McCollum, 1957) and ending with the discovery of threonine in 1935 (McCoy, Meyer & Rose, 1935). The realisation that certain amino acids were essential was a process intimately associated with the move from dogs to rodents as experimental animals. Willcock & Hopkins, describing their early work on amino acid essentiality, justify this change to rodents:

> Whatever objection may be attached to the use of small animals in metabolism experiments, they do not apply to observations of the simple kind necessary for the enquiry at its present stage. The observations are of a strictly comparative kind .... It was desirable at the outset of the enquiry to work with a large number of individuals, and the difficulty and expense of preparing the special constituents of the dietary for use of numerous large animals would be almost prohibitive. The general result being established, it will be possible to study the detailed effects of the dietary upon a smaller number of larger animals (1906, p. 92).

Such remarks notwithstanding, dogs were used in some of the early work. Abderhalden fed dogs a mixture of 16 different amino acids instead of whole protein, and, while they did not accept the food enthusiastically and experienced some vomiting and diarrhoea, dogs maintained nitrogen equilibrium for 6–8 days (Abderhalden, 1912). Moreover, he found that a proteolytic product of whole protein supported growth of dogs (Abderhalden & Hirsch, 1912).

In the 1920s and 1930s, Rose and coworkers conducted a series of experiments attempting to support growth in rats using purified amino acids as the sole source of nitrogen, experiments which culminated in 1935 with the discovery of threonine (McCoy, Meyer & Rose 1935). In 1938, Rose published a paper in which he characterised the amino acids as either nutritionally indispensable or dispensable for rats (Rose, 1938). A year later, Rose & Rice presented evidence that the same scheme could be applied to dogs as well (Rose & Rice, 1939).

It thus became clear that dietary protein is essential first because it contains amino acids which cannot be synthesised by the organism, and secondly to meet the physiological requirement for nitrogen. This nitrogen is used to synthesise dispensable amino acids as well as other nitrogen-containing compounds, such as nucleic acids, purines, pyrimidines, and neurotransmitters. Strictly, animals do not have a 'protein requirement' as such since it can be met by a combination of a source of essential amino acids and a non-protein source of nitrogen. However, protein is the most effective means of meeting these two requirements[1].

Until recently, most experiments designed to study protein requirements have used intact proteins of varying quality. However, a more effective approach is to feed a diet which exceeds the minimal requirements for all essential amino acids and then to evaluate what additional amount of nitrogen results in optimal performance of the animal. Now that purified essential amino acids are readily available and there are some estimates of the requirements for them of both dogs and cats, such studies can, and have been done.

Factors affecting nitrogen balance and protein requirement estimates

The dietary requirement for protein may be defined as the minimal intake that promotes optimal performance of the animal. Performance has been evaluated in many ways, but the most frequently used criteria are nitrogen balance (at all ages) and growth rate (of the young). Figure 1 illustrates the application of these two response variables in an experiment to determine the nitrogen requirement of the kitten (Smalley *et al.*, 1985). The requirement is the dietary level of protein at which the response just reaches a plateau.

As mentioned earlier, nitrogen balance is the more reliable index of the two, the drawback of bodyweight changes being that body composition changes are not taken into account. However, many factors besides the level of dietary protein can influence nitrogen balance, and therefore the experimentally determined protein requirement. Among these are: the amino acid composition of the protein and its digestibility, energy intake, activity level, and the physiological state of the animal, including prior nutritional status. It is clearly important to understand the possible impact of these confounding factors, to control for them as far as possible in any experiment, and to keep them in mind when evaluating published work. Therefore the effects of the most important factors are discussed below.

Protein quality

Nitrogen balance is so responsive to protein quality that it is the basis of the Biological Value techniques for measuring protein quality (Allison, 1964). Many investigators have demonstrated the effect of protein quality on nitrogen balance and growth in both dogs and cats. For example,

---

[1]In practice nutritionists measure dietary nitrogen (usually by the Kjeldahl method) and estimate the protein content of the diet by assuming a fixed nitrogen:protein ratio. In this paper, we follow the convention of converting dietary nitrogen concentration to crude protein by assuming 16% nitrogen in protein (i.e. protein = N × 6.25), such values being quoted as crude protein equivalents (CPE).

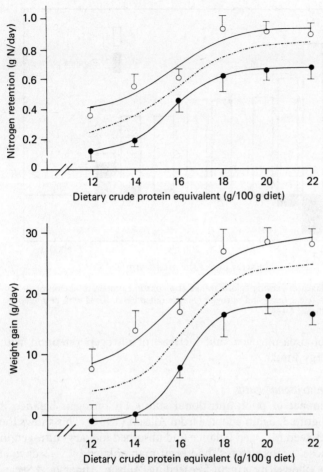

Fig. 1. Mean daily weight gain and nitrogen retention of male (open circles) and female (filled circles) kittens. Mean of male and female results shown as dotted line. Vertical bars represent the standard error of individual means. Curves represent calculated non-linear regressions. (Used with permission from Smalley *et al.*, 1985. Copyrighted by and reprinted with permission of Cambridge University Press.)

Mabee & Morgan (1951) fed puppies a number of proteins at the same concentration in isocaloric diets. While all diets promoted positive nitrogen balance, magnitude of nitrogen retention varied widely with protein source.

Energy intake

The effect of energy intake on nitrogen balance is clearly illustrated in Figure 2, adapted from Allison (1956). Adult fox terriers were fed the same amount of nitrogen, and energy intake was varied by adding or removing carbohydrate. The nitrogen-sparing effect of energy can be seen in

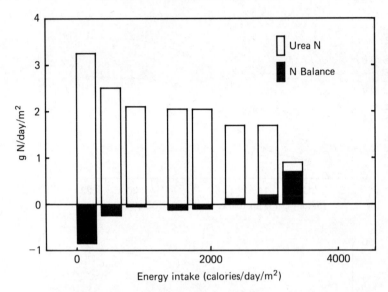

Fig. 2. Effect of varying caloric intake at a constant protein intake on urea-nitrogen (clear bars) and nitrogen balance (filled bars). (Used with permission from Allison, 1956.)

the decrease of urea-nitrogen and increase of nitrogen retention with increasing energy intake.

Prior nutritional status

The impact of prior nutritional status on nitrogen balance is illustrated in Figure 3, again adapted from Allison (1956). This shows the relationship between nitrogen balance and absorbed (dietary) nitrogen in adult dogs whose 'protein stores' had been manipulated by 2–3 days of plasmapheresis followed by protein-free feeding (Allison, Anderson & Seeley 1946). Dogs were fed diets containing various levels of high-quality protein. Dog A was well nourished and had full protein stores, dog D was severely protein depleted and B and C were intermediate. The amount of absorbed nitrogen required *to produce nitrogen equilibrium* (*zero* balance) depended on the degree of protein depletion: the most depleted dog required the *least* absorbed nitrogen. If nitrogen equilibrium was being used to define dietary nitrogen requirement of these adult dogs, the requirement of the most depleted dog (D) would be assessed at one-quarter of the well-nourished dog (A). Clearly, prior nutritional state influences the outcome of protein requirement determinations. However, the greater slope of the line for dog D illustrates the 'increased anabolic activity associated with the depleted state' (Allison, 1956). Arguably, therefore, this animal's need for nitrogen is not being met by only 1 g/day/m² absorbed nitrogen, even though nitrogen equilibrium is achieved.

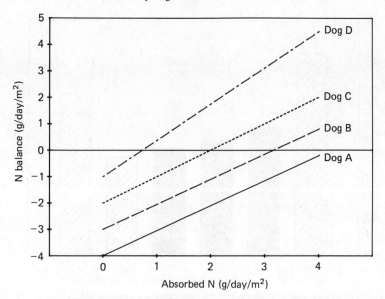

Fig. 3. The effect of prior protein nutriture on the response of nitrogen balance to nitrogen absorption. Dog A had full protein stores, dog D had severely depleted protein stores, and dogs B and C were intermediate. (Used with permission from Allison, 1956.)

Dietary methionine concentration

Feeding a dog a protein-free diet produces a negative nitrogen balance (Fig. 3); however, supplementing with small amounts of methionine significantly reduces the extent of nitrogen loss, as illustrated in Figure 4, adapted from Allison, Anderson & Seeley (1947). These workers fed adult dogs a protein-free diet for 4 days and then supplemented the diet with a small amount of methionine. Urinary nitrogen excretion decreased immediately and remained low for several days after removal of methionine from the diet. This action of methionine in a protein-free diet is associated with a reduction in the excretion of urea, with little or no change in excretion of nitrogen as ammonia (Allison *et al.*, 1947). This effect of methionine varies with the 'protein stores', being less marked when they are depleted (Miller, 1944). Neither the mechanism by which methionine supplementation of a protein-free diet improves nitrogen balance, nor its implications, have been investigated. However, since methionine is known to be involved in initiation of protein synthesis, lipid utilisation and secretion of some hormones, this is a fertile field for research.

Feeding pattern

Feeding pattern of dogs and cats should also be considered as a factor which might affect nitrogen balance, and hence protein requirement.

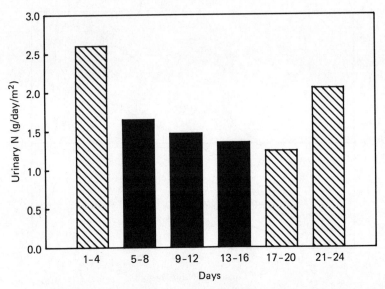

Fig. 4. Urinary nitrogen concentration in dogs fed a protein-free diet (hatched bars), and the same diet with an additional 0.24 g of D,L-methionine nitrogen/ day/m² body surface (filled bars). (Used with permission from Allison, Anderson & Seeley, 1947.)

There is evidence that in rats decreasing meal frequency increases nitrogen excretion (Fabry, 1967). Rats allowed free access to food adopt a nibbling pattern, taking numerous meals throughout the 24-hour period, but especially at night. Rats tube-fed in two meals the amount of food consumed by a matched, *ad libitum* fed group in 24 hours, have a higher urinary nitrogen output (Cohn, Joseph, Bell & Oler, 1963), an effect not related to the intubation process itself. Moreover, rats consuming one 2-hour meal a day had a higher percentage of body fat than *ad libitum* fed controls (Leveille, 1972).

Like rats, dogs and cats are naturally 'nibblers' rather than gorgers. Mugford (1977) and Mugford & Thorne (1980) have shown that, while there is considerable breed variation, both dogs and cats generally consume 10–13 meals in a day. Kane *et al.* (1981, 1986) confirmed that cats eat 12–18 meals per day about equally spaced between the light and dark periods. Dogs, however, consume most of their meals in the light cycle (Mugford & Thorne, 1980). Romsos *et al.* (1978), have examined the effect of feeding dogs isoenergetically either one or four meals in 48 hours. Although nitrogen balance was not measured, the difference in meal frequency did not affect body composition or bodyweight. Direct comparisons of the effect of meal pattern on nitrogen balance in cats have not been made, but if, as with the rat, limiting the number of meals decreases nitrogen retention and utilisation, the apparent protein requirement may be higher with meal feeding than with free-feeding.

Age and growth rate

Growth rate decreases with age, even during the stage of very rapid growth. This may be illustrated by expressing growth rate as a function of metabolic body size (e.g. kg$^{\frac{3}{4}}$). Growth rate of half-grown puppies, for example, is approximately half that of puppies in early weaning (Fig. 5) and protein requirement consequently is lower (Payne, 1965). Even relatively small differences in age could account for some differences in experimental results. For instance, in studies by Heiman (1947), pups of 22–28 weeks of age did not grow well on diets of 17% mixed protein (w/w), whereas by 29–35 weeks of age it appeared adequate.

Dietary fat content (protein–energy ratio)

Ontko, Wuthier & Phillips (1957) showed that a diet containing 25% crude protein equivalent (CPE) from mixed protein supported maximal growth of weanling pups when the dietary fat content was 20% (w/w), but 29% CPE was required when the fat content was increased to 30%. In other words, as energy density of the diet increased, a higher concentration of protein was required for maximal nitrogen retention. Evaluations of different studies on protein and amino acid requirements must therefore take account of differences in the fraction of dietary energy contributed by protein, the protein:energy, or P:E ratio.

Protein requirements of the dog and cat

With the preceding caveats in mind, we summarised in Table 1 the principal estimates of protein and amino acid requirements of adult dogs and cats and of growing animals in Table 2. When adequate information was available in the original report, we expressed requirements either as P:E ratios (assuming an ME of crude protein of 4.0 kcal/g), or as CPE, or both.

Theoretical values for protein requirements were calculated for dogs and cats at different stages of life by Payne (1965) and Greaves (1965), respectively, and their estimates are shown in Figures 5 and 6. The basis for these requirements, and the factorial method used to derive them, is discussed in chapter 6. They are expressed in terms of a hypothetical fully utilisable reference protein called by Payne Net Dietary-protein (ND-p), and the requirements are therefore minima, from which experimentally determined values will differ, depending on protein quality and other factors.

Protein requirement for adult maintenance

Adult dog

Figures 5 and 6, and Table 1, show clearly that the adult animal requires less protein for maintenance than does the growing animal for maximal growth.

Experimental estimates of the protein requirement for adult dogs are for the most part based on nitrogen balance studies, and there have been many

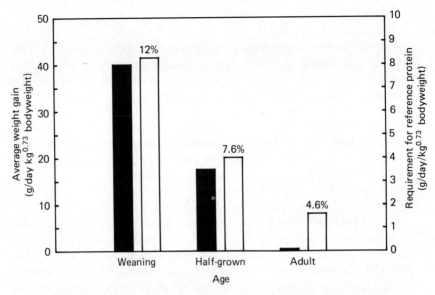

Fig. 5. Effect of age on weight gain (filled bars) and on protein requirement (clear bars) in dogs. Percentages in bars show fraction of energy which should be provided as fully utilisable reference protein (NDpCal%). (Adapted from Payne, 1965.)

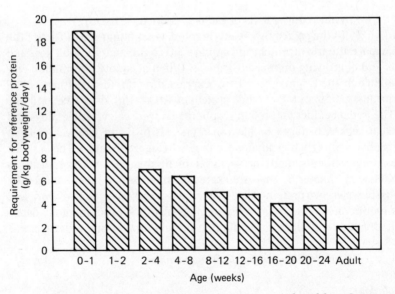

Fig. 6. Effect of age on protein requirement in cats. (Adapted from Greaves, 1965.)

more such studies in adult dogs than in adult cats or the young of either species. When high-quality protein (such as lactalbumin or methionine-supplemented casein) is fed, estimates of the P:E ratio required by adult dogs to maintain nitrogen balance range are from 4.3–6.9% (Table 1). Casein without supplemental methionine, its first limiting amino acid, is required in slightly greater quantity, at P:E ratios from 6.4 to 9.4% (Table 1). These values approximate the theoretical values arrived at by Payne (1965) and Kendall, Blaza & Holme (1982). According to Payne's calculations, the protein requirement of the adult dog should be met by a diet with 4.6% of the energy provided by reference protein. Kendall, Blaza & Holme (1982) found that a minimum of 273 mg $N/W^{0.75}$ or 1.73 g reference protein/$W^{0.75}$ was required to replace total endogenous nitrogen output. This minimum protein requirement is met when a diet with a P:E of 5.2% (reference protein) is consumed at an energy intake equal to the NRC (1974) requirement (132 kcal/$W^{0.75}$/day).

While as little as 5% of the energy as a high-quality protein meets the maintenance requirement, it has been suggested by Romsos & Ferguson (1983) that adult dogs choose diets with P:E ratios of approximately 30%, but the results of this study are ambiguous. Dogs were given choices either between diets with P:E ratios of 20 and 46%, or those with 25 and 46%. If dogs ate randomly from each pair of diets presented, the average P:E ratios would have been 33 and 35%, respectively. To demonstrate a positive choice of 30% of energy as protein, P:E ratios of diets should be such that random choice would not result in values close to 30%.

Diets containing protein just at the level required for nitrogen balance may not promote optimal performance of the adult animal. Dogs fed at that level may be more susceptible to toxicity of some drugs (Allison, Wannemacher & Migliarese, 1954), and higher levels may be needed to maintain optimal protein reserves (Wannemacher & McCoy, 1966). It may therefore be desirable to feed higher levels of protein to adult dogs under some kinds of stress; however, caution should be exercised in this regard (see later section on Protein Excess).

Adult cat

Far less work has been done to evaluate the protein requirement of adult cats, but all evidence indicates that it exceeds that of dogs by a considerable margin. Estimates, shown in Table 1, range over P:E ratios of 10 to 19%. The exceptional value of 10% (Burger *et al.*, 1984) was determined using nitrogen balance in cats fed a soy-protein based diet supplemented with amino acids sufficient to meet all the amino acid requirements of growing kittens. It is doubtful that 10% of any *high*-quality protein without these levels of amino acids would support nitrogen equilibrium in all adult cats.

We suggest that the minimum protein requirements of adult cats and dogs are about 15 and 5% of energy, respectively, as high-quality proteins. If

Table 1. *Protein requirement for adult maintenance*

Authors	Requirement (% of energy)	Criterion	Comments
Dogs			
High-quality protein			
Arnold & Schad, 1954	4.5	N balance	Casein + methionine
Kade, Phillips & Phillips, 1948	4.3	N balance	Lactalbumin or casein + methionine
Melnick & Cowgill, 1937	6.9	N balance	Lactalbumin
Payne, 1965	4.6 (NDpCal%)	endogenous N loss	Based on summarised calculation for inevitable N loss, on estimated caloric requirements, and assuming a reference protein of 100% NPU.
Kendall, Blaza & Holme, 1982	5.2	endogenous N loss	Based on measured N loss on protein-free diet and expressed in terms of NRC, 1974 caloric requirement of 132 kcal/kg$^{0.75}$ and reference protein of 100% NPU.
National Research Council, 1962	6.3	N balance	High-quality protein
National Research Council, 1974	21.8	N balance (under stress)	High-quality protein
National Research Council, 1985	9.3	weight gain	High-quality protein
Low-quality protein			
Arnold & Schad, 1954	6.6	N balance	Unsupplemented casein
Kade, Phillips & Phillips, 1948	6.4	N balance	Unsupplemented casein, whole or hydrolyzed
Melnick & Cowgill, 1937	9.4	N balance	Unsupplemented casein
Wannemacher & McCoy, 1966	21.1	N balance	Gliadin
	6.3	N balance	Casein
Cats			
Greaves & Scott, 1960	18.7	N balance	Fish and liver
Burger et al., 1984	10.0	N balance	Soy protein and amino acids
Allison et al., 1956	12.0	N balance	Casein and liver powder

protein is supplemented with essential amino acids to meet published kitten requirements, the protein requirement of the adult cat is apparently about 10% of energy, but there is no evidence that comparable amino acid supplementation reduces requirements in adult dogs.

Protein requirement for growth

Growing dogs

Investigations into the protein requirement of puppies in the 1940s and 1950s generally tested diets containing mixed proteins and used pups of various breeds of 6–8 weeks of age. Requirements derived from the data ranged over P:E ratios from 17–22% (Table 2).

Recently Burns, LeFaivre & Milner (1982) concluded that the protein requirement of puppies is considerably lower than these earlier estimates. They based this conclusion on response of weight gain to dietary protein concentration, but an analysis of their nitrogen balance measurements is more consistent with earlier work. Burns, LeFaivre & Milner (1982) fed the high-quality protein lactalbumin at levels of 0–20% crude protein, and both weight gain and nitrogen balance were evaluated in 8 to 10 week-old and 13 to 17 week-old pups after 2–6 weeks of feeding. On the basis of their data, the investigators calculated that 11.3% CPE (P:E of 11.0%) was required to maximise weight gain in 13 to 17 week-old pups. For 8 to 10 week-old pups with a slightly greater growth rate, it was estimated that 15% CPE (P:E of 14.6%) maximised growth. If the results from these experiments are graphed (Fig. 7), weight gain of 13 to 17 week-old pups plateaus at approximately 11–12% CPE (10.7–11.7% P:E), but nitrogen retention responds linearly over the whole range of intakes. In the 8 to 10 week-old pups, there was no plateau, but a linear response occurred for both variables. Our interpretation of these data is that at least 20% CPE (19.5% P:E), and possibly more, is needed to maximise nitrogen retention of puppies.

This conclusion is supported by results from another study by Milner (1981) in which puppies were fed diets containing two levels of nitrogen from purified amino acids and several levels of lysine. We have plotted data from that study in Figure 8. When the dietary amino acids were doubled from 14% w/w (about 13.1% CPE and P:E) to 28% w/w, nitrogen retention increased significantly at the highest level of lysine fed. This indicated that the 14% amino acid diet did not provide enough protein for maximal nitrogen retention, although weight gain was maximal at that level.

Growing cats

Whereas estimates for the protein requirement of the puppy range from P:E of 12 to 22%, those for the kitten are from 16 to greater than 30% (Table 2). Some earlier estimates are based on anecdotal evidence, or on experiments

Table 2. *Protein requirement for growth*

Authors	Requirement (% of energy)	Criterion	Comments
Dogs			
Heiman, 1947	20	Weight gain	Mixed protein, 8 week-old Cocker Spaniels, 32 wk feeding
Gessert & Phillips, 1956	17.2	Weight gain	Mixed protein, 7 wk-old, mixed breed, 14 wk feeding
Ontko, Wuthier & Phillips, 1957	22	Weight gain	Mixed protein, 6 wk-old Beagles, Shepherds, Collies, 10 wk feeding
Burns, LeFaivre & Milner, 1982	≥20	N balance and weight gain	Lactalbumin, 8- to 10 wk-old Beagles, 2–6 wk feeding
	≥20	N balance	Lactalbumin, 13- to 17 wk-old Beagles, 2 wk feeding
	12.2	Weight gain	Lactalbumin, 13- to 17 wk-old Beagles, 2 wk feeding
National Research Council, 1962	12	Weight gain	High-quality protein
National Research Council, 1974	20.3	N balance (allowance)	
National Research Council, 1985	11.5	Weight gain N balance	High-quality protein Lactalbumin or equivalent
Cats			
Dickinson & Scott, 1956	>30	Weight gain	White fish, herring and liver
Miller & Allison, 1958	>25	Weight gain	Casein
Greaves, 1965	≥30	Weight gain	Review of literature
Jansen et al., 1975	28	Weight gain	Casein
Smally et al., 1985	16	N balance and weight gain (allowance)	Amino acid diet and casein + amino acids
National Research Council, 1978	28	Weight gain	High-quality protein

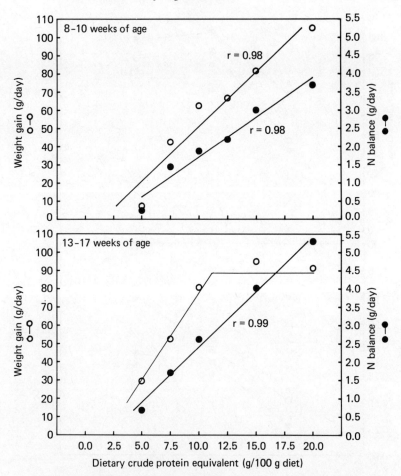

Fig. 7. Effect of dietary protein concentration (CPE) on weight gain (open circles) and N balance (filled circles) in 8–10 and 13–17 week-old pups. (Data from Burns, LeFaivre & Milner, 1982.)

using mixed protein with weight gain as the critical variable measured. Probably the low protein diets used would not have been palatable to cats, and the consequently reduced food intake could have impaired growth, confounding interpretation of results. Another complicating factor is the higher requirements of kittens for certain amino acids, unknown at the time of the earliest studies. For example, if casein is the sole source of protein, it can be calculated that, even if digestibility were 100%, arginine and the sulphur amino acids become limiting in diets containing less than 28 and 24% CPE, respectively. Therefore, at casein concentrations below these, growth would be limited by essential amino acid (EAA) concentration rather than by nitrogen, and requirements for the latter could not be

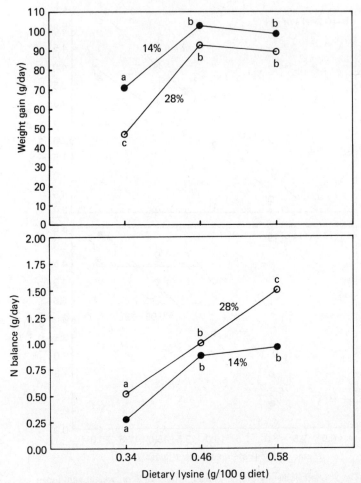

Fig. 8. Effect of dietary lysine concentration on weight gain and N balance in pups fed 14% amino acids (filled circles) or 28% amino acids (open circles). (Data from Milner, 1981. Reproduced with permission of Instituto de Nutricion de Centro America Y Panama.)

accurately estimated. Since the EAA requirements of the kitten have been determined during the last five years, this need no longer be a complicating factor.

In a study to determine the minimum protein requirement of the kitten, Smalley *et al.* (1985) fed diets designed to provide EAA *in excess* of requirements, even at the lowest level of protein. The nitrogen was provided either by crystalline amino acids alone or by a mixture of casein and crystalline amino acids. We have pooled the data from these experiments and show response curves for bodyweight and nitrogen retention in Figure 9. No clear plateau in either variable is reached even at 22% w/w. However, experiments done in this laboratory with diets providing 23.2% CPE yield nitrogen retention of 0.85 g/day, essentially identical to that obtained by Smalley *et al.* with 22% CPE. Data for higher protein diets are not available. Thus the protein requirement of the kitten, based on nitrogen retention, is probably not much higher than 22% CPE in a diet containing about 5 kcal/g (i.e. a P:E of 17.6%).

This estimate is lower than others in the literature, but differences in experimental designs and diets may partially account for these discrepancies. Other studies used intact proteins, and as implied above, if the protein is not supplemented with its limiting amino acid, the requirement for that amino acid is in effect being determined rather than the nitrogen requirement. Also, the nitrogen requirement itself may in fact be higher if EAA are provided *just* at their required level than if they are in *excess* (unpublished data, Rogers & Morris). The former is more likely with intact, unsupplemented protein (e.g. unsupplemented casein, as used in the work of Jansen *et al.*, 1975). Using crystalline amino acids to supplement intact proteins, Smalley *et al.* (1985) ensured that EAA requirements were exceeded; this could account in part for their low estimate.

We conclude that the minimum protein requirement of growing kittens is a P:E of about 18% if all EAA are present in excess of their requirements.

Obligatory nitrogen loss in the cat is greater than that in the dog and rat; this can be inferred by comparing the proportion of ingested nitrogen that is

Fig. 9. Effect of dietary protein concentration on weight gain (open circles) and nitrogen retention (filled circles) of kittens (N = 32). Dissimilar symbol represents N retention data from a separate experiment. (Adapted from Smalley *et al.*, 1985.)

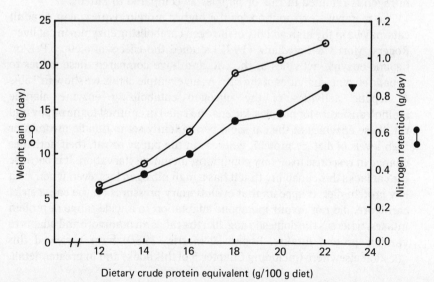

Table 3. *Per cent ingested protein (nitrogen) retained by kittens, puppies and rats fed various quantities of protein*

Dietary protein (% of energy)	Kitten[1] $n=24$	Puppies[2] $n=5$	Rats[2] $n=10$
5.0		25	65
7.5		70	72
9.6	34		
10.0		62	72
11.2	38		
12.5		59	71
12.8	41		
14.4	47		
15.0		66	71
16.0	48		
17.6	47		
20.0		59	63

Notes:
[1] Calculated from Smally *et al.*, 1985. Kittens were fed amino acids or casein supplemented with amino acids.
[2] Calculated from Burns, LeFaivre & Milner, 1982. Puppies and rats were fed lactalbumin.

retained in the young of these species (Table 3). The lower efficiency of utilisation of nitrogen by the kitten is illustrated most clearly at low protein intakes where, for example, at 8–10% dietary protein, about twice as much nitrogen is retained in rats or puppies as compared to kittens.

The metabolic explanation for the higher protein requirement of adult cats may lie in the high activity of nitrogen catabolising enzymes in cat liver. Rogers, Morris & Freedland (1977) studied the effects of dietary P:E on hepatic enzyme activities in the cat, and have compared these values to hepatic enzyme activities of the rat. (As an example of this we show in Table 4 hepatic activities of the nitrogen catabolising enzyme alanine aminotransferase for puppies, kittens and rats.) In contrast to the puppy and rat, these enzymes in the cat seem permanently set to handle medium or high levels of dietary protein, whereas in the puppy or rat, their activities change in response to dietary conditions, including starvation. It is because the cat lacks this capability that it has a high nitrogen loss, even when fed a low protein diet. It appears that evolutionary pressures on the cat, a strict carnivore, did not favour metabolic adaptation to a wide range of protein intakes, whereas the domestic dog, like the rat, is an omnivore and adapts to variable protein intakes. Rogers & Morris (1980) have reviewed this question elsewhere (including Chapter 5 of this book) and in greater detail.

Table 4. *Effect of dietary protein on hepatic alanine amino transferase activities*

Treatment	μ mols/min/g (mean ± SEM)		
	Kittens[1]	Puppies[2]	Rats[1]
Standard protein	51 ± 2 (8)	17 ± 1 (7)	15 ± 2 (3)
High protein	63 ± 3 (8)	31 ± 2 (7)	41 ± 1 (3)
Food deprived (5 days)	51 ± 4 (3)*	8 ± 1 (5)	17[+]

Notes:
[1] Taken from unpublished work of Rogers, Tews, Morris & Freedland. Numbers in parentheses indicate size of group.
[2] Rogers, Morris & Freedland, unpublished.
* Adult cats.
[+] Calculated from % change when three days food deprived.

Amino acid requirements of the dog and cat

Although the nine amino acids essential for the rat and the dog were identified 50 years ago (Rose, 1938; Rose & Rice 1939), until recently virtually no published data existed on the essentiality of specific amino acids in cats, nor on the amounts required for optimal growth in either dogs or cats. In 1979, Rogers & Morris (1979) and Milner (1979a, b) demonstrated that the same ten amino acids (the nine required by the adult plus arginine) essential for the growing rat were indispensable in the diet for growing cats and dogs. Since then, quantitative estimates of most EAA requirements for growing cats and dogs have been published, although more work is still needed on the requirements for maintenance or reproduction.

Criteria for requirements

The requirement of an EAA is generally defined as the minimum dietary concentration that maximises (or minimises) a specific measurable response in the animal. Nitrogen balance and weight gain are most often measured as response variables. They generally give similar requirement values, although nitrogen balance is often the more sensitive index, in that a somewhat higher dietary concentration of an amino acid is required to maximise nitrogen retention than to maximise weight gain. Figure 10 illustrates how the histidine requirement of the kitten is determined using these variables. Both plateau at approximately the same dietary concentration, a histidine requirement just above 2.0 g per kg diet. These patterns of response are typical of other EAA and, of course, resemble those to changes in dietary protein concentrations shown in Figure 1.

Other physiological variables have been studied as possible indices for

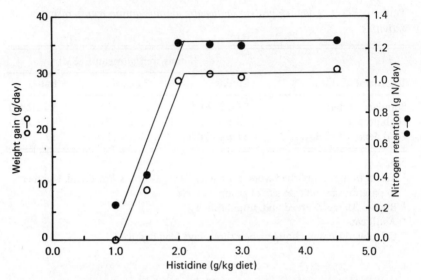

Fig. 10. Effect of dietary histidine concentration on weight gain (open circles) and nitrogen retention (filled circles) of kittens. (Adapted with permission from Quam, Morris & Rogers, 1986.)

estimating EAA requirements. The response of plasma level of an EAA to changes in its dietary concentration has been examined, but the response has proved too variable to be consistently useful for predicting requirements. In the kitten, for example, plasma leucine concentration increases linearly with increasing concentrations of leucine in the diet (Hargrove, Rogers & Morris, 1984). On the other hand, the response of plasma methionine to increasing dietary concentrations can best be described as exponential (Schaeffer, Rogers & Morris, 1982), while the response curve for histidine is S-shaped, plateauing at very high (3 g/kg) and very low (1.5 g/kg) dietary concentrations.

Because histidine is present in high concentrations in haemoglobin, blood haemoglobin concentration has proved the most sensitive index for determining the histidine requirement of kittens (Quam, Morris & Rogers, 1986). The dose–response curves for dietary histidine concentration v. haematocrit (PCV) and haemoglobin concentration for the kitten are shown in Figure 11. To maximise both PCV and blood haemoglobin, 3 g histidine/kg diet were required. This was also the minimal dietary concentration required by kittens to prevent cataract formation in long-term experiments (Quam, Morris & Rogers, 1986). Therefore, even though weight gain and nitrogen retention plateau at 2 g histidine/kg diet, the requirement should be taken as 3 g.

Urinary orotic acid may be a good indicator of arginine requirement. Figure 12 illustrates how, in the dog, this is minimised at a dietary arginine

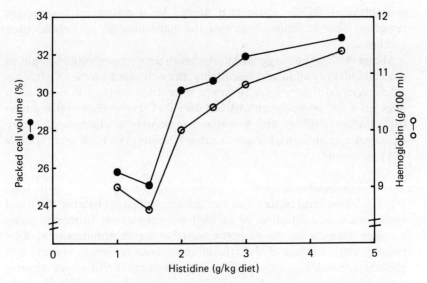

Fig. 11. Effect of dietary histidine concentration on packed cell volume (closed circles) and haemoglobin (open circles) concentration of kittens. (Adapted with permission from Quam, Morris & Rogers, 1986.)

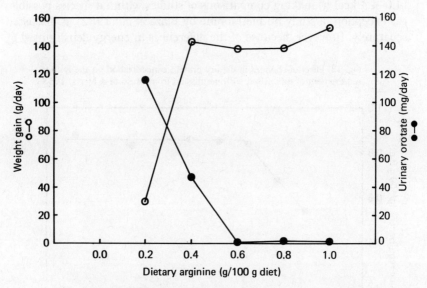

Fig. 12. Response of weight gain (open circles) and orotic acid excretion (closed circles) to changes in dietary arginine concentration in dogs. (Data from Czarnecki & Baker, 1984.)

concentration slightly above that needed for a plateau in bodyweight response. This minimum indicates the normalisation of hepatic urea synthesis.

Milner (1979a) has suggested that urinary urea concentrations might be a useful index of amino acid essentiality, since elevations were seen in dogs fed diets devoid of some EAA. However, since urinary urea also increased in dogs fed a diet adequate in EAA but devoid of several dispensable amino acids (Milner, 1979a), and it was not responsive to changes in dietary isoleucine concentrations (Burns, Garton & Milner, 1984), its general use is not appropriate.

Confounding factors

The several factors that can influence nitrogen balance discussed earlier must be controlled, or at least accommodated, in studies using nitrogen balance as the response variable for determination of EAA requirements. In view of the critical importance of dietary energy, it is probably desirable to express EAA requirements, like those for protein, relative to dietary energy. However, little work has been published on this, and we have therefore expressed EAA requirements as grams per 100 g diet. The energy densities of the diets used in studies on cats have been fairly similar (4.5–5.0 kcal/g), as, with one exception, have those used for dogs (4.0–4.4 kcal/g) making comparisons of studies within a species possible; the exception, a study on methionine by Blaza *et al.* (1982), is discussed separately. However, because of the differences in energy density used in

Fig. 13. Effect of changes in dietary protein concentration on the lysine requirement of rats. (Used with permission from Bressani & Mertz, 1958.)

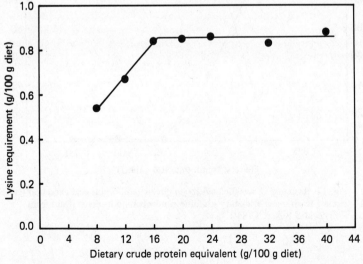

studies of the two species, we show comparative requirements in Table 5, corrected on the assumption they *are* related to energy density.

Another factor which can influence nitrogen balance is the protein concentration of the diet. For some amino acids and in some species, the requirement increases as the protein concentration of the diet increases from subadequate to adequate levels, until the requirement for protein is met, at which point no further increase in amino acid requirement can be observed. Figure 13 clearly illustrates this for the lysine requirement of the rat. A similar response has been observed for lysine in growing pigs (Baker, Katz & Easter, 1975; Brinegar *et al.*, 1950), for tryptophan in chicks (Boomgaardt & Baker, 1971) and pigs (Boomgaardt & Baker, 1973), and for isoleucine in swine (Becker *et al.*, 1957), but it has not been demonstrated for dogs or cats and does not occur for every amino acid. However, in order to ensure that the true amino acid requirement is being assessed, the protein content of the diet should meet or slightly exceed the requirement. With these provisos in mind, it is now possible to review the literature on EAA requirements for cats and dogs.

Amino acid requirements of growing dogs

Threonine, tryptophan, histidine, leucine, isoleucine, valine, phenylalanine and tyrosine
In a series of experiments designed to estimate the requirements of threonine, tryptophan, histidine, leucine, isoleucine, and valine, Burns & Milner (1982) and Burns, Garton & Milner (1984) measured food intake, growth rate, nitrogen balance and urinary urea concentrations in young Beagle dogs fed various levels of these acids in diets containing about 13% CPE (12.7% P:E). The levels shown to optimise performance are shown in Table 5. With the exception of isoleucine, where the opposite was true, nitrogen balance was a more sensitive index of performance than growth. Urinary urea concentration in most cases was minimised by dietary concentrations similar to those that maximised nitrogen retention. Isoleucine again was the exception, in that urinary urea excretion did not appear responsive to dietary isoleucine.

As noted earlier, haemoglobin is generally a useful response parameter for histidine. However, although one of the deficient diets resulted in significantly lower haemoglobin levels in the blood, there was, in this experiment, no consistent response of haemoglobin to dietary histidine concentration. The explanation may be related to the relatively short duration (15 days). In much longer experiments with adult dogs, Cianciaruso, Jones & Kopple (1981) found a decrease in haemoglobin when a histidine-deficient diet was fed.

An estimate of the requirement for tryptophan was also reported by Czarnecki & Baker (1982). They found that the concentration of tryptophan

Table 5. *Current estimates of amino acid requirements of growing kittens and puppies*

| Amino acid | Requirement (g/100 g diet*) | | References | |
	Kitten	Puppy	Kitten	Puppy
Arginine	1.05 (0.92)	0.60	Anderson, Baker & Corbin, 1979; Costello, Morris & Rogers, 1980	Ha, Milner & Corbin, 1978; Czarnecki & Baker, 1984
Histidine	0.30 (0.26)	0.21	Anderson et al., 1980a; Quam, Morris & Rogers, unpublished	Burns & Milner, 1982
Isoleucine	0.50 (0.44)	0.40	Anderson et al., 1980b; Hargrove, Rogers & Morris, 1984	Burns, Garton & Milner, 1984
Leucine	1.20 (1.06)	0.65	Anderson et al., 1980b; Hargrove, Rogers & Morris, 1984	Burns, Garton & Milner, 1984
Lysine	0.80 (0.70)	0.65	Anderson, Baker & Corbin, 1979; Milner, 1981; O'Donnell, Rogers & Morris, unpublished	Milner, 1981
Methionine with excess cys	0.40 (0.35)	0.3	Smalley, Rogers & Morris, 1983;	Blaza et al., 1982; Hirakawa & Baker, 1985; Burns & Milner, 1981
total sulphur AA	0.75 (0.66)	0.6	Teeter, Baker & Corbin, 1978; Schaeffer, Rogers & Morris, 1982	
Phenylalanine with excess tyr	0.40 (0.35)	0.40	Anderson et al., 1980a	Milner, Garton & Burns, 1984

			References	
total aromatic	0.90 (0.79)	0.80	Milner, Garton & Burns, 1984	Williams, Rogers & Morris, unpubl.
Threonine	0.70 (0.61)	0.52	Burns & Milner, 1982	Anderson et al., 1980b; Rogers & Morris, 1979; Titchenal et al. 1980
Tryptophan	0.13 (0.11)	0.15	Czaranecki & Baker, 1982; Burns & Milner, 1982	Anderson et al., 1980a; Hargrove, Rogers & Morris, 1983
Valine	0.60 (0.53)	0.45	Burns, Garton & Milner, 1984	Anderson et al., 1980b; Hardy, Morris & Rogers, 1977

* Energy densities of diets vary: for puppies, range is 4.0–4.4 kcal/g; for kittens, range is 4.5–5.0 kcal/g. Numbers in parentheses are the requirements for kittens corrected to densities used for the puppies (see text).

in a diet of 14.7% CPE (13.5% P:E) required to maximise growth of English Pointer puppies varied from less than 0.12% for pups of 12–14 weeks of age, to at least 0.16% at 6–10 weeks old. The authors calculated the requirement for maximal growth to be 0.15%, an estimate similar to that of Burns & Milner (1982). Czarnecki & Baker also found that, unlike the rat, which is a very efficient user of D-tryptophan (Ohara *et al.*, 1980), the pup can use D-tryptophan with only about 36% efficiency.

Aromatic amino acid requirements were determined by Milner *et al.* (1984). Using diets containing about 13% CPE (12.6% P:E), they found that 0.8% phenylalanine resulted in maximal growth and nitrogen balance and that about half of the phenylalanine requirement could be met by tyrosine. Neither plasma nor urinary urea were useful as predictors of the phenylalanine requirement.

Lysine
Milner (1981) estimated the lysine requirement of male and female Beagle pups using a diet containing 14.0% (w/w) amino acids (13.1% CPE and P:E). He found that 0.58% lysine resulted in maximal weight gain and nitrogen balance, and suggested this was the requirement of the immature dog. However, dietary concentrations of lysine of 0.69% or above were used in three of Milner's experiments and, if the results are plotted and a break point determined for both weight gain and nitrogen retention, two of them give break points at about 0.65% lysine for both variables. It has been shown in several species that lysine is one of the amino acids for which requirement increases as dietary protein concentration is increased from subadequate to adequate (e.g. see Fig. 13). Milner (1981) also studied the effect of doubling the dietary total amino acid concentration on the response to graded concentrations of dietary lysine. As Figure 8 shows, there was a significant decrease in growth rate seen in dogs fed 28% amino acids and 0.35% lysine compared to those fed the same concentration of lysine but with 14% amino acids. This could be the result of an amino acid imbalance, created by the high concentrations of non-limiting amino acids; certainly the significant reduction in food intake found in that group is a characteristic of this. Although nitrogen retention appears low in all groups in this experiment, the results indicate there was a higher lysine requirement when 28% amino acids were fed. It is possible that requirements of other amino acids determined with diets containing a protein content of 12–15% CPE have been underestimated.

Arginine
Arginine occupies a unique position in amino acid nutrition, for, unlike all other EAA, many mammals, when fed a diet devoid of arginine, continue to grow slowly and maintain a positive nitrogen balance. However, arginine is

required in the diet of the young of most species for *optimal* nitrogen retention and growth rates, and, at least during the period of rapid growth, to prevent abnormalities of intermediary nitrogen metabolism.

In a study of young of a number of species, Milner, Prior & Visek (1975) showed that young dogs fed an arginine-free diet for three days exhibited marked increases in urinary orotic and citric acid concentrations, decreases in urinary urea concentrations, and a decrease in food intake; nitrogen balance was not determined. The arginine-free diet also caused emesis occurring shortly after eating. The authors proposed that arginine deficiency decreases the capacity of the urea cycle to detoxify ammonia, which is subsequently shunted into pyrimidine synthesis because of the excess formation of carbamyl phosphate. The result is an increase in urinary orotic acid, a product of pyrimidine metabolism. These sequelae to the consumption of an arginine-free diet are similar to those reported in the kitten by Morris & Rogers (1978a). Increased urinary orotic acid with arginine deficiency occurred with such reliability across species in the experiments of Milner *et al.* (1975), that the authors suggested that it should be considered as an indicator of arginine deficiency in ureotelic mammals.

Ha, Milner & Corbin (1978) also conducted studies to determine the arginine requirement of immature dogs, using weight gain and orotic acid excretion as response variables. Labrador Retriever and Beagle pups (weighing 3 kg – age not given) were fed amino acid diets of varying nitrogen content for periods of 4 or 9 days. With an arginine-free gel diet, food intake was depressed, urinary orotic acid elevated, pups lost weight and experienced episodes of emesis. With 0.56% of arginine in the diet, although urinary orotic acid was normalised, growth of the Labrador pups was suboptimal, and blood ammonia and citrate levels were higher than control levels. Interestingly, therefore, orotic acid excretion seems to have been the least sensitive index of requirement. The authors concluded that the puppy's requirement for arginine is greater than 0.56% in a purified amino acid diet of 14% CPE (14% P:E). An interesting observation in this study was that when the arginine-free diet was force-fed rather than being fed as a gel diet, dogs exhibited more severe emesis, as well as muscle tremors and hypersalivation, an observation confirmed in adult dogs in a subsequent paper from the same group (Burns, Milner & Corbin, 1981). The difference in the amount of food ingested per unit of time, and therefore presumably in the rate of nitrogen absorption, may well account for the differences in severity of clinical signs between the two methods of feeding.

Another estimate of the dietary arginine requirement of the rapidly growing dog has been made by Czarnecki & Baker (1984). English Pointer puppies were fed diets containing about 15% amino acids (about 14.8% CPE, 13.6% P:E) for 10 days. The authors found that 0.4% arginine was sufficient to maximise weight gain in 11 week-old pups but, in contrast to

Table 6. *Methionine requirement of the puppy*

	TSAA requirement as methionine (g/100 g diet)	mg per 100 kcal ME	Breed	Age (wks)	Weeks of feeding	Protein in diet (g/100g)	Fat in diet (g/100g)	Response variables
Burns & Milner, 1981	0.39	95	Beagles	8–14	1 or 2	15.6[a]	15	weight gain, N balance
Hirakawa & Baker, 1985	0.50	125	Pointers	4	3 or 4	15.6[a]	15	weight gain
	req ≤ 0.62	≤127	Beagles	7	10	22.1[a]	25	weight gain
Blaza et al., 1982	0.62 < req ≤ 0.97	127–200	Labs	7	10	22.1[a]	25	weight gain, N balance
	0.58 < req ≤ 0.76	124–162	Beagles, Labs	7	12	22.5[b]	25	weight gain, N balance

Notes:
[a] As crystalline amino acids
[b] As soy protein

Ha, Milner & Corbin (1978), they found a higher level of 0.6% was required to minimise urinary orotic acid. The requirement was calculated by breakpoint analysis of urinary orotic acid response to be 0.53%

From the above studies, it can be concluded that the arginine requirement of the immature dog is at least 0.56%, and is probably about 0.6% of the diet.

Sulphur amino acids

Milner demonstrated that, for the growing dog, methionine is a dietary essential (Milner, 1979*b*), while cystine is dispensable (Milner, 1979*a*). Moreover, as for almost all mammalian species studied, approximately half of the methionine requirement of the puppy can be met by cystine (Burns & Milner, 1981). Hence, investigators generally refer to a total sulphur amino acid (TSAA) requirement rather than a requirement for methionine *per se*.

Several groups of investigators have attempted to determine the TSAA requirement of puppies with some conflicting results. Many experimental variables differ among these studies, including duration of feeding, levels of TSAA fed, age and breed of pups used, source and level of dietary nitrogen, and energy density of the diet. Table 6 shows the levels of TSAA (as methionine equivalents calculated on an iso-sulphurous basis) determined by each group to maximise the response variables measured and some of the important experimental conditions used. The marked differences between the studies cannot be attenuated by relating the results to the energy density of the diet.

The diets used in the studies of Burns & Milner (1981) and Hirakawa & Baker (1985) contained approximately the same level of nitrogen as crystalline amino acids and the same level of fat. The pups in the latter study were considerably younger, and the study period was twice as long; both factors could account for the higher estimate made by Hirakawa & Baker. Requirements for most nutrients decrease with age, and the potential effects of small age differences on requirements are illustrated by their impact on tryptophan requirements (see above). The importance of duration of experiment was shown in the work of Blaza *et al.*, (1982) in which differences in growth rates in response to dietary level of methionine were not evident after two weeks of feeding the test diets, but were observable after four weeks.

The intervals between methionine concentrations fed by Blaza *et al.* (1982) were large and the levels fed were few; it is therefore difficult to estimate the methionine requirement from their data. However, it is clear from their data that neither Beagle nor Labrador Retriever pups had maximal growth or nitrogen retention when fed levels of methionine approximating those recommended by Burns & Milner (1981) and Hirakawa & Baker (1985). The equivalent of 0.58% TSAA in soy protein diets was inadequate for both breeds of dogs. Although 0.62% TSAA in amino acid diets supported maximal weight gain in Beagle pups, it was inadequate for Labradors.

Three unique aspects of the experimental design of Blaza *et al.* (1982) may account for these different results: the duration of feeding, the level of protein fed, and the source of dietary nitrogen. The feeding period of 10–12 weeks evidently allowed the appearance of effects that were not evident in the shorter term. The relatively unexplored effect of methionine on nitrogen balance on a protein-free diet discussed previously and illustrated in Figure 4 is one possible factor influencing results in these experiments. It is possible that, even in protein-containing diets, there is a transitory improvement in nitrogen balance, and hence perhaps weight gain, attributable to this action of methionine. If so, experiments to determine methionine requirements would have to be longer than those of Milner *et al.* This is suggested by the work of Blaza *et al.* (1982), where dogs on three different levels of methionine apparently grew at the same rate for the first two weeks; only thereafter did growth of the pups fed the lowest level slow markedly.

The possible effect of the dietary concentration of protein on amino acid requirements was discussed above and may have effected the results in these studies, as it does in the chick (Almquist, 1949). The protein requirement should have been met by the diets fed by Blaza *et al.* which contained about 22% (w/w) of either soy protein or amino acids, but it is possible that the approximately 14.5% protein (CPE) fed by Burns & Milner (1981) and Hirakawa & Baker (1985) limited growth and hence generated an apparently lower methionine requirement. Differences in ages of pups were probably too small to account for the higher estimates of Blaza *et al.*; at the time of their nitrogen balance determinations the pups were 17–19 weeks old, not substantially different from those in the study of Burns & Milner (16 weeks at end of test).

Availability of dietary sulphur amino acids from the nitrogen source used could also influence the determined requirement. TSAA availability in crystalline amino acid diets would be 100% but is low in many protein preparations such as the soy-isolate used in some of the work by Blaza *et al.* (1982). If availability is low, more TSAA would be required to meet the apparent requirement.

Although in the long-term study of Blaza *et al.* the TSAA requirement may have exceeded 0.76% for growing Labradors, all three groups of investigators estimated TSAA requirements of 0.62% or less for either Beagle or Pointer pups. Therefore, we conclude that the TSAA requirement for puppies is approximately 0.6%, and that 50% of that requirement can be met with cysteine.

An important contribution was made by Burns & Milner (1981) in their examination of the utilisation of methionine analogues, isomers, and derivatives in meeting the requirement of Beagle pups for methionine. They fed diets containing 0.15% cystine and with methionine-like compounds at levels iso-sulphurous to 0.185% methionine. This level of sulphur amino acids would have been just adequate to support maximal growth and nitrogen balance in their experiments, if the methionine-like compounds were 100% utilised. D-methionine, D,L-methionine, N-acetyl-L-methionine and calcium hydroxy-methionine were all as efficacious as L-methionine in supporting optimal performance, whereas the utilisation of N-acetyl-D-methionine appeared to be negligible.

In contrast to the kitten (*vide infra*), no evidence exists suggesting an essential role for dietary taurine in canine nutrition. Beagle pups nourished intravenously with taurine-free parenteral nutrition formulas for 14 days were able to maintain plasma, cerebral and hepatic pools of taurine, and showed no detrimental effects of the regimen (Malloy *et al.*, 1981). The significance of the decrease in taurine conjugated bile acids in gall bladder bile which occurred on this regimen is uncertain. This is in contrast to the importance of taurine in feline nutrition.

Amino acid requirements of adult dogs

Far less work has been done on the EAA requirements of adult dogs, although the pattern of essentiality appears to be identical to that for immature dogs, and includes arginine and histidine. Clearly the amounts of EAA required for adult maintenance would be expected to be less than concentrations required for growth, but data are sparse.

In 1939, Rose & Rice had reported that for nitrogen balance and growth arginine was dispensable from the diet of the adult dog. Burns, Milner & Corbin (1981) have since re-examined this matter measuring urinary orotic acid as an additional response variable. Adult female Pointers were force-fed for four days a purified diet devoid in arginine and containing about 13% CPE as crystalline amino acid. They exhibited the same signs as did puppies force-fed such diets – emesis, muscle tremors, slight frothing at the mouth. An increase in urinary orotate and citrate and a decrease in urinary urea were also seen. The authors observed, like Rose & Rice in 1939, that the arginine diet did not effect nitrogen balance, and they suggested that the decreases in bodyweight observed were probably due to the vomiting. When the arginine-free diet was fed as a gel, severe emesis and orotic aciduria occurred by the second day of feeding, although no muscle tremor or frothing were evident. Arginine at the level of 0.28% of the diet was sufficient to prevent any significant alterations in intermediary metabolism, and at that level, food intake and weight were not different from control. Burns *et al.* therefore concluded that arginine is an EAA for the mature dog and that, for mature dogs at least, urinary orotic acid is a more sensitive index of arginine status than nitrogen balance.

Cianciarus, Jones & Kopple (1981) have studied the essentiality of histidine for adult dogs. They tube-fed adult female mongrel dogs either a histidine-free diet or one with 0.33% histidine. Whereas the dogs fed the complete diet adapted quickly to the tube-feeding process and would lick the bowl containing the remaining slurry of food, the histidine-deficient dogs showed persistent aversion to tube feeding, never ceasing to resist it, and refusing to lick the feeding bowl. Clinical signs of deficiency, however, did not appear until 40–60 days after the start of feeding, when dogs fed the histidine-free diet had significant weight loss and decreased hematocrit, serum albumin, and total protein. Feeding the histidine-free diet to one dog for 72 days resulted in its death. The authors suggest the delayed onset of deficiency symptoms might be due to decreases in haemoglobin synthesis, histidine degradation, or histidine excretion. Alternatively there may have been production of histidine in the kidney, possibly from the histidine-containing dipeptide carnosine. No estimate of the level of histidine required was made in this experiment.

Amino acid requirements of growing cats

There are a number of interesting aspects to the amino acid nutrition of the growing kitten; they will be listed only briefly here, however, since this topic has been reviewed recently by MacDonald, Rogers & Morris (1984). The requirements as summarised by those authors are presented in slightly modified form in Table 5.

Histidine, leucine, isoleucine, valine, lysine, phenylalanine and tyrosine
The histidine requirement listed (0.3%) is the level required to maximise haemoglobin and PCV (Fig. 11) and to prevent cataracts in long-term feeding. As discussed on page 178, this is higher than the level needed to maximise nitrogen retention. Marginal levels of dietary tryptophan have also recently been observed to result in cataract formation in kittens (Morris & Rogers, unpublished). Few long-term experiments have been reported in the cat or dog with other EAA, so it is not clear whether cataracts or other defects might occur with feeding marginal levels of other EAA over prolonged periods.

Isoleucine and valine requirements are similar in the cat, rat, and dog, whereas the cat's leucine requirement is somewhat higher (Hargrove, Rogers & Morris, 1984). Branched-chain amino acid (BCAA) antagonism is much less severe in the kitten than in the rat; feeding 10% of any of the three BCAA did not decrease the growth rate of kittens fed a low-protein diet that otherwise met the BCAA requirement (Hargrove *et al.*, 1988).

The total phenylalanine plus tyrosine requirement of the kitten is about 0.8–0.9% of the diet, about 60% of which can be met with tyrosine, a dispensable amino acid (Rogers & Morris, 1979).

Quantitatively, the cat's tryptophan requirement is similar (Anderson *et al.*, 1980*a*; Hargrove, Rogers & Morris, 1983) to several other species. However, as discussed in Chapter 11, there is negligible flow of tryptophan through the quinolinic acid pathway and therefore the cat does not synthesise a significant quantity of niacin (Leklem *et al.*, 1971; Ikeda *et al.*, 1965; Suhadolnik *et al.*, 1957). Nothing unique has been reported for lysine deficiency in the cat. Dietary asparagine may be required for maximal growth at least in the very young kitten, e.g., after early weaning (Kamikawa, Morris & Rogers, unpublished), but this work has yet to be verified.

Arginine

The arginine requirement listed in Table 5 is 1.05%, the level required to prevent the orotic aciduria which occurs when an arginine-deficient diet is fed (Costello, Morris & Rogers, 1980). However, only 0.8–0.9% is required to produce maximal weight gain.

The response to arginine deficiency is the most dramatic deficiency response in the cat and is similar to, but more extreme than, the response of the dog. The feeding of a single meal of an arginine-free diet may result in hyperammonaemia in less than one hour. The hyperammonaemia is progressive, and, by 2–5 hours, some cats exhibit severe signs of ammonia intoxication which include: lethargy, emesis, vocalisation, hypersalivation, hyperactivity, hyperesthesia, ataxia, emprosthotonos, extended limbs and exposed claws. In the most severely affected cats there is hypothermia, bradypnea, cyanosis, and some die (Morris & Rogers, 1978*a*, *b*; Morris *et al.*, 1979; Rogers & Morris, 1980). These observations raise the question of why the cat is so much more sensitive to arginine deficiency than other mammals. It is known that animals in the post-absorptive state are more sensitive to ammonia load (Wergedal & Harper, 1964) and that administration of arginine greatly decreases the ammonia intoxication under these conditions (Greenstein *et al.*, 1956). The generally accepted reason for the prevention of hyperammonaemia by arginine is that it acts anaplerotically to stimulate urea synthesis by providing an intermediate of the urea cycle, since the intermediates are depleted during the post-absorptive state. Some anaplerotic stimulation of the urea cycle occurs indirectly through other urea intermediates, but the cat's ability to keep the urea cycle 'filled' with intermediates is minimal. There appears to be at least two reasons for this: the cat is apparently less able to synthesise ornithine, and may be less able to provide citrulline to the kidney for conversion to arginine.

Several lines of evidence point to an impairment in ornithine synthesis in cats. Dietary ornithine completely prevents the hyperammonaemia of arginine deficiency in cats, although animals still lose about the same amount of weight as they would if any other EAA was omitted from the diet (Morris & Rogers, 1978*b*; Morris *et al.*, 1979; Rogers & Morris, 1979). Stewart *et al.* (1981) measured ornithine, citrulline and arginine in the livers of cats fed an arginine-free diet. They confirmed that cats which showed severe hyperammonaemia were those which were most severely depleted of liver ornithine.

One pathway of *de novo* synthesis of ornithine is from glutamic acid through the action of the enzyme pyrroline-5-carboxylate synthase (P-5-C-syn). This is the principal route of ornithine synthesis in the small intestine of the rat (Ross, Dunn & Jones, 1978; Henslee & Jones, 1982). However, Rogers & Phang (1985) have shown that P-5-C-syn in cat small intestinal mucosa has only about 5% as much activity/kg bodyweight as is found in the rat, thus verifying that the cat has a severe limitation in the synthesis of ornithine from glutamate. Citrulline can completely replace arginine in the diet of the cat (Morris *et al.*, 1979), but a higher dietary concentration of citrulline than arginine is needed for an equivalent growth rate (Johannsen, Rogers & Morris, unpublished). In the rat, *de novo* citrulline synthesis occurs primarily in the intestine (Windmueller & Spaeth, 1981), and citrulline released from the intestine is converted in the kidney to arginine for use by other tissues (Featherstone, Rogers & Freedland, 1973; Rogers, Freedland & Symmons, 1972). Since citrulline is made from ornithine, the low intestinal P-5-C synthase in cat precludes citrulline synthesis from glutamate. In addition, there may be other metabolic defects in the cat, such as a lack of carbamyl phosphate synthase and ornithine transcarbamylase in the intestine, which would further reduce the ability of the cat to synthesise citrulline. This would explain why the level of citrulline in cat plasma is normally quite low (Rogers & Morris, 1980; Morris, 1985).

Teleologically it is attractive to postulate that the cat, a strict carnivore, has never had a diet limiting in protein (and therefore in arginine or ornithine), but much of the excess of amino acids is needed for gluconeogenesis (since carbohydrate is low in animal tissues). Thus, there is little need to make ornithine when nitrogen intake is high (since arginine from dietary protein would provide the ornithine needed anaplerotically in the urea cycle as well as the arginine needed for protein synthesis). However, *during the post-absorptive state*, there is a metabolic benefit to ornithine depletion since this would reduce urea synthesis and thereby prevent excessive catabolism of amino acids. Work done in the mink and ferret (Deshmukh & Shope, 1983; Leoschke & Elvehjem, 1959) indicates that arginine metabolism in the cat might not be unique but is representative of strict carnivores (Morris, 1985 and Chapter 5 of this book).

Sulphur amino acids
There are several unique aspects of sulphur amino acid nutrition and metabolism in the cat. First, the dietary requirement is considerably higher than in most other mammals. Secondly, the cat excretes a unique branched-chain sulphur amino acid, felinine, in the urine. Third, dietary taurine is required to prevent feline central retinal degeneration (FCRD). Finally, methionine is the most toxic of amino acids for kittens with reductions in food intake and weight gain beginning at about 1.5% of diet.

This high total sulphur amino acid (TSAA) requirement results in TSAA being the most limiting in most cat diets. Assuming 100% bioavailability and that the TSAA requirement is about 0.75% of the diet, about 19% animal protein is needed to meet the TSAA requirement of the growing kitten. Thus, the requirement is just barely met by a meat diet providing enough protein to meet the nitrogen requirement (Morris & Rogers, 1982). Cystine will spare about one-half of the methionine requirement in the cat, as in other species (Smalley, Rogers & Morris, 1983; Teeter, Baker & Corbin, 1978). Therefore, part of the higher TSAA requirement appears to be for cystine and part for methionine as such.

The reason for the high TSAA requirement of the cat is unclear. Part may be due to the use of cysteine for felinine synthesis, and part to the need for the cat's thick coat of hair. The cat normally eats a high fat diet, and there may be an increased need for S-adenosylmethionine for methylation reactions involved in the synthesis of phospholipids necessary for fat absorption and transport, which would be reflected in an increased TSAA requirement. A high constitutive level of methionine transaminase would also result in an increased need for dietary methionine. An examination of the activities of the enzymes in the various pathways should help clarify the reason for this anomaly.

Felinine was first identified by Datta & Harris (1951) and isolated and characterised by Westall (1953). Because it is not available commercially, very little work has been done on felinine. Roberts (1963) examined several aspects of felinine metabolism and reported that felinine was highest in adult male cat urine, quite low in female cat urine, and was detectable in kittens of either sex. He found about a four-fold variation in felinine excretion among a group of 19 male cats. He also showed that giving an adult female cat testosterone for one week increased feline excretion whereas the administration of oestradiol to an adult male cat for two weeks had no effect. Isotopic incorporation methods have been used to show that leucine, mevalonic acid, and acetate can serve as precursors to felinine (Roberts, 1963; Avizonis & Wriston, 1959; Wang, 1964). However, attempts to identify the site and pathway of felinine biosynthesis have largely been unsuccessful.

The metabolic role of felinine is still a matter for speculation. It has been suggested that it may be a urinary component for territorial marking or perhaps somehow involved in the regulation of sterol metabolism (Shapiro, 1962).

The total quantity of dietary sulphur amino acids required to provide for the synthesis of felinine which is excreted in the urine of a male cat should not exceed 50–75 mg (0.05% of the diet). It is therefore doubtful that felinine synthesis alone is the reason for the high TSAA requirement of the cat (especially of the kitten).

Taurine deficiency in the cat has been shown to cause feline central retinal degeneration (FCRD) (Hayes, Carey & Schmidt, 1975; Hayes, Rabin & Berson, 1975). Taurine is a beta-amino sulphonic acid (2-amino ethanesulphonic acid) and, as such, is not present in protein. It is made from cysteine in the liver (and some other tissues, e.g. brain) of mammals (Hayes & Sturman, 1981). Except for its conjugation to bile acids (Haslewood, 1964; Vessey, 1978) and its presence in a few peptides such as the hormone glutarine (gamma-L-glutamyl-taurine) (Feuer *et al.*, 1978; Furka *et al.*, 1980) taurine is not further metabolised, and is eventually excreted in the urine.

FCRD had been described before Hayes and coworkers' discovery of the nutritional cause of the disease (Bellhorn, Aguirre & Bellhorn, 1974; Bellhorn & Fischer, 1970; Morris, 1965; Rubin, 1963; Rubin & Lipton, 1973; Scott, Greaves & Scott, 1964). The discovery of taurine deficiency as the cause of FCRD has led to an examination of taurine synthesis and metabolism in the cat, and the role of taurine in FCRD and its function in general have been extensively reviewed (Barbeau & Huxtable, 1978; Hayes, 1976; Hayes & Sturman, 1981; Sturman & Hayes, 1980). The length of time it takes to produce FCRD varies from 23 weeks to more than a year, and the incidence under laboratory conditions varies from 12–100% (National Research Council, 1981; Anderson *et al.*, 1979; Barnett & Burger, 1980; Hayes, Carey & Schmidt, 1975). Since several workers have reproduced FCRD under laboratory conditions, it is clear that some cats need taurine in the diet. However, the taurine requirement has not been experimentally defined.

Part of the uncertainty is caused by unknown dietary or environmental factors which affect the requirement. Morris & Rogers (1981) have examined the importance of type of dietary fat, level of dietary protein, presence of dietary fibre, and the level of dietary sulphur amino acids on the incidence and severity of FCRD in cats, and conclude that only the level of sulphur amino acids (SAA) has an effect. Weanling kittens fed a taurine-free diet and with SAA included near the requirement, all developed FCRD within 5–6 months (Anderson *et al.*, 1979). But, when twice the SAA requirement was fed, no FCRD was seen in one year (Morris & Rogers, 1981). Therefore, it appears that the problem of FCRD in kittens fed commercial cat food containing 200–300 mg taurine/kg diet is that the total 'available' SAA were near the requirement. But, even when kittens were fed a high level of SAA with no dietary taurine, a few developed FCRD. Therefore, the recommendation made by the 1981 NRC committee still appears valid; that is, that the taurine requirement (when SAA requirement is just met) is between 250 mg and 1000 mg/kg of dry diet. An interpolation of existing data puts the minimal requirement about 350 mg/kg diet. A more accurate estimate may not be known for some time since studies at the lower levels should probably be extended to 3–5 years before those concentrations can be considered safe.

Threonine
Anderson *et al.* (1980*b*) selected 0.8% of the diet as the threonine requirement for the kitten. However, a close examination of their results indicates that kittens grew just as well at 0.7% threonine as at 0.8% threonine. Titchenal *et al.* (1980) and Rogers & Morris (1979) found maximal weight gain at 0.7% of the diet as threonine and somewhat less than maximal growth at a dietary level of threonine of 0.6%. Therefore the minimal threonine requirement appears to be about 0.7% of the diet. Kittens fed 0.4% threonine developed neurological problems as early as the fifth day, even though they were still growing (Titchenal *et al.*, 1980). Signs included tremor, ataxia, inco-ordination, dysequilibrium and defective righting reflex. Neurological signs, which were progressive but completely reversible, were indicative of cerebellar dysfunction. In addition to the neurological signs, several kittens developed thoracic and pelvic limb lameness associated with carpal deviation and stiff movement of the hind limbs. They seemed reluctant to walk or run, but no visible pain or discomfort could be induced by vigorous palpation of the affected limbs and joints. These problems were completely reversed by feeding adequate threonine. These interesting observations have not been studied further.

In a report to the Pet Food Institute, Morris & Rogers (1984) described experiments in which they fed kittens diets containing all essential amino acids at concentrations of either 75, 100, or 125% of the levels listed in Table 5. All diets contained 24% protein (CPE). Growth was poor when 75% of requirements was fed. Although gain of the kittens consuming 125% essential amino acids was somewhat greater than that of kittens receiving 100%, the differences were not statistically significant. Performance of kittens given the 100% diet was similar to that of kittens given an iso-nitrogenous diet of whole protein (casein and soy). This work substantiates the view that the estimates listed in Table 5 closely approximate the true requirements of growing kittens.

Dietary concentrations of the essential amino acids required for maintenance of adult cats have not been determined.

Protein excess
While there is some evidence that feeding adult dogs levels of protein higher than those required to maintain nitrogen balance may be beneficial in certain cases (Allison *et al.*, 1954; Wannemacher & McCoy, 1966), caution should be exercised in feeding excess protein. Some investigators are of the opinion that feeding high levels of protein over long periods of time may be detrimental to kidney function (Brenner, Mayer & Hostetter, 1982; Murphy, 1983). Romsos *et al.* (1976) fed intermediate and high levels of protein (e.g. 20% and 48%) to young dogs, and saw no influence of dietary protein level on energy intake, nitrogen balance, body-

weight gain, body fat, fat-free mass, and most serum metabolites. However, plasma concentration of urea was significantly higher in dogs fed the higher levels of protein. Since kidney function often declines with age in dogs, and the kidney is responsible for urea excretion, it would appear prudent not to feed adult dogs a great excess of protein.

The cat's response to high-protein diets is variable, and depends in part on the source of protein and the physical form of the diet. Hills, Morris & Rogers (1982) fed kittens either 33 or 68% soy-protein diets, and found that, while the higher concentration of protein increased the potassium requirement, kittens fed 68% protein grew slightly better than those fed 33% protein when the potassium requirement was met, indicating no adverse effect of high dietary protein for kittens. Cook *et al.* (1985) found that kittens avoided high-protein diets based on casein but did not avoid high-protein diets based on isolated soy. Aversion to high-casein diets can be overcome on a short-term basis by feeding the diet as a gel (Hargrove, Rogers & Morris, unpublished). A gel-like consistency presumably increases the palatability of such diets. However, rate of nutrient absorption may be affected by the physical form of the diet, and rate of nitrogen absorption could conceivably affect blood ammonia concentrations and hence food intake. On a long-term basis, high concentrations of dietary casein could be aversive to cats by producing metabolic acidosis (Cook, Rogers & Morris, unpublished).

REFERENCES

Aberhalden, E. (1912). Futterungsversuche mit vollstandig abgebauten Nahrungsstoffen Losung des Problems der kunstlichen Darstellung der Nahrungsstoffe. *Hoppe-Seyler's Zeitschrift fur Physiologische Chemie*, **77**, 22–58.

Abderhalden, E. & Hirsch, P. (1912). Futterungsversuche mit Gelatine, Ammonsalzen, vollstandig abgebautem Fleisch und einem aus allen bekannten Aminosauren bestehenden Gemisch ausgefuhrt an jungen Hunden. *Hoppe-Seyler's Zeitschrift fur Physiologische Chemie*, **81**, 323–8.

Allison, J.B. (1956). Optimal nutrition correlated with nitrogen retention. *American Journal of Clinical Nutrition*, **4**, 662–72.

Allison, J.B. (1964). The nutritive value of dietary proteins. In *Mammalian Protein Metabolism*, vol. II, p. 47, ed. H.N. Munro & J.B. Allison, New York: Academic Press.

Allison, J.B., Anderson, J.A. & Seeley, R.D. (1946). The determination of the nitrogen balance index in normal and hypoproteinaemic dogs. *Annals of the New York Academy of Sciences*, **47**, 245–71.

Allison, J.B., Anderson, J.A. & Seeley, R.D. (1947). Some effects of methionine on the utilization of nitrogen in adult dogs. *Journal of Nutrition*, **33**, 361–70.

Allison, J.B., Miller, S.A., McCoy, J.R. & Brush, M.K. (1956). Studies on the protein nutrition of the cat. *The North American Veterinarian*, **37**, 38–43.

Allison, J.B., Wannemacher, R.W. & Migliarese, J.F. (1954). Diet and the metabolism of 2-aminofluorene. *Journal of Nutrition*, **52**, 415–25.

Almquist, H.J. (1949). Amino acid balance at super-normal dietary levels. *Proceedings of the Society for Experimental Biology and Medicine*, **72**, 179–80.

Anderson, P.A., Baker, D.H. & Corbin, J.E. (1979). Lysine and arginine requirements of the domestic cat. *Journal of Nutrition*, **109**, 1368–72.

Anderson, P.A., Baker, D.H., Corbin, J.E. & Helper, L.C. (1979). Biochemical lesions associated with taurine deficiency in the cat. *Journal of Animal Science*, **49**, 1227–34.

Anderson, P.A., Baker, D.H., Sherry, P.A. & Corbin, J.E. (1980a). Histidine, phenylalanine-tyrosine and tryptophan requirements for growth of the young kitten. *Journal of Animal Science*, **50**, 479–83.

Anderson, P.A., Baker, D.H., Sherry, P.A., Teeter, R.G. & Corbin, J.E. (1980b). Threonine, isoleucine, valine and leucine requirements of the young kitten. *Journal of Animal Science*, **50**, 266–71.

Arnold, A. & Schad, J.S. (1954). Nitrogen balance studies with dogs on casein or methionine-supplemented casein. *Journal of Nutrition*, **53**, 265–73.

Avizonis, P.V. & Wriston, J.C. (1959). On the biosynthesis of felinine. *Biochimica and Biophysica Acta*, **34**, 279–81.

Baker, D.H., Katz, R.S. & Easter, R.A. (1975). Lysine requirement of growing pigs at two levels of dietary protein. *Journal of Animal Science*, **40**, 851–6.

Barbeau, A. & Huxtable, R.J. (1978). *Taurine and Neurological Disorders*. New York: Raven.

Barnett, K.C. & Burger, I.H. (1980). Taurine deficiency retinopathy in the cat. *Journal of Small Animal Practice*, **21**, 521–34.

Becker, D.E., Jensen, A.H., Terrill, S.W., Smith, I.D. & Norton, H.W. (1957). The isoleucine requirement of weanling swine fed two protein levels. *Journal of Animal Science*, **16**, 26–34.

Belavady, B., Madhavan, T.V. & Gopalan, C. (1967). Production of nicotinic acid deficiency (Black tongue) in pups fed diets supplemented with leucine. *Gastroenterology*, **53**, 749–53.

Bellhorn, R.W., Aguirre, G.D. & Bellhorn, M.B. (1974). Feline central retinal degeneration. *Investigative Ophthalmology*, **13**, 608–16.

Bellhorn, R.W. & Fischer, C.A. (1970). Feline central retinal degeneration. *Journal of the American Veterinary Medical Association*, **157**, 842–9.

Blaza, S.E., Burger, I.H., Holme, D. W. & Kendall, P.T. (1982). Sulphur-containing amino acid requirements of growing dogs. *Journal of Nutrition*, **112**, 2033–42.

Boomgaardt, J. & Baker, D.H. (1971). Tryptophan requirement of growing chicks as affected by dietary protein level. *Journal of Animal Science*, **33**, 595–9.

Boomgaardt, J. & Baker, D.H. (1973). Tryptophan requirement of growing pigs at three levels of dietary protein. *Journal of Animal Science*, **36**, 303–6.

Brenner, B.M., Meyer, T.W. & Hostetter, H.H. (1982). Dietary protein intake and the progressive nature of kidney disease. *New England Journal of Medicine,* **307,** 652–9.

Bressani, R. & Mertz, E.T. (1958). Relationship of protein level to the minimum lysine requirement of the rat. *Journal of Nutrition,* **65,** 481–91.

Brinegar, M.J., Williams, H.H., Ferris, F.H., Loosli, J.K. & Maynard, L.A. (1950). The lysine requirement for the growth of swine. *Journal of Nutrition,* **42,** 129–38.

Burger, I.H., Blaza, S.E., Kendall, P.T. & Smith, P.M. (1984). The protein requirement of adult cats for maintenance. *Feline Practice,* **14,** 8–14.

Burns, R.A., Garton, R.L. & Milner, J.A. (1984). Leucine, isoleucine and valine requirements of immature beagle dogs. *Journal of Nutrition,* **114,** 204–9.

Burns, R.A., LeFaivre, M.H. & Milner, J.A. (1982). Effects of dietary protein quantity and quality on the growth of dogs and rats. *Journal of Nutrition,* **112,** 1843–53.

Burns, R.A. & Milner, J.A. (1981). Sulphur amino acid requirements of immature beagle dogs. *Journal of Nutrition,* **111,** 2117–24.

Burns, R.A. & Milner, J.A. (1982). Threonine, tryptophan and histidine requirements of immature beagle dogs. *Journal of Nutrition,* **112,** 447–52.

Burns, R.A., Milner, J.A. & Corbin, J.E. (1981). Arginine: an indispensable amino acid for mature dogs. *Journal of Nutrition,* **111,** 1020–4.

Cianciarus, B., Jones, M.R. & Kopple, J.D. (1981). Histidine: an essential amino acid for adult dogs. *Journal of Nutrition,* **111,** 1074–84.

Cohn, C., Joseph, D., Bell, L. & Oler, A. (1963). Feeding frequency and protein metabolism. *American Journal of Physiology,* **205,** 71–8.

Cook, N.E., Kane, E., Rogers, Q.R. & Morris, J.G. (1985). Self-selection of dietary casein and soy-protein by the cat. *Physiology and Behaviour,* **34,** 583–94.

Costello, M.J., Morris, J.G. & Rogers, Q.R. (1980). Effect of dietary arginine level on urinary orotate and citrate excretion in growing kittens. *Journal of Nutrition,* **110,** 1204–8.

Czarnecki, G.L. & Baker, D.H. (1984). Urea cycle function in the dog with emphasis on the role of arginine. *Journal of Nutrition,* **114,** 581–90.

Czarnecki, G.L. & Baker, D.H. (1982). Utilization of D- and L-tryptophan by the growing dog. *Journal of Animal Science,* **55,** 1405–10.

Czarnecki, G.L., Hirakawa, D.A. & Baker, D.H. (1985). Antagonism of arginine by excess dietary lysine in the growing dog. *Journal of Nutrition,* **115,** 743–52.

Datta, S.P. & Harris, H. (1951). A convenient apparatus for paper chromatography. Results of a survey of the urinary amino-acid patterns of some animals. *Journal of Physiology*, 114, 39P–41P.

Deady, J.E., Anderson, B., O'Donnell, J.A. III, Morris, J.G. & Rogers, Q.R. (1981). Effect of level of dietary glutamic acid and thiamin on food intake, weight gain, plasma amino acids, and thiamin status of growing kittens. *Journal of Nutrition*, 111, 1568–79.

Deady, J.E., Rogers, Q.R. & Morris, J.G. (1981). Effect of high dietary glutamic acid on the excretion of ^{35}S-thiamin in kittens. *Journal of Nutrition*, 111, 1580–85.

Deshmukh, D.R. & Shope, T.C. (1983). Arginine requirement and ammonia toxicity in ferrets. *Journal of Nutrition*, 113, 1664–7.

Dickinson, C.D. & Scott, P.P. (1956). Nutrition of the cat. 2. Protein requirements for growth of weanling kittens and young cats maintained on a mixed diet. *British Journal of Nutrition*, 10, 311–16.

Fabry, P. (1967). Metabolic consequences of the pattern of food intake. In: *Handbook of Physiology, section 6, Alimentary Canal, Vol. 1*, ed. C.F. Code. Washington, DC: American Physiological Society.

Fau, D., Smalley, K.A., Rogers, Q.R. & Morris, J.G. (1983). Effects of excess dietary methionine in the kitten. *Federation Proceedings*, 42, 542, Abstract #1469.

Featherstone, W.R., Rogers, Q.R. & R.A. Freedland (1973). Relative importance of kidney and liver in synthesis of arginine by the rat. *American Journal of Physiology*, 224, 127–9.

Feuer, L., Torok, L.J., Kapa, E. & Csaba, G. (1978). The effect of gamma-L-glutamyl-taurine (litoralon) on the amphibian metamorphosis. *Comparative Biochemistry and Physiology*, 61C, 67–71.

Furka, A., Sebestyen, F., Feuer, L., Horvath, A., Hercsel, J., Ormai, S. & Banyai, B. (1980). Isolation of gamma-glutamyl taurine from the protein-free aqueous extract of bovine parathyroid powder. *Acta Biochimica et Biophysica, Academy of Sciences Hungary*, 15, 39–47.

Gessert, C.F. & Phillips, P.H. (1956). Protein in the nutrition of the growing dog. *Journal of Nutrition*, 58, 415–21.

Greaves, J.P. (1965). Protein and calorie requirements of the feline. In *Canine and Feline Nutritional Requirements*, ed. O. Graham-Jones. Oxford: Pergamon Press.

Greaves, J.P. & Scott, P.P. (1960). Nutrition of the cat. 3. Protein requirements for nitrogen equilibrium in adult cats maintained on a mixed diet. *British Journal of Nutrition*, 14, 361–9.

Greenstein, J.P., Winitz, M., Gullino, P., Birnbaum, S.M. & Otey, M.C. (1956). Studies on the metabolism of amino acids and related compounds *in vivo*. III. Prevention of ammonia toxicity by arginine and related compounds. *Archives of Biochemistry and Biophysics*, 64, 342–54.

Ha, Y.H., Milner, J.A. & Corbin, J.E. (1978). Arginine requirements in immature dogs. *Journal of Nutrition*, 108, 203–10.

Hardy, A.J., Morris, J.G. & Rogers, Q.R. (1977). Valine requirement of the growing kitten. *Journal of Nutrition*, 107, 1308–12.

Hargrove D.M., Rogers, Q.R., Calvert, C.C. & Morris, J.G. (1988). Effects of dietary excess of the branched-chain amino acids on growth, food intake and plasma amino acid concentrations of kittens. *Journal of Nutrition*, **118**, 311–20.

Hargrove, D.M., Rogers, Q.R. & Morris, J.G. (1983). The tryptophan requirement of the kitten. *British Journal of Nutrition*, **50**, 487–93.

Hargrove, D.M., Rogers, Q.R. & Morris, J.G. (1984). Leucine and isoleucine requirements of the kitten. *British Journal of Nutrition*, **52**, 595–605.

Haslewood, G.A.D. (1964). The biological significance of chemical differences in bile salts. *Biological Review*, **39**, 537–74.

Hayes, K.C. (1976). A review on the biological function of taurine. *Nutrition Reviews*, **34**, 161–5.

Hayes, K.C., Carey, R.E. & Schmidt, S.Y. (1975). Retinal degeneration associated with taurine deficiency in the cat. *Science*, **188**, 949–51.

Hayes, K.C., Rabin, A.R. & Berson, E.L. (1975). An ultrastructural study of nutritionally induced and reversed retinal degeneration in cats. *American Journal of Pathology*, **78**, 505–24.

Hayes, K.C. & Sturman, J.A. (1981). Taurine in metabolism. *Annual Review of Nutrition*, **1**, 401–25.

Heiman, V. (1947). The protein requirements of growing puppies. *Journal of the American Veterinary Medical Association*, **111**, 304–8.

Henslee, J.G. & Jones, M.E. (1982). Ornithine synthesis from glutamate in rat small intestinal mucosa. *Archives of Biochemistry and Biophysics*, **49**, 186–97.

Hills, D.L., Morris, J.G. & Rogers, Q.R. (1982). Potassium requirement of kittens as affected by dietary protein. *Journal of Nutrition*, **112**, 216–22.

Hirakawa, D.A. & Baker, D.H. (1985). Sulphur amino acid nutrition of the growing puppy: determination of dietary requirements for methionine and cystine. *Nutrition Research*, **5**, 631–42.

Ikeda, M., Tsuji, H., Nakamura, S., Ichiyama, A. Nishizuka, Y. & Hayaishi, O. (1965). Studies on the biosynthesis of nicotinamide adenine dinucleotide. II. A role of picolinic carboxylase in the biosynthesis of nicotinamide adenine dinucleotide from tryptophan in mammals. *Journal of Biological Chemistry*, **240**, 1395–401.

Jansen, G.R., Deuth, M.A., Ward, G.M. & Johnson, D.E. (1975). Protein quality studies in growing kittens. *Nutrition Reports International*, **11**, 525–36.

Kade, C.F., Phillips, J.H. & Phillips, W.A. (1948). The determination of the minimum nitrogen requirement of the adult dog for maintenance of nitrogen balance. *Journal of Nutrition*, **36**, 109–21.

Kane, E., Leung, P.M.B., Rogers, Q.R. & Morris, J.G. (1986). Diurnal feeding and drinking patterns in adult cats as affected by changing the level of fat in the diet. *Appetite* (in press).

Kane, E., Rogers, Q.R., Morris, J.G. & Leung, P.M.B. (1981). Feeding behaviour of the cat fed laboratory and commercial diets. *Nutrition Research*, **1**, 499–507.

Kendall, P.T., Blaza, S.E. & Holme, D.W. (1982). Assessment of endogenous nitrogen output in adult dogs of contrasting size using a protein-free diet. *Journal of Nutrition,* **112,** 1281–6.

Leklem, J.E., Brown, R.R. Hankes, L.V. & Schmaeler, M. (1971). Tryptophan metabolism in the cat: A study with carbon-14-labelled compounds. *American Journal of Veterinary Research,* **32,** 335–44.

Leoschke, W.L. & Elvehjem, C.A. (1959). The importance of arginine and methionine for the growth and fur development of mink fed purified diets. *Journal of Nutrition,* **69,** 147–50.

Leveille, G.A. (1972). The long-term effects of meal-eating on lipogenesis, enzyme activity and longevity in the rat. *Journal of Nutrition,* **102,** 549–56.

Mabee, D.M. & Morgan, A.F. (1951). Evaluation by dog growth of egg yolk protein and six other partially purified protein, some after heat treatment. *Journal of Nutrition,* **43,** 261–79.

MacDonald, M.L., Rogers, Q.R. & Morris, J.G. (1984). Nutrition of the domestic cat, a mammalian carnivore. *Annual Review of Nutrition,* **4,** 521–62.

Magendie, M.F. (1816). Sur les proprietes nutritives des substances qui ne contiennent pas d'azote. *Annales de Chimie et de Physique,* **3,** 66–77.

Malloy, M.H., Rassin, D.K., Gaull, G.E. & Heird, W.C. (1981). Development of taurine metabolism in beagle pups: effects of taurine-free total parenteral nutrition. *Biology of the Neonate,* **40,** 1–8.

Manson, J.A. & Carpenter, K.J. (1978). The effect of a high level of dietary leucine on the niacin status of dogs. *Journal of Nutrition,* **108,** 1889–98.

McCollum, E.V. (1957). *A History of Ideas in Nutrition. The Sequence of Ideas in Nutrition Investigations.* Boston: Houghton-Mifflin.

McCoy, R.H., Meyer, C.E. & Rose, W.C. (1935). Feeding experiments with mixtures of highly purified amino acids. VIII. Isolation and identification of a new essential amino acid. *Journal of Biological Chemistry,* **112,** 283–302.

Melnick, D. & Cowgill, G.R. (1937). The protein minima for nitrogen equilibrium with different proteins. *Journal of Nutrition,* **13,** 401–24.

Miller, L.L. (1944). The metabolism of dl-methionine and l-cystine in dogs on a very low protein diet. *Journal of Biological Chemistry,* **152,** 603–11.

Miller, S.A. & Allison, J.B. (1958). The dietary nitrogen requirements of the cat. *Journal of Nutrition,* **64,** 493–501.

Milner, J.A. (1979a). Assessment of indispensable and dispensable amino acids for the immature dog. *Journal of Nutrition,* **109,** 1161–7.

Milner, J.A. (1979b). Assessment of the essentiality of methionine, threonine, tryptophan, histidine and isoleucine in immature dogs. *Journal of Nutrition,* **109,** 1351–7.

Milner, J.A. (1981). Lysine requirements of the immature dog. *Journal of Nutrition,* **111,** 40–5.

Milner, J.A., Garton, R.L. & Burns, R.A. (1984). Phenylalanine and tyrosine requirements of the immature Beagle dog. *Journal of Nutrition*, **114**, 2212–16.

Milner, J.A., Prior, R.L. & Visek, W.J. (1975). Arginine deficiency and orotic aciduria in mammals. *Proceedings of the Society for Experimental Biology and Medicine*, **150**, 282–8.

Morris, J.G. (1985). Nutritional and metabolic responses to arginine deficiency in carnivores. *Journal of Nutrition*, **115**, 524–31.

Morris, J.G. & Rogers, Q.R. (1978*a*). Ammonia intoxication in the near-adult cat as a result of a dietary deficiency of arginine. *Science*, **199**, 431–2.

Morris, J.G. & Rogers, Q.R. (1978*b*). Arginine: An essential amino acid for the cat. *Journal of Nutrition*, **108**, 1944–53.

Morris, J.G. & Rogers, Q.R. (1981). Report to the Pet Food Institute Feline Task Force, 2nd Annual Pet Food Institute Technical Seminar, September 18, 1981.

Morris, J.G. & Rogers, Q.R. (1982). Do cats really need more protein? In *Waltham Symposium, No. 4, Recent Advances in Feline Nutrition*, ed. A.T.B. Edney, *Journal of Small Animal Practice*, **23**, 521-32.

Morris, J.G. & Rogers, Q.R. (1984). Amino acid requirements of kittens for growth. In *5th Annual Pet Food Institute Technical Symposium*, Nashville, Tennessee.

Morris, J.G., Rogers, Q.R., Winterrowd, D.L. & Kamikawa, E.M. (1979). The utilization of ornithine and citrulline by the growing kitten. *Journal of Nutrition*, **109**, 724–9.

Morris, M.L. (1965). Feline degenerative retinopathy. *Cornell Veterinarian*, **55**, 295–308.

Mugford, R.A. (1977). External influences on the feeding of carnivores. In *The Chemical Senses and Nutrition*, pp. 25–50, ed. M.R. Kare & O. Maller. New York: Academic Press.

Mugford, R.A. & Thorne, C. (1980). Comparative studies of meal patterns in pet and laboratory housed dogs and cats. In *Nutrition of the Dog and Cat*, ed. R.S. Anderson. Oxford: Pergamon Press.

Munro, H.N. (1964). Historical introduction: the origin and growth of our present concepts of protein metabolism. In *Mammalian Protein Metabolism*, vol. 1, ed. H.N. Munro & J.B. Allison. New York: Academic Press.

Murphy, D.H. (1983). Too much of good thing: protein and a dog's diet. *International Journal for the Study of Animal Problems*, **4**, 101–7.

National Research Council (1962). *Nutrient Requirements of Dogs*. Washington, DC: National Academy Press.

National Research Council (1974). *Nutrient Requirements of Dogs*. Washington, DC: National Academy Press.

National Research Council (1985). *Nutrient Requirements of Dogs*. Washington, DC: National Academy Press.

National Research Council (1978). *Nutrient Requirements of Cats*. Washington, DC: National Academy Press.

National Research Council (1981). *Taurine Requirement of the Cat*, pp. 1–4, Washington, DC: National Academy Press.

National Research Council (1985). *Nutrient Requirements of Dogs.* Washington, DC: National Academy Press.

Ohara, I., Otsuka, S-I., Yugari, Y. & Ariyoshi, S. (1980). Comparison of the nutritive values of L-, DL- and D-tryptophan in the rat and chick. *Journal of Nutrition,* **110**, 634–40.

Ontko, J.A., Wuthier, R.E. & Phillips, P.H. (1957). The effect of increased dietary fat upon the protein requirement of the growing dog. *Journal of Nutrition,* **62**, 163–9.

Payne, P.R. (1965). Assessment of the protein values of diets in relation to the requirements of the growing dog. In *Canine and Feline Nutritional Requirements,* ed. O. Graham-Jones. Oxford: Pergamon Press.

Quam, D.D., Morris, J.G. & Rogers, Q.R. (1986). Histidine requirement of kittens for growth, haematopoiesis and prevention of cataracts. *British Journal of Nutrition* (in press).

Roberts, R.N. (1963). *A Study of Felinine and its Excretion by the Cat.* PhD Thesis. State University of New York, Buffalo.

Rogers, Q.R., Freedland, R.A. & Symmons, R.A. (1972). *In vivo* synthesis and utilization of arginine in the rat. *American Journal of Physiology,* **223**, 236–40.

Rogers, Q.R. & Morris, J.G. (1978). Amino acid nutrition and metabolism in the cat. In *Proceedings of the Kal Kan Symposium for the Treatment of Dog and Cat Diseases,* pp. 6–12. Kal Kan Foods, Inc.: Vernon, CA.

Rogers, Q.R. & Morris, J.G. (1979). Essentiality of amino acids for the growing kitten. *Journal of Nutrition,* **109**, 718–23.

Rogers, Q.R. & Morris, J.G. (1980). Why does the cat require a high protein diet? In *Nutrition of the Dog and Cat,* ed. R.S. Anderson. Oxford: Pergamon Press.

Rogers, Q.R., Morris, J.G. & Freedland, R.F. (1977). Lack of hepatic enzyme adaptation to low and high levels of dietary protein in the adult cat. *Enzyme,* **22**, 348–56.

Rogers, Q.R. & Phang, J.M. (1985). Deficiency of pyrroline-5-carboxylate synthase in the intestinal mucosa of the cat. *Journal of Nutrition,* **115**, 146–50.

Romsos, D.R., Belo, P.S. Bennink, M.R., Bergen, W.G. & Leveille, G.A. (1976). Effects of dietary carbohydrate fat and protein on growth, body composition and blood metabolite levels in the dog. *Journal of Nutrition,* **106**, 1452–64.

Romsos, D.R., Belo, P.S., Bergen, W.G. & Leveille, G.A. (1978). Influence of meal frequency on body weight, plasma metabolites, and glucose and cholesterol metabolism in the dog. *Journal of Nutrition,* **108**, 238–47.

Romsos, D.R. & Ferguson, D. (1983). Regulation of protein intake in adult dogs. *Journal of the American Veterinary Medical Association,* **182**, 41–3.

Rose, W.C. (1938). The nutritive significance of the amino acids. *Physiological Reviews,* **18**, 109–36.

Rose, W.C. & Rice, E.E. (1939). The significance of the amino acids in canine nutrition. *Science*, **90**, 186–7.

Ross, G., Dunn, D. & Jones, M.E. (1978). Ornithine synthesis from glutamate in rat intestinal mucosa homogenates: Evidence for the reduction of glutamate to gamma-glutamyl semialdehyde. *Biochemistry and Biophysics Research Communications*, **83**, 140–7.

Rubin, L.F. (1963). Atrophy of rods and cones in the cat retina. *Journal of the American Veterinary Medical Association*, **142**, 1415–20.

Rubin, L.F. & Lipton, D.E. (1973). Retinal degeneration in kittens. *Journal of the American Veterinary Medical Association*, **162**, 467–9.

Schaeffer, M.C., Rogers, Q.R. & Morris, J.G. (1982). Methionine requirement of the growing kitten, in the absence of dietary cystine. *Journal of Nutrition*, **112**, 962–71.

Scott, P.P., Greaves, J.P. & Scott, M.G. (1964). Nutritional blindness in the cat. *Experimental Eye Research*, **3**, 357–64.

Shapiro, I.L. (1962). *In Vivo Studies on the Metabolic Relationship Between Felinine and Serum Cholesterol in the Domestic Cat*. PhD Thesis. University of Delaware, Newark.

Smalley, K.A., Rogers, Q.R. & Morris, J.G. (1983). Methionine requirement of kittens given amino acid diets containing adequate cystine. *British Journal of Nutrition*, **49**, 411–17.

Smalley, K.A., Rogers, Q.R., Morris, J.G. & Eslinger, L.L. (1985). The nitrogen requirement of the weanling kitten. *British Journal of Nutrition*, **53**, 501–12.

Stewart, P.M., Batshaw, M., Valle, D. & Walser, M. (1981). Effects of arginine-free meals on ureagenesis in cats. *American Journal of Physiology*, **241**, E310–15.

Sturman, J.A. & Hayes, K.C. (1980). The biology of taurine in nutrition and development. *Advances in Nutrition Research*, **3**, 231–99.

Sturman, J.A., Rassin, D.K., Hayes, K.C. & Gaull, G.E. (1978). Taurine deficiency in the kitten: exchange and turnover of [^{35}S] taurine in brain, retina, and other tissues. *Journal of Nutrition*, **108**, 1462–76.

Suhadolnik, R.J., Stevens, C.O., Decker, R.H., Henderson, L.M. & Hankes, L.V. (1957). Species variation in the metabolism of 3-hydroxyanthranilate to pyridine carboxylic acids. *Journal of Biochemistry*, **228**, 973–82.

Teeter, R.G., Baker, D.H. & Corbin, J.E. (1978). Methionine and cystine requirements of the cat. *Journal of Nutrition*, **108**, 291–5.

Titchenal, C.A., Rogers, Q.R., Indrieri, R.J. & Morris, J.G. (1980). Threonine imbalance, deficiency and neurologic dysfunction in the kitten. *Journal of Nutrition*, **110**, 2444–59.

Vessey, D.A. (1978). The biochemical basis for the conjugation of bile acids with either glycine or taurine. *Biochemical Journal*, **174**, 621–6.

Wang, T.M. (1964). *Further Studies on the Biosynthesis of Felinine*. PhD Thesis. University of Delaware, Newark.

Wannemacher, R.E. & McCoy, J.R. (1966). Determination of optimal dietary protein requirements of young and old dogs. *Journal of Nutrition*, **88**, 66–74.

Wergedal, J.E., & Harper, A.E. (1964). Metabolic adaptations in higher animals IX. Effect of high protein intake on amino nitrogen catabolism *in vivo. Journal of Biological Chemistry,* **239,** 1156–63.

Westall, R.G. (1953). The amino acids and other ampholytes of urine. 2. The isolation of a new sulphur-containing amino acid from cat urine. *Biochemical Journal,* **55,** 244–8.

Willcock, E.G. & Hopkins, F.G. (1906). The importance of individual amino acids in metabolism. *Journal of Physiology,* **35,** 88–102.

Windmueller, H.G. & Spaeth, A.E. (1981). Source and fate of circulating citrulline. *American Journal of Physiology,* **241,** E473–80.

11

Tryptophan metabolism in the cat

J.R. MERCER AND S.V.P.S. SILVA

Introduction

Tryptophan is an essential amino acid consisting of a benzene and a pyrrole ring to which is attached a non-polar side chain. It was discovered by Hopkins & Cole in 1901 and was the first amino acid shown to be essential in the diet of animals. Although tryptophan is an essential amino acid it is unlikely to be first limiting in any diet. Lysine or the sulphur amino acids are usually first limiting and the nutritional interest in tryptophan stems largely from the key functions of a number of its metabolic products.

Tryptophan participates in many biochemical processes and is important in the regulation of protein turnover and of some aspects of brain metabolism. Some of the key enzymes of tryptophan metabolism in the rat have been intensively studied. In particular, much is known of the regulation of the glutarate pathway and the synthesis of nicotinate in the liver, and of the regulation of brain serotonin synthesis. However, most of the information has been obtained from work with the rat and there is a paucity of information on the nutritional biochemistry of tryptophan in other animals.

Estimates of the tryptophan requirement of the growing cat range from 1.1 g/kg diet (Hargrove, Rogers & Morris, 1983) to 1.5 g/kg diet (Anderson *et al.*, 1980) depending on the growth rate, and the requirement is comparable with that of other animals at similar stages of growth. The rate of tryptophan catabolism in cat liver is much lower than in the rat and the activities of the key enzymes regulating its catabolism are generally less responsive to the factors which alter their activities in the rat. The net result of these differences is that the cat cannot synthesise nicotinate from tryptophan and excretes much smaller quantities of the cancer-promoting metabolites of tryptophan in its urine. Although tryptophan may only perform a relatively minor direct role in the regulation of gluconeogenesis in the cat liver, it probably participates in glucose homeostasis indirectly through its effect on appetite and food intake. This paper deals with some of these unusual features of tryptophan metabolism in the cat.

207

Tryptophan metabolism

In the mammal, tryptophan metabolism occurs by four pathways, the essential features of which are shown in Figure 1. The oxidation of tryptophan by the kynurenine–glutarate pathway in the liver is quantitatively the most important (Young, St Arnaud-McKenzie & Sourkes, 1978) accounting for about 99% of whole body tryptophan metabolism (Bender, 1982). In the rat liver at least, tryptophan metabolism occurs exclusively by the kynurenine-glutarate pathway (Smith, Carr & Pogson, 1980). In the rat brain, the 5-hydroxyindole pathway, which accounts for about 1% of total body tryptophan catabolism (Bender, 1982) is responsible for the synthesis of the neurotransmitter serotonin. The transamination of tryptophan to indol-3-ylpyruvate (indole pathway) and the decarboxylation to tryptamine (tryptamine pathway) are only of minor interest in mammals. Although tryptamine and indol-3-ylacetate occur in the urine, most of these are probably produced by microflora in the gut (Chen, Gholson & Raica, 1974).

In the rat, under normal conditions, little transamination of kynurenine or 3-hydroxykynurenine occurs. However, in vitamin B_6 deficiency, the urinary excretion of xanthurenate and kynurenate increases (Lepkovsky, Roboz & Haagen-Smit, 1943). Both kynureninase and kynurenine aminotransferase require vitamin B_6 as a coenzyme. However, kynureninase only occurs in the cytoplasm but kynurenine aminotransferase also occurs in the mitochondria and the mitochondrial membrane probably protects the aminotransferase from being depleted of its cofactor (Ogasawara, Hagino & Kotake, 1962). Xanthurenate is diabetogenic as it binds to insulin and reduces its action but, *in vitro* at least, zinc can displace the xanthurenate and restore insulin action (Ikeda & Kotake, 1984).

Metabolic roles of tryptophan

The concentration of tryptophan in the proteins of mammals is much lower than that of other amino acids and it was suggested some years ago that it may play a special role in the regulation of protein synthesis at the translational step (Fleck, Shepherd & Munro, 1965). It was subsequently shown that tryptophan exerted the cytoplasmic effect on protein synthesis in rat liver by initiating ribosome aggregation (Sidransky *et al.*, 1968) as a result of an increased transport of m-RNA into the cytoplasm (Murty & Sidransky, 1972), and it was also shown to have an effect in the nucleus by stimulating the synthesis of DNA-dependent RNA polymerase thereby initiating transcription (Henderson, 1970). Tryptophan and other amino acids are known to inhibit lysosomal protein degradation in rat liver cells (Hopgood, Clark & Ballard, 1977) and we have shown similar effects in cat liver cells (Silva & Mercer, 1986). Recently Grinde (1984) showed that, in

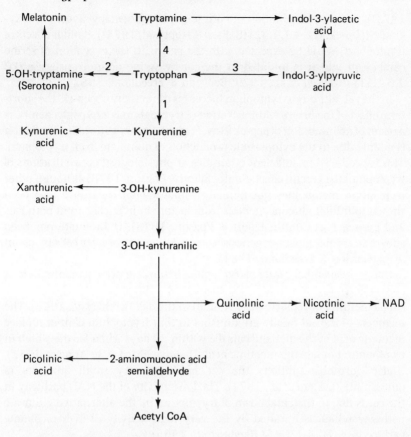

1   kynurenine–glutarate pathway
2   5–hydroxy–indole pathway
3   indole pathway
4   tryptamine pathway

Fig. 1. Metabolic fates of tryptophan.

rat liver cells, the tryptophan metabolite kynurenine was a more potent inhibitor of lysosomal protein degradation than tryptophan itself.

In the rat, tryptophan has been shown to have important stimulatory effects on the activities of several enzymes not involved in tryptophan degradation. The effect occurs by an increased synthesis of the enzyme protein (e.g. serine dehydratase (EC 4.2.1.13) (Jost, Khairallah & Pitot, 1968)) or by stabilising the enzyme protein thereby decreasing the rate of its degradation (e.g. ornithine aminotransferase (EC 2.6.1.13) (Chee & Swick, 1976), tyrosine aminotransferase (EC 2.6.1.5) (Čihák, Lamar & Pitot,

1973)), or by a combination of both (e.g. phosphoenolpyruvate carboxy-kinase (pep-ck) (EC 4.1.1.32) (Ballard & Hopgood, 1973)). Similar effects of tryptophan would be expected with the cat, with the exception of serine catabolism which is initiated in the cat by serine aminotransferase (EC 2.6.1.51) (Rowsell *et al.*, 1979; Beliveau & Freedland, 1982).

In the rat, although tryptophan increases the activity of pep-ck, the short-term effect of the *in vivo* administration of tryptophan is a hypoglycaemia as a result of an inhibition of pep-ck (Ray, Foster & Lardy, 1966). The inhibition is mainly due to the cytoplasmic production of quinolate from tryptophan. This is produced in sufficient quantities at physiological concentrations of tryptophan to exert an effect (Smith, Elliott & Pogson, 1978) although other tryptophan metabolites not formed in mammalian cells have also been shown to inhibit gluconeogenesis from lactate in liver cells from both rats and guinea pigs (Smith, Elliott & Pogson, 1979). Quinolinate has been shown to inhibit gluconeogenesis from pyruvate in cat liver cells by about 40% (Gordon & Freedland, 1981).

One of the metabolic fates of tryptophan in the liver of most mammals is its conversion to nicotinate (Nishizuka & Hayaishi, 1963) thereby supplementing or replacing the vitamin in the diet (Krehl *et al.*, 1946). The members of the cat family are unusual in that tryptophan cannot replace nicotinic acid in the diet and cats die within 20 days when fed diets high in tryptophan but lacking nicotinic acid (Da Silva, Fried & De Angelis, 1952). Under normal conditions the cat produces very small amounts of quinolinate (Leklem *et al.*, 1971). The low activity of the NAD pathway in the cat is due to the catabolism of tryptophan by the alternative glutarate pathway which is initiated by the very high activity of liver picolinate carboxylase (EC 4.1.1.45) (Ikeda *et al.*, 1965).

Tryptophan is the precursor of serotonin (5-hydroxytryptamine), the neurotransmitter which influences the response of the brain to a variety of inputs, and which has been shown to be involved in appetite control, behaviour and mood, in sleep–awake rhythms and in other cyclic neural events (Wurtman, 1980). Serotonin in turn is the precursor of melatonin which is produced in the pineal gland and facilitates the seasonality and sensitivity of breeding to day-length (Reiter & Richardson, 1980).

Picolinate, a product of the glutarate pathway of tryptophan metabolism is necessary for the intestinal absorption of zinc (Evans & Johnson, 1979). Picolinate produced in the liver is excreted in the urine as the glycine conjugate (Mehler & May, 1956) but it is also produced in the exocrine pancreas and secreted into the small intestine where it forms a complex with zinc and facilitates its absorption (Evans, 1980).

Tryptophan occurs in the plasma both free and non-covalently bound to albumin, 80–90% being in the bound form (McMenamy *et al.*, 1961). Non-esterified fatty acids displace tryptophan from plasma albumin (Curzon, Friedel & Knott, 1973) although they appear to have different binding sites

(McMenamy, 1965). Only the unbound tryptophan can be transported into the brain (Etienne, Young & Sourkes, 1976) and the liver (Smith & Pogson, 1980) so the level of binding to albumin is likely to influence brain serotonin synthesis and liver catabolism of tryptophan.

Regulation of tryptophan dioxygenase

The cytosolic kynurenine–glutarate pathway which is responsible for the regulation of the concentration of circulating tryptophan, also leads to the production of nicotinate although not, as stated previously, in the cat. The first enzyme in the pathway is tryptophan 2,3-dioxygenase (tryptophan pyrrolase) (EC 1.13.11.11), a haemoprotein which catalyses the oxidative cleavage of the pyrrole ring of tryptophan to produce N-formylkynurenine. Its functions and the regulation of its activity have often been reviewed (Čihák, 1979; Badawy, 1984).

In most mammals, tryptophan 2,3-dioxygenase (TDO) occurs in at least two forms; an inactive haem-free apoenzyme and an active reduced holoenzyme. It occurs in both forms in the rat, mouse, pig, turkey, chicken and man but the cat, in common with the guinea pig, ruminants, rabbit, gerbil, hamster and frog, does not apparently possess an apoenzyme (Badawy & Evans, 1976).

In animals which possess both apo- and holo-forms of the enzyme, the activity is regulated in a number of ways. Hormonal control has been demonstrated by the increased synthesis of the apoenzyme which accompanies the administration of glucocorticoids to rats (De Lap & Feigelson, 1978). Tryptophan itself participates in regulation in at least two ways. Tryptophan occupies a catalytic and a regulatory site on the enzyme (Schimke, Sweeney & Berlin, 1965) and it also increases holoenzyme activity by stimulating the synthesis or availability of haem (Greengard & Feigelson, 1961; Badawy, 1984). The overall effect of tryptophan administration is an increased saturation of the apoenzyme with haem and a decrease in the degradation rate of the enzyme protein. There is good evidence that in rat liver, TDO is allosterically inhibited by nicotinamide adenine dinucleotides, particularly when in the reduced state (Cho-Chung & Pitot, 1967). Experiments *in vivo* using ethanol and various drugs have also provided indirect evidence that the reduced nucleotides inhibit TDO (Badawy, 1977). However, Pogson *et al.* (1984) have shown that the large changes in redox state induced by ethanol and xylitol have little effect on TDO flux in rat liver cells, and they concluded that the *in vivo* concentration of free NADH in the cytosol would be too low to be effective as an inhibitor of the enzyme.

We have conducted experiments on the regulation of TDO in isolated cat liver cells. Previous experiments by Leklem, Woodford & Brown (1969) and Badawy & Evans (1976) showed that the activity of TDO in cat liver homogenates was about 50% of that in the rat liver, and the enzyme was not induced by the administration of cortisol (Leklem, Woodford & Brown,

1969). We have shown that TDO flux in liver cells from 16 h fasted cats (Silva & Mercer, 1985) is no more than about 25% of that in rat liver cells (Smith, Carr & Pogson, 1980) and it is not increased by the administration of dexamethasone (Silva & Mercer, 1986). We have also shown that the enzyme is responsive to an increased supply of protein. In experiments using liver cells from cats fed 17% or 70% protein diets, the flux of TDO measured at physiological concentrations of tryptophan was three times higher in cells from high protein than from low protein fed animals (Silva & Mercer, 1985). The increase in flux with tryptophan concentration pointed to an elevated synthesis of enzyme protein without any change in catalytic rate and therefore conforms with the belief that tryptophan acts directly through stabilisation of the holoenzyme or indirectly through a higher rate of synthesis resulting from an increased synthesis or availability of haem.

The low activity of TDO in the cat is surprising in view of the high tryptophan intakes which would accompany the consumption of a high protein diet, and particularly in wild cats which feed infrequently, there would seem to be a physiological requirement for the rapid adaptation of regulatory enzymes such as TDO. To clarify this, we have examined some aspects of the control of TDO in liver cells from 48 h fasted cats which had previously been fed a commercial dry food to appetite. The flux rate of TDO was measured at a tryptophan concentration of 0.1 mM in Krebs–Henseleit bicarbonate buffer containing 2% bovine serum albumin using a radiochemical technique (Smith & Pogson, 1980) and the results are shown in Table 1. The true concentration of tryptophan was about 0.05 mM after allowing for it being bound to the albumin (Smith & Pogson, 1980). The control rate of 2.87 nmol/mg dw/h was about 70% of that reported in 24 h starved rats by Smith, Carr & Pogson (1980). The incubation of cells with 0.5 mM 5-amino-laevulinate for 30 min resulted in an average increase ($P < 0.10$) of 20% in TDO flux rate. Badawy & Evans (1975) observed an increase in TDO activity in rats following an intraperitoneal injection of 5-aminolaevulinate. One hour after the injection, total enzyme activity was unchanged but holoenzyme activity was increased by 53% indicating an increased saturation of the apoenzyme with haem. In the cat, however, only holoenzyme would be assayed, so our results point to an activation of the enzyme by an increased availability of haem. Cat liver TDO probably also participates in the regulation of the free-haem pool since incubating the liver cells with 2-allyl-2-isopropylacetamide for 30 min caused a decrease of 46% in TDO flux. Allylisopropylacetamide is a porphyrinogenic agent which increases the activity of 5-aminolevulinate synthase, the first and 'rate-limiting' enzyme in haem biosynthesis. The induction is largely due to the enhanced degradation of hepatic haem caused by allylisopropylacetamide (Bonkowsky et al., 1980). Badawy & Morgan (1980) found that a single subcutaneous injection of allylisopropylacetamide to 48 h starved rats

Table 1. *Factors affecting tryptophan dioxygenase flux in cat liver cells*

Treatment	Flux rate (SEM) nmol/mg dw/h
Control	2.87
5-aminolaevulinate (0.5 mM)	3.45 (0.16)
2-Allyl-2-isopropylacetamide (1 mM)	1.83* (0.17)
Ethanol (10 mM)	2.42* (0.10)
Nicotinate (0.5 mM)	2.25 (0.06)
Pyrazinamide (1 mM)	2.54* (0.08)

Note:
Each value is the mean of five animals with the SEM from the analysis of variance in parentheses. Significant treatment differences ($P < 0.05$) from the control are indicated by an asterisk. Treatment additions were made 30 min before the L-[ring-2-^{14}C] tryptophan.

caused a reduction of 59% in holoenzyme activity within 15 min without any effect on total enzyme activity.

TDO in the rat is controlled to some extent by end-product inhibition through the pyridine nucleotides. Cho-Chung & Pitot (1967) showed that nicotinate and a number of its derivatives, particularly NAD(P)H inhibited the activity of purified TDO. As shown in Table 1, nicotinate inhibited TDO flux in cat liver cells by about 22%. Ethanol increases the reductive potential of the cytosol and the lower NAD/NADH ratio inhibited TDO flux by 16%. Although this was statistically significant, it is unlikely to be of biological significance because the high concentrations of the nucleotides needed to cause a significant inhibition (Cho-Chung & Pitot, 1967) are unlikely to occur *in vivo* (Pogson *et al.*, 1984).

Our results with cat liver cells point to a degree of control of tryptophan degradation by the availability of haem to TDO and, because of the rapid rate of turnover of the enzyme, to a role of TDO in the regulation of haem metabolism in the liver. However, relative to the rat, the cat liver enzyme is rather insensitive to factors known to influence its activity. In animals such as the guinea pig in which TDO occurs only as the holoenzyme and is not induced by glucocorticoids, a tryptophan load is toxic. However, as shown by Leklem, Woodford & Brown (1969), large doses of tryptophan are not toxic to cats.

When pyrazinamide, an anti-tuberculosis drug, is injected into the rat, it is metabolised to a potent inhibitor of picolinate carboxylase and results in increased liver NAD levels (Nasu *et al.*, 1981). The inhibition of picolinate carboxylase which occurs when rat liver cells are incubated with pyrazinamide also causes an accumulation of quinolinate with a resultant inhibition of gluconeogenesis (Cook & Pogson, 1983). Cat liver cells were incubated with pyrazinamide to see whether an inhibition of picolinate carboxylase would result in end-product inhibition of TDO by NAD(P)H but only a small (but statistically significant) reduction occurred (Table 1).

Picolinate carboxylase and nicotinic acid synthesis

The high activity of picolinate carboxylase in cat liver, which is believed to be responsible for the inability of the cat to synthesise nicotinate (Ikeda *et al.*, 1965) is intriguing. Although Suhadolnik *et al.* (1957) were the first to show that the activity was much higher in the cat than in the rat, Ikeda *et al.* (1965) showed that the activity in cat liver was much higher than in the livers of a wide variety of mammals and birds. Poston & Combs (1980) found that the activity in the livers of various salmonid fishes was relatively high, particularly in lake trout and atlantic salmon, but was still only about 55% of the activity in cat liver. The ability of animals to synthesise nicotinate from tryptophan is directly related to the ratio of the activities of 3-hydroxyanthranilate oxygenase (EC 1.13.11.6) and picolinate carboxylase (Henderson & Swan, 1971) and the high activity of picolinate carboxylase activity in salmonid livers together with the low activity of 3-hydroxyanthranilate oxygenase means that they are apparently less able to synthesise nicotinate than the cat (Poston & Combs, 1980). There is some evidence, however, that the activity of cat liver picolinate carboxylase *in vivo* may be much lower and more similar to that in rat liver (Suhadolnik *et al.*, 1957; Leklem *et al.*, 1971). The activity of the enzyme in rat liver is increased by alloxan-induced diabetes (Mehler, McDaniel & Hundley, 1958) and by a high dietary protein intake (Johnson & Evans, 1984).

We designed an experiment both to examine the possibility that the high picolinate carboxylase activity in the cat is due in part to its high protein intake, and at the same time, to see whether a low intake of nicotinamide was reflected in the concentration of pyridine nucleotides in the liver. The two isoenergetic (22.8 MJ GE/kg) diets which were fed contained 23% or 46% protein (Table 2) and were based on isolated soy protein supplemented with amino acids, minerals and vitamins (with the exception of nicotinate) according to Hargrove *et al.* (1983). Twelve cats of mixed sex (500–1000 g), immunised against panleucopaenia and respiratory disease and treated for worms, were assigned to the two diets. Within each group, there were two levels of nicotinamide which was administered as a subcutaneous injection in sterile normal saline solution; three cats in each group received 2 mg/day and three received 10 mg/day. The lower level of nicotinamide was selected so as to just prevent a deficiency and the upper level as the optimum intake (Da Silva, Fried & de Angelis, 1952). After at least two weeks, cats were anaesthetised with 'Saffan' (alphaxalone and alphadolone) and liver removed for enzyme and metabolite assays.

Table 2. *Composition of diets (g/kg diet) used to examine the effect of protein intake on liver picolinate carboxylase activity in the cat*

Protein content (%)	23	46
Beef tallow	265.2	290.0
Sucrose	169.5	125.0
Corn starch	248.2	97.0
Isolated soy protein[a]	231.0	462.0
Arginine hydrochloride	0.92	1.84
Cystine	0.23	0.46
Methionine	5.89	11.78
Threonine	2.01	4.02
Glycine	10.70	21.40
Taurine	0.75	0.75
Choline chloride (50%)	6.92	6.92
Mineral mix[b]	57.5	57.5
Vitamin mix[c]	2.18	2.18

Notes:

[a] Supro 620. Ralston Purina

[b] Mineral mix. Hargrove *et al.* (1983)

[c] Vitamin mix (mg/kg diet): retinyl palmitate 80, cholecalciferol 5, DL-α-tocopherol 640, menadione 15, thiamine hydrochloride 25, riboflavin 10, pyridoxine 10, calcium pantothenate 20, myoinositol 200, folic acid 10, cobalamin 50, biotin 1, ascorbic acid 400.

The growth rate of the cats was related to protein intake, being approximately 11 g/day and 21 g/day on the 23% and 46% protein diets respectively. The minimum protein requirement of growing kittens is 19–21% (Rogers & Morris, 1982) and, although the cats in this experiment did not grow maximally on the 23% protein diet, they were conventional cats not specifically bred as laboratory animals. There was no effect of nicotinamide dose on growth within either of the two diets. As shown in Table 3, there was no effect of protein intake or nicotinamide level on the activity of liver picolinate carboxylase. This is in contrast to the situation in the rat where the activity in normal rats was 25% lower on a low protein diet, while, in diabetic rats, the activity was 38–58% lower depending on the level of intake of minerals (Johnson & Evans, 1984). Table 3 also shows that there were no significant treatment effects on the ratio or the concentrations of the pyridine nucleotides; 2 mg nicotinamide/d was obviously adequate. The higher concentration of NADH on the high protein diet was reflected in a lower total NAD/NADH ratio and the high free mitochondrial NAD/NADH ratio on the high protein diet was due to a lower concentration of β-hydroxybutyrate.

Table 3. *Effect of protein and nicotinamide intake on picolinate carboxylase activity and NADH concentration in the liver of the cat*

Protein (%)	23	23	46	46	SEM
Nicotinamide (mg/day)	2	10	2	10	
Picolinate carboxylase (μmol/g wet/h)	43.4	44.3	53.8	41.1	4.96
NAD (μmol/g wet)	0.656	0.625	0.637	0.626	0.028
NADH (μmol/g wet)	0.142	0.135	0.186	0.192	0.016
NAD/NADH free cytoplasmic	432	186	301	355	93
NAD/NADH free mitochondrial	4.9	4.8	8.1	9.3	1.7

Although the cat cannot synthesise nicotinate from tryptophan it possesses the enzymes to do so. Ikeda *et al.* (1965) showed that quinolinate phosphoribosyl transferase (EC 2.4.2.19), which is believed to be rate-limiting for the NAD pathway, is present in cat liver at an activity similar to that in rat liver. Quinolinate phosphoribosyl transferase in rat liver has a high K_m and a low V_{max} (Bender, Magboul & Wynick, 1982) so considerable changes in flux rate would be expected with variation in the supply of quinolinate as could occur from an increased intake of tryptophan. However, Satyanarayana & Narasinga Rao (1977) showed that the activity of the enzyme in rat liver was inversely related to the protein content of the diet. Even if quinolinate phosphoribosyl transferase in the cat was sensitive to protein intake, the high picolinate carboxylase activity would prejudice the likelihood of quinolinate being produced and converted to pyridine nucleotides.

The major source of pyridine nucleotides in the cat is probably from dietary NADH and NADPH much of which is hydrolysed in the intestine, absorbed as nicotinamide and converted to NAD in the liver. The hepatic NAD glycohydrolase (EC 3.2.2.5) can then release the pyridine nucleus as nicotinamide which is transported to other tissues and resynthesised as NAD. The NAD occurs in a functional form bound to enzymes and so acts in redox reactions, but it also occurs in a free or storage form which can be acted on by NAD glycohydrolase producing nicotinamide, which can be excreted in the urine as N^1-methylnicotinamide, and also releasing ADP-ribose. Poly(ADP-ribose) has an important function in DNA replication and the repair of damaged DNA, and therefore in differentiation, mutation and oncogenesis (Henderson, 1983). Although urine collections were not made in our study, we did sample urine from the bladder of the anaesthetised cats before the liver samples were taken. The ratio of the concentrations of N^1-methylnicotinamide to creatinine was not affected by protein intake and only reflected the nicotinamide dose, the mean values (with standard errors) being 2.1 ± 0.3 and 6.2 ± 0.9 μg N^1-methylnicotinamide μmol creatinine^{-1}

on the 2 and 10 mg nicotinamide/d treatments respectively. In the rat, Bender, Magboul & Wynick (1982) concluded that tissue pyridine nucleotide concentrations are controlled by the rate of hydrolysis of NAD to nicotinamide rather than by synthesis from quinolinate and this control is likely to be even more rigid in the cat given that very little quinolinate is likely to be produced.

Tryptophan metabolism and gluconeogenesis

The regulation of gluconeogenesis in the cat is effected partly through the redox potential of the liver mitochondria. Cat liver contains cytosolic and mitochondrial forms of pep-ck (Hanson & Garber, 1972) each of which accounts for about 50% of the phosphoenolpyruvate formation (Gordon & Freedland, 1981). Gluconeogenesis from lactate in the isolated perfused liver from cats fasted for 48 h has been shown to be inhibited by 60% in the presence of octanoate, an effect attributed to the lowered mitochondrial NAD/NADH ratio which accompanies fatty acid oxidation (Arinze & Hanson, 1973). When liver cells from cats fasted for 72 h were incubated in the presence of 2 mM octanoate, gluconeogenesis from pyruvate and lactate was inhibited by 60% and 80% respectively (Gordon & Freedland, 1981). These effects, which are similar to those observed in other animals which possess both forms of pep-ck, such as the guinea pig and man, point to an important role of mitochondrial redox state on gluconeogenesis in the cat liver.

The hypoglycaemia which follows the administration of tryptophan to rats is partly due to an accumulation of quinolinate which inhibits cytosolic pep-ck, but in animals which also possess a mitochondrial pep-ck there is no relationship between the intracellular distribution of pep-ck and the inhibitory effect of quinolinate on gluconeogenesis (Smith, Elliott & Pogson, 1979). We have conducted experiments on the effect of tryptophan and its metabolites on gluconeogenesis in cat liver cells. In line with the effect of tryptophan in the rat and the lack of an effect in the guinea pig, it was not surprising that tryptophan itself had no effect on glucose synthesis from lactate in the cat liver cell. The effects of a number of tryptophan metabolites on gluconeogenesis was examined in liver cells from cats fasted for 72 h in experiments similar to those reported by Smith, Elliott & Pogson (1979) on the rat and guinea pig; their results and those from the cat are shown in Table 4.

The different effects of the various metabolites were not consistent with the intracellular distribution of pep-ck. This was best illustrated by the effects of kynurenine and 3-hydroxyanthranilate in the three different species. Smith, Elliott & Pogson (1979) argued that if the differences in the effects of the tryptophan metabolites were due entirely to the pattern of pep-ck distribution, then all metabolites in the pathway from tryptophan to quinolinate should show weak effects compared to those of quinolinate

Table 4. *The effect of tryptophan metabolites and pyrazinamide on gluconeogenesis from lactate in liver cells*

Treatment	Glucose output (% of control)		
	rat[a]	guinea pig[a]	cat
Kynurenine (0.5 mM)	46*	147*	124
3-Hydroxyanthranilate (0.5 mM)	75	115*	85
Quinolinate (5 mM)	19[b]	61[b]	32*
Nicotinate (0.5 mM)	81	95	76*
Picolinate (0.5 mM)	62*	101	40*
Pyrazinamide (1 mM)	—	—	84
5-Hydroxytryptophan (0.5 mM)	80	81	81*
5-Hydroxytryptamine (0.5 mM)	28*	56*	60
5-Hydroxyindole-3-acetate (0.5 mM)	100	85	86
Tryptamine (0.5 mM)	4*	14*	24*
Indole 3-acetate (0.5 mM)	62*	13*	79*

Notes:

[a] Smith *et al.* (1979). Glucose production rates with 10 mM lactate alone were (mean $\pm$ SE) $231 \pm 32(4)$ and $192 \pm 13(3)$ nmol/mg dw/h for the rat and guinea pig respectively.

[b] Quinolinate results from Elliott, Pogson & Smith (1977) using the same experimental conditions.

Each value for the cat is the mean of five experiments and the control value was 95 ± 5 nmol/mg dw/h. Significant differences ($P < 0.05$) from the control are indicated by an asterisk.

itself. However, this was not so as kynurenine stimulated gluconeogenesis in liver cells from the guinea pig (Smith, Elliott & Pogson, 1979) and cat. Further support for the view that quinolinate is not entirely responsible for the inhibition of gluconeogenesis by tryptophan was obtained from work with the gerbil (Muñoz-Clares *et al.*, 1981). Although pep-ck in the gerbil is almost entirely cytosolic, neither tryptophan nor quinolinate were effective inhibitors of gluconeogenesis from lactate in gerbil liver cells.

As shown in Table 4, picolinate was a strong inhibitor of gluconeogenesis in the cat liver cell but its importance *in vivo* is likely to be less significant in view of the apparently low production (Suhadolnik *et al.*, 1957) and urinary excretion (Leklem *et al.*, 1971), which contrast with both its high production *in vitro* (Suhadolnik *et al.*, 1957) and its urinary excretion as the glycine conjugate when injected intraperitoneally (Mehler & May, 1956). Pyrazinamide, which causes an inhibition of gluconeogenesis in rat liver cells as a result of the accumulation of quinolinate (Cook & Pogson, 1983), had a similar effect in cat liver cells. The means by which picolinate carboxylase activity in cat liver is regulated, if at all, would seem a fruitful area of interest. The intermediates of the minor pathways of tryptophan metabolism had similar effects to those observed in the guinea pig by Smith, Elliott & Pogson (1979) although the effect of indole-3-acetate was more similar to the effect in the rat.

The nature of glucose metabolism in the cat departs in some important aspects from that in other animals. The glucose entry (irreversible loss) rate and the body glucose mass of fasted cats fed a high protein diet (63%) were shown by Kettelhut, Foss & Migliorini (1980) to be 3.1 mg/kg/min and 244 mg/kg respectively. The rate of irreversible loss was only 47% of that in fasted rats (6.5 mg/kg/min) fed the same diet. It was however, about 50% greater than in non-pregnant non-lactating sheep (2.1 mg/kg/min) (Leng, 1970), an animal, which like the cat, is largely dependent on gluconeogenesis for its glucose requirements. The energy requirements of the ruminant are largely satisfied by the metabolism of ruminally produced acetate, and glucose plays a smaller role than for most non-ruminants. The traditional high fat diet of the cat may mean that much of its energy requirements are obtained from the oxidation of fatty acids. However, ketone body production from oleate in cat liver cells is lower than in the rat (Gordon & Freedland, 1981) but the rates of glucose production in the cat are also generally low (Kettelhut, Foss & Migliorini, 1980) and the rates of gluconeogenesis from a variety of substrates by cat liver cells are similar to those in liver cells from both the rat (Gordon & Freedland, 1981) and sheep (Clark, Filsell & Jarrett, 1976).

The relatively low rate of irreversible loss of glucose demonstrated by Kettelhut, Foss & Migliorini (1980) has also been shown indirectly in glucose tolerance studies (Middleton, 1984). The decline in serum glucose concentration following an intravenous glucose load in normal cats occurred at 1.5% min^{-1} which was somewhat lower than in the dog (Mattheeuws *et al.*, 1984) and man (Vermeulen, Daneels & Thiery, 1970) where the fractional disappearance rate was about 2.6. Values less than 1.2 have been used as an index of abnormality and diagnostic of diabetes mellitus in man (Reaven & Olefsky, 1974), and although the insulin response in some of her experimental cats was slow, Middleton (1984) concluded it unlikely that normal cats were in a state of latent diabetes, particularly as overt diabetes mellitus in the cat is rare. In view of the dependence of the cat on the *in vivo* synthesis of glucose, albeit at an apparently low rate, quantitative studies on glucose kinetics and the contribution of amino acids to glucose and energy metabolism may help in understanding the high protein requirement of the cat.

Tryptophan and appetite regulation

Although direct evidence is lacking, there are good reasons for believing that tryptophan may play an important role in appetite regulation in the cat. High protein diets or diets containing deficiencies or excesses of essential amino acids reduce feed intake of rats (Harper, Benevenga & Wohlhueter, 1970) but amino acids also participate in the regulation of food intake under normal dietary conditions and this aspect has been well reviewed (Li & Anderson, 1983). Anderson and his co-workers have shown that, when given a choice, rats regulate protein intake at a constant proportion of the energy intake and also compensate for dietary dilution to maintain an adequate intake of protein. This area of nutrition in the cat has been neglected.

In rats offered a choice, protein intake has been shown to be inversely related to the ratio of the plasma concentration of tryptophan to the sum of the concentrations of the other large neutral amino acids (^{trp}/naa), viz, leucine, isoleucine, valine, phenylalanine and tyrosine (Ashley & Anderson, 1975). This work complemented the earlier results of Fernstrom & Wurtman (1972) who showed that the large neutral amino acids competed for the same transport system for entry into the brain. Elevated levels of brain serotonin depress food intake in rats (Barrett & McSharry, 1975) while the serotonin antagonist cyproheptadine induces hyperphagia (Noble, 1969), and it is generally believed that serotonin participates in the hypothalamic satiety response and in the regulation of feeding behaviour (Bender, 1982). There is evidence, for example, that the serotonergic system is important in regulating the proportions of protein and carbohydrate (or the proportion of energy as protein) consumed by rats when in a free choice situation. A high carbohydrate diet is accompanied by an increased uptake of branched chain amino acids by peripheral tissues and an increased plasma tryptophan concentration following the insulin response and this results in an increased transport of tryptophan into the brain. The consumption of a high protein diet results in a higher concentration of all branched chain amino acids leading to a situation less favourable to the transport of tryptophan (Li & Anderson, 1983). Although it seems likely therefore that serotonin is involved in the regulation of food intake, much of the supporting evidence is based on the results of pharmacological experiments or the use of extreme diets. In this regard, Ashley, Leathwood & Moennoz (1984) showed that although the plasma ^{trp}/naa ratio in rats was increased after a short fast, there was no effect on brain tryptophan or serotonin concentrations, and although the consumption of diets of increasing protein content was accompanied by a lowering of the ^{trp}/naa ratio, it was not reflected by brain serotonin levels. Nevertheless, the possible role of tryptophan in food intake regulation under normal feeding conditions is worthy of attention, particularly in the cat which consumes a high protein–low carbohydrate diet in the wild, but as a domestic animal often consumes a high protein–high carbohydrate diet.

Summary

In addition to its important role as an essential amino acid, in the rat, tryptophan is involved in the initiation of protein synthesis and is the precursor of serotonin, melatonin and nicotinate, and a number of its metabolites participate in the regulation of gluconeogenesis and protein degradation.

In the cat, some of these effects of tryptophan are attenuated, partly because of the lower activity of tryptophan 2,3-dioxygenase and its resistance to factors which are potent modifiers of the enzyme in rat liver. The means by which tryptophan catabolism is regulated in the liver of the cat is an open question but it is unlikely to be controlled solely by the concentration of tryptophan in the liver. Many enzyme systems in cat liver have been shown to be unresponsive to the level of dietary protein (Rogers, Morris & Freedland, 1977), and this area of intermediary metabolism and the ways in which it is influenced by nutritional status should be a fruitful area of research.

The regulation of tryptophan metabolism in the cat is very likely important in the low incidence of cancer of the bladder, in mood and behaviour and in the regulation of food intake. But the main conundrum in cat liver tryptophan metabolism is the high activity of picolinate carboxylase which may be responsible for the very low urinary excretion of the metabolites of tryptophan which promote cancer of the bladder in man. However, picolinate carboxylase cannot be entirely responsible because only small quantities of tryptophan metabolites appear in the urine and, in particular, very small quantities of intermediates preceding 3-hydroxyanthranilate are excreted. Nevertheless, picolinate carboxylase is generally believed to protect the cat against the toxic effects of the high intake of tryptophan which accompanies the carnivorous diet. On the other hand, it is apparently because of the high activity of picolinate carboxylase that the cat cannot synthesise nicotinate from tryptophan even though the enzymes necessary for its synthesis are present at levels similar to those in the rat. Furthermore, although picolinate is a strong inhibitor of gluconeogenesis in the liver of the cat, it is apparently produced in very small quantities since only small amounts are excreted in the urine.

REFERENCES

Anderson, P.A., Baker, D.H., Sherry, P.A. & Corbin, J.E. (1980). Histidine, phenylalanine, tyrosine and tryptophan requirements for growth of the young kitten. *Journal of Animal Science*, **50**, 479–83.

Arinze, I.J. & Hanson, R.W. (1973). Mitochondrial redox state and the regulation of gluconeogenesis in the isolated, perfused cat liver. *FEBS Letters*, **31**, 280-82.

Ashley, D.V.M. & Anderson, G.H. (1975). Correlation between the plasma tryptophan to neutral amino acid ratio and protein intake in the self-selecting weanling rat. *Journal of Nutrition*, **105**, 1412–21.

Ashley, D.V.M., Leathwood, P.D. & Moennoz, D. (1984). Carbohydrate meal increases brain 5-hydroxytryptamine synthesis in the adult rat only after prolonged fasting. In *Progress in Tryptophan and Serotonin Research*, ed. H.G. Schlossberger, W. Kochen, B. Linzen & H.

Steinhart, pp. 591–4. Berlin: Walter de Gruyter.

Badawy, A.A.-B. (1977). The functions and regulation of tryptophan pyrrolase. *Life Sciences*, 21, 755–68.

Badawy, A.A.-B. (1984). The functions and regulation of tryptophan pyrrolase. In *Progress in Tryptophan and Serotonin Research*, ed. H.G. Schlossberger, B. Linzen & H. Steinhart, pp. 641–50. Berlin: Walter de Gruyter.

Badawy, A.A.-B. & Evans, M. (1975). The regulation of rat liver tryptophan pyrrolase by its cofactor haem. Experiments with haematin and 5-aminolaevulinate and comparison with the substrate and hormonal mechanisms. *Biochemical Journal*, 150, 511–20.

Badawy, A.A.-B. & Evans, M. (1976). Animal liver tryptophan pyrrolases. Absence of apoenzyme and of hormonal induction mechanism from species sensitive to tryptophan toxicity. *Biochemical Journal*, 158, 79-88.

Badawy, A.A.-B. & Morgan, C.J. (1980). Tryptophan pyrrolase in haem regulation. The relationship between the depletion of rat liver tryptophan pyrrolase haem and the enhancement of 5-aminolaevulinate synthase activity by 2-allyl-2-isopropylacetamide. *Biochemical Journal*, 186, 763–72.

Ballard, F.J. & Hopgood, M.F. (1973). Phosphopyruvate carboxylase inhibition by L-tryptophan. Effects on synthesis and degradation of the enzyme. *Biochemical Journal*, 136, 259–64.

Barrett, A.M. & McSharry, L. (1975). Inhibition of drug-induced anorexia in rats by methysergide. *Journal of Pharmacy and Pharmacology*, 27, 889–95.

Beliveau, G.P. & Freedland, R.A. (1982). Metabolism of serine, glycine and threonine in isolated cat hepatocytes *Felis domestica*. *Comparative Biochemistry and Physiology*, 71B, 13–18.

Bender, D.A. (1982). Biochemistry of tryptophan in health and disease. *Molecular Aspects of Medicine*, 6, 101–97.

Bender, D.A., Magboul, B.I. & Wynick, D. (1982). Probable mechanisms of regulation of the utilization of dietary tryptophan, nicotinamide and nicotinic acid as precursors of nicotinamide nucleotides in the rat. *British Journal of Nutrition*, 48, 119–27.

Bonkowsky, H.L., Healey, J.F., Sinclair, P.R., Mayer, Y.P. & Erny, R. (1980). Metabolism of hepatic haem and green pigments in rats given 2-allyl-2-isopropylacetamide and ferric citrate. A new model for hepatic haem turnover. *Biochemical Journal*, 188, 289–95.

Chee, P.Y. & Swick, R.W. (1976). Effect of dietary protein and tryptophan on the turnover of rat liver ornithine aminotransferase. *Journal of Biological Chemistry*, 251, 1029–34.

Chen, N.C., Gholson, R.K. & Raica, N. (1974). Isolation and identification of indole-3-carboxaldehyde: a major new urinary metabolite of D-tryptophan. *Biochimica et Biophysica Acta*, 343, 167–72.

Cho-Chung, Y.S. & Pitot, H.C. (1967). Feedback control of rat liver tryptophan pyrrolase. I. End product inhibition of tryptophan pyrrolase activity. *Journal of Biological Chemistry,* **242,** 1192–8.

Čihák, A. (1979). L-tryptophan action on hepatic RNA synthesis and enzyme induction. *Molecular and Cellular Biochemistry,* **24,** 131–42.

Čihák, A., Lamar, C. & Pitot, H.C. (1973). L-tryptophan inhibition of tyrosine aminotransferase degradation in rat liver *in vivo. Archives Biochemistry Biophysics,* **156,** 188–94.

Clark, M.G., Filsell, O.H. & Jarrett, I.G. (1976). Gluconeogenesis in isolated intact lamb liver cells. Effects of glucagon and butyrate. *Biochemical Journal,* **156,** 671–80.

Cook, J.S. & Pogson, C.I. (1983). Tryptophan and glucose metabolism in rat liver cells. The effects of DL-6-chlorotryptophan, 4-chloro-3-hydroxyanthranilate and pyrazinamide. *Biochemical Journal,* **214,** 511–16.

Curzon, G., Friedel, J. & Knott, P.J. (1973). The effect of fatty acids on the binding of tryptophan to plasma protein. *Nature,* **242,** 198–200.

Da Silva, A.C., Fried, R. & De Angelis, R.C. (1952). The domestic cat as a laboratory animal for experimental nutrition studies. III. Niacin requirements and tryptophan metabolism. *Journal of Nutrition,* **46,** 399–409.

De Lap, L. & Feigelson, P. (1978). Effect of cycloheximide on the induction of tryptophan oxygenase mRNA by hydrocortisone *in vivo. Biochemical Biophysical Research Communications,* **82,** 142–9.

Elliott, K.R.F., Pogson, C.I. & Smith, S.A. (1977). Permeability of the liver cell membrane to quinolinate. *Biochemical Journal,* **164,** 283–6.

Etienne, P., Young, S.N. & Sourkes, T.L. (1976). Inhibition by albumin of tryptophan uptake by rat brain. *Nature,* **262,** 144–5.

Evans, G.W. (1980). Normal and abnormal zinc absorption in man and animals: The tryptophan connection. *Nutrition Reviews,* **38,** 137–41.

Evans, G.W. & Johnson, P.E. (1979). Purification and characterization of a zinc binding ligand in human milk. *Federation Proceedings,* **38,** 703 (abs. 2501).

Fernstrom, J.D. & Wurtman, R.J. (1972). Brain serotonin content: physiological regulation by plasma neutral amino acids. *Science,* **178,** 414–16.

Fleck, A., Shepherd, J. & Munro, H.N. (1965). Protein synthesis in rat liver: Influence of amino acids in diet on microsomes and polysomes. *Science,* **150,** 628–9.

Gordon, S. & Freedland, R.A. (1981). Metabolic capacities of isolated hepatocytes from the cat *Felis domestica. Comparative Biochemistry and Physiology,* **69B,** 257–63.

Greengard, O. & Feigelson, P. (1961). The activation and induction of rat liver tryptophan pyrrolase *in vivo* by its substrate. *Journal of Biological Chemistry,* **236,** 158–61.

Grinde, B. (1984). Effect of amino acid metabolites on lysosomal protein degradation. A regulatory role for kynurenine? *European Journal of Biochemistry,* **145,** 623–7.

Hanson, R.W. & Garber, A.J. (1972). Phosphoenolpyruvate carboxykinase. I. Its role in gluconeogenesis. *American Journal of Clinical Nutrition*, 25, 1010–21.

Hargrove, D.M., Rogers, Q.R. & Morris, J.G. (1983). The tryptophan requirement of the kitten. *British Journal of Nutrition*, 50, 487–93.

Harper, A.E., Benevenga, N.J. & Wohlhueter, R.M. (1970). Effect of ingestion of disproportionate amounts of amino acids. *Physiological Reviews*, 50, 428–58.

Henderson, A.R. (1970). The effect of feeding with a tryptophan-free amino acid mixture on rat liver magnesium ion-activated deoxyribonucleic acid-dependent ribonucleic acid polymerase. *Biochemical Journal*, 120, 205–14.

Henderson, L.M. (1983). Niacin. In *Annual Review of Nutrition*, vol. 3, ed. W.J. Darby, H.P. Broquist & R.E. Olson, pp. 289–307. Palo Alto, California: Annual Reviews Inc.

Henderson, L.M. & Swan, P.B. (1971). Picolinic acid carboxylase. In *Methods in Enzymology*, vol. 18, ed. D.B. McCormick & L.D. Wright, pp. 175–80. New York: Academic Press.

Hopgood, M.F., Clark, M.G. & Ballard, F.J. (1977). Inhibition of protein degradation in isolated rat hepatocytes. *Biochemical Journal*, 164, 399–407.

Hopkins, F.G. & Cole, S.W. (1901). A contribution to the chemistry of proteins. Part I. A preliminary study of a hitherto undescribed product of tryptic digestion. *Journal of Physiology (London)*, 27, 418–28.

Ikeda, S. & Kotake, Y. (1984). Urinary excretion of xanthurenic acid and zinc in diabetes. In *Progress in Tryptophan and Serotonin Research*, ed. H.G. Schlossberger, W. Kochen, B. Linzen & H. Steinhart, pp. 355–8. Berlin: Walter de Gruyter.

Ikeda, M., Tsuji, H., Nakamura, S., Ichiyama, A., Nishizuka, Y. & Hayaishi, O. (1965). Studies on the biosynthesis of nicotinamide adenine dinucleotide. II. A role of picolinic carboxylase in the biosynthesis of nicotinamide adenine dinucleotide from tryptophan in mammals. *Journal of Biological Chemistry*, 240, 1395–401.

Johnson, W.T. & Evans, G.W. (1984). Effects of the interrelationship between dietary protein and minerals on tissue content of trace metals in streptozotocin-diabetic rats. *Journal of Nutrition*, 114, 180–90.

Jost, J-P., Khairallah, E.A. & Pitot, H.C. (1968). Studies on the induction and repression of enzymes in rat liver. *Journal of Biological Chemistry*, 243, 3057–66.

Kettelhut, I.C., Foss, M.C. & Migliorini, R.H. (1980). Glucose homeostasis in a carnivorous animal (cat) and in rats fed a high-protein diet. *American Journal of Physiology*, 239, R437–44.

Krehl, W.A., Sarma, P.S., Teply, L.J. & Elvehjem, C.A. (1946). Factors affecting the dietary niacin and tryptophan requirement of the growing rat. *Journal of Nutrition*, 31, 85–106.

Leklem, J.E., Brown, R.R., Hankes, L.V. & Schmaeler, M. (1971).

Tryptophan metabolism in the cat: A study with carbon-14-labelled compounds. *American Journal of Veterinary Research,* **32**, 335–44.

Leklem, J.E., Woodford, J. & Brown, R.R. (1969). Comparative tryptophan metabolism in cats and rats: Differences in adaptation of tryptophan oxygenase and *in vivo* metabolism of tryptophan, kynurenine and hydroxykynurenine. *Comparative Biochemistry and Physiology,* **31**, 95–109.

Leng, R.A. (1970). Glucose synthesis in ruminants. In *Advances in Veterinary Science,* **14**, 209–60.

Lepkovsky, S., Roboz, E. & Haagen-Smit, A.J. (1943). Xanthurenic acid and its rôle in the tryptophane metabolism of pyridoxine-deficient rats. *Journal of Biological Chemistry,* **149**, 195–201.

Li, E.T.S. & Anderson, G.H. (1983). Amino acids in the regulation of food intake. *Nutrition Abstracts and Reviews,* **53A**, 169–81.

Mattheeuws, D., Rottiers, R., Kaneko, J.J. & Vermeulen, A. (1984). Diabetes mellitus in dogs: Relationship of obesity to glucose tolerance and insulin response. *American Journal of Veterinary Research,* **45**, 98–103.

McMenamy, R.H. (1965). Binding of indole analogues to human serum albumin. Effects of fatty acids. *Journal of Biological Chemistry,* **240**, 4235–43.

McMenamy, R.H., Lund, C.C., Van Marike, J. & Oncley, J.L. (1961). The binding of L-tryptophan in human plasma at 37°C. *Archives Biochemistry Biophysics,* **93**, 135–9.

Mehler, A.H. & May, E.L. (1956). Studies with carboxyl-labelled 3-hydroxyanthranilic and picolinic acids *in vivo* and *in vitro. Journal of Biological Chemistry,* **223**, 449–55.

Mehler, A.H., McDaniel, E.G. & Hundley, J.M. (1958). Changes in the enzymatic composition of liver. I. Increase of picolinic carboxylase in diabetes. *Journal of Biological Chemistry,* **232**, 323–30.

Middleton, D.J. (1984). Aspects of feline hepatic and pancreatic disorders. PhD thesis, University of Sydney, pp. 391.

Muñoz-Clares, R.A., Lloyd, P., Lomax, M.A., Smith, S.A. & Pogson, C.I. (1981). Tryptophan metabolism and its interaction with gluconeogenesis in mammals: Studies with the guinea pig, mongolian gerbil, and sheep. *Archives of Biochemistry and Biophysics,* **209**, 713–17.

Murty, C.N. & Sidransky, H. (1972). The effect of tryptophan on messenger RNA of the livers of fasted mice. *Biochimica et Biophysica Acta,* **262**, 328–35.

Nasu, S., Yamaguchi, K., Sakakibara, S., Imai, H. & Ueda, I. (1981). The effect of pyrazines on the metabolism of tryptophan and nicotinamide adenine dinucleotide in the rat. Evidence of the formation of a potent inhibitor of aminocarboxy-muconate-semialdehyde decarboxylase from pyrazinamide. *Biochimica et Biphysica Acta,* **677**, 109–19.

Nishizuka, Y. & Hayaishi, O. (1963). Enzymic synthesis of niacin nucleotides from 3-hydroxyanthranilic acid in mammalian liver.

Journal of Biological Chemistry, **238**, 483–85.

Noble, R.E. (1969). Effect of cyproheptadine on appetite and weight gain in adults. *Journal of the American Medical Association*, **209**, 2054–5.

Ogasawara, N., Hagino, Y. & Kotake, Y. (1962). Kynurenine transaminase and the increase of xanthurenic acid excretion. *Journal of Biochemistry (Tokyo)*, **52**, 162–6.

Pogson, C.I., Muñoz-Clares, R.A., Cook, J.S. & Smith, S.A. (1984). Tryptophan metabolism and its control in mammalian liver. In *Progress in Tryptophan and Serotonin Research*, ed. H.G. Schlossberger, W. Kochen, B. Linzen & H. Steinhart, pp. 625–32. Berlin: Walter de Gruyter.

Poston, H.A. & Combs, G.F. (1980). Nutritional implications of tryptophan catabolizing enzymes in several species of trout and salmon. *Proceedings of the Society for Experimental Biology and Medicine*, **163**, 452–4.

Ray, P.D., Foster, D.O. & Lardy, H.A. (1966). Paths of carbon in gluconeogenesis and lipogenesis. IV. Inhibition of L-tryptophan of hepatic gluconeogenesis at the level of phosphoenolpyruvate formation. *Journal of Biological Chemistry*, **241**, 3904–8.

Reaven, G.M. & Olefsky, J.M. (1974). Relationship between insulin response during the intravenous glucose tolerance test, rate of fractional glucose removal and the degree of insulin resistance in normal adults. *Diabetes*, **23**, 454–9.

Reiter, R.J. & Richardson, B.A. (1980). The physiology of melatonin. In *Biochemical and Medical Aspects of Tryptophan Metabolism*, ed. O. Hayaishi, Y. Ishimura & R. Kido, pp. 247–56. Amsterdam: Elsevier.

Rogers, Q.R. & Morris, J.G. (1982). Do cats really need more protein? *Journal of Small Animal Practice*, **23**, 521–32.

Rogers, Q.R., Morris, J.G. & Freedland, R.A. (1977). Lack of hepatic enzyme adaptation to low and high levels of dietary protein in the adult cat. *Enzyme*, **22**, 348–56.

Rowsell, E.V., Carnie, J.A., Wahbi, S.D., Al-Tai, A.H. & Rowsell, K.V. (1979). L-serine dehydratase and L-serine-pyruvate aminotransferase activities in different animal species. *Comparative Biochemistry and Physiology*, **63B**, 543–55.

Satyanarayana, U. & Narasinga Rao, B.S. (1977). Effect of dietary protein level on some key enzymes of the tryptophan-NAD pathway. *British Journal of Nutrition*, **38**, 39–45.

Schimke, R.T., Sweeney, E.W. & Berlin, C.M. (1965). Studies of the stability *in vivo* and *in vitro* of rat liver tryptophan pyrrolase. *Journal of Biological Chemistry*, **240**, 4609–20.

Sidransky, H., Sarma, D.S.R., Bongiorno, M. & Verney, E. (1968). Effect of dietary tryptophan on hepatic polyribosomes and protein synthesis in fasted mice. *Journal of Biological Chemistry*, **243**, 1123–32.

Silva, S.V.P.S. & Mercer, J.R. (1985). Effect of protein intake on amino acid catabolism and gluconeogenesis by isolated hepatocytes from the cat (*Felis domestica*). *Comparative Biochemistry and Physiology*, **80B**, 603–7.

Silva, S.V.P.S. & Mercer, J.R. (1986). Protein degradation in cat liver cells. *Biochemical Journal*, **240**, 843–6.

Smith, S.A., Carr, F.P.A. & Pogson, C.I. (1980). The metabolism of L-tryptophan by isolated rat liver cells. Quantification of the relative importance of, and the effect of nutritional status on, the individual pathways of tryptophan metabolism. *Biochemical Journal*, **192**, 673–86.

Smith, S.A., Elliott, K.R.F. & Pogson, C.I. (1978). Differential effects of tryptophan on glucose synthesis in rats and guinea pigs. *Biochemical Journal*, **176**, 817–25.

Smith, S.A., Elliott, K.R.F. & Pogson, C.I. (1979). Inhibition of hepatic gluconeogenesis by tryptophan metabolites in rats and guinea pigs. *Biochemical Pharmacology*, **28**, 2145–8.

Smith, S.A. & Pogson, C.I. (1980). The metabolism of L-tryptophan by isolated rat liver cells. Effect of albumin binding and amino acid competition on oxidation of tryptophan by tryptophan 2,3-dioxygenase. *Biochemical Journal*, **186**, 977–86.

Suhadolnik, R.J., Stevens, C.O., Decker, R.H., Henderson, L.M. & Hankes, L.V. (1957). Species variation in the metabolism of 3-hydroxyanthranilate to pyridinecarboxylic acids. *Journal of Biological Chemistry*, **228**, 973–82.

Vermeulen, A., Daneels, R. & Thiery, M. (1970). Effects of oral contraceptives on carbohydrate metabolism. *Diabetologia*, **6**, 519–23.

Wurtman, R.J. (1969). Time-dependent variations in amino acid metabolism: mechanism of the tyrosine transaminase rhythm in rat liver. *Advances in Enzyme Regulation*, **7**, 57–67.

Wurtman, R.J. (1980). Nutritional control of brain tryptophan and serotonin. In *Biochemical and Medical Aspects of Tryptophan Metabolism*, ed. O. Hayaishi, Y. Ishimura & R. Kido, pp. 31–46. Amsterdam: Elsevier.

Young, S.N., St Arnaud-McKenzie, D. & Sourkes, T.L. (1978). Importance of tryptophan pyrrolase and aromatic amino acid decarboxylase in the catabolism of tryptophan. *Biochemical Pharmacology*, **27**, 763–7.

12

Is carbohydrate essential for pregnancy and lactation in dogs?

SANDRA E. BLAZA, DEREK BOOLES AND IVAN H. BURGER

Introduction

Recent work (Romsos *et al.*, 1981) has suggested that pregnant bitches require a dietary source of carbohydrate in order to whelp and rear healthy puppies. Bitches fed a carbohydrate-free diet became hypoglycaemic, hypoalanaemic and ketotic towards the end of gestation. Only 63% of the puppies from these bitches were alive at birth and there was a high mortality rate immediately after birth, although the losses were confined to a sub-population of the test group. However, there was no similar failure of lactation.

These findings are difficult to reconcile with the evolutionary development of the dog as a hunter, since the body of prey would have supplied only a little available carbohydrate. In addition, there is now good evidence to show that dogs, unlike many other mammals of omnivorous habit, are able to maintain normal plasma glucose levels and develop only very moderate ketosis in the face of both total starvation (de Bruijne *et al.*, 1981) and extremely high levels of energy expenditure when fed a carbohydrate-free diet (Kronfeld *et al.*, 1977).

The aim of this study was to examine the effect of feeding either a diet which supplied no available carbohydrate, or one in which carbohydrate supplied 11% of the estimated metabolisable energy (ME), to bitches through pregnancy and lactation. Two breeds of dogs (Beagle and Labrador Retriever) were used to assess the possibility of breed differences.

Materials and methods

Diets

Two diets were fed, a control diet containing starch (CC) and a test diet (LC) which was a commercial canned dog food based on meat, poultry, offals, bone grits and some soya protein, but having no measurable available carbohydrate. Specific analyses confirmed this. Seven proximate analyses were performed which gave carbohydrate by difference as 1.5% and this was

229

Table 1. *Nutrient content of LC diet*

	%
Moisture	77.9
Protein	12.2
Fat	5.2
Ash	3.1
Predicted ME	96 kcal (402 kJ)/100 g

Essential fatty acids per 100 g (3 analyses): linoleic acid, 0.28 g; linolenic acid, 0.04 g; arachidonic acid, 0.04 g.

Minerals per 100 g (3 analyses): calcium, 0.58 g; phosphorus, 0.44 g; sodium, 0.30 g; potassium, 0.24 g; magnesium, 27 mg; iron, 4.2 mg; copper, 0.29 mg; manganese, 0.8 mg; zinc, 1.4 mg.

Vitamins per 100 g (3 analyses): vitamin A, 2580 IU; vitamin E, 6.3 IU; thiamin, 0.20 mg; riboflavin, 0.26 mg; pantothenic acid, 0.66 mg; niacin, 2.19 mg; pyridoxine, 0.09 mg; folic acid, 9 μg; biotin, 11 μg; vitamin B12, 5 μg; choline, 70 mg.

Amino acids per 100 g (3 analyses): arginine, 0.71 g; methionine, 0.18 g; cystine, 0.10 g; lysine, 0.65 g; phenylalanine, 0.47 g; tyrosine, 0.31 g; histidine, 0.25 g; threonine, 0.42 g; leucine, 0.85 g; isoleucine, 0.33 g; valine, 0.50 g; serine, 0.53 g; glutamine, 2.11 g; proline, 0.78 g; glycine, 1.18 g; alanine, 0.76 g; aspartic acid, 0.82 g. Tryptophan not analysed.

CC Diet

As above with addition of 3% (by weight) food grade starch (fortified with minerals and vitamins to maintain above profile).

largely accounted for by the presence of gelling agents which supply little available carbohydrate (Leibetseder, 1984). Full nutrient analysis (Table 1) shows that this food contained more than the recommended nutrient allowances for dogs and exceeded published amino acid allowances for adult maintenance and puppy growth (NRC, 1985).

The control diet (CC) was based on the LC diet with the addition of 3% (by weight) food grade maize starch (Laing-National Ltd, Manchester, UK). A vitamin and mineral supplement was mixed with the starch to ensure that the final levels of these nutrients were similar in both regimes. In this diet, 11% of the total available energy came from carbohydrate (calculated using modified Atwater factors of 4, 9 and 4 kcal ME/g for protein, fat and carbohydrate respectively).

Husbandry

Animals

Seven Labrador Retriever and five Beagle bitches were fed the test and control diets through two parities in a cross-over design. Age at the start of the study ranged from 2 to 7 years (Labradors) and 2 to 5 years (Beagles). Each breed group included two maiden bitches, the remainder having successfully reared at least one litter.

Bitches were housed in pairs in pens measuring 1.7 × 1.2 m (Beagles) and 2.6 × 1.2 m (Labradors) until three weeks before the expected whelping date. They were then moved to an adjacent unit in the same building where they were housed individually in pens of 1.5 × 1.6 m (Beagles) and 3.7 × 1.6 m (Labradors). Both units were maintained at 20°C and bitches had access to outside runs at all times, through flap doors.

Bitches were vaccinated against canine parvovirus (Kavak, Duphar Veterinary Ltd, Southampton, UK) at mating and against hardpad, distemper, hepatitis and leptospirosis (Canilep DD, Glaxovet Ltd, Uxbridge, UK) at one week and seven weeks after mating. Puppies received intranasal vaccination against kennel cough (Intrac I, Mycofarm Ltd, Weybridge, UK) at two weeks of age and were treated with Coopane (Wellcome Foundation, Crewe, UK) at three weeks.

Feeding and measurements
Six bitches (four Labrador, two Beagle) were assigned to the test diet (LC) for the first parity, six (three Labrador, three Beagle) to the control diet (CC). The foods were fed solus from the first day of mating (11–15 days after oestrus observed) until weaning. Food allocations for pregnancy and lactation were based on a theoretical scale (NRC, 1985). For the first four weeks of pregnancy the adult maintenance allowance was given and this was then increased by 10% (compounded) per week. Food allocation was never below the theoretical amount but additional food was fed if needed during lactation, as assessed by the individual animal's bodyweight and condition.

All bitches were weighed weekly throughout the trial, puppies twice weekly. A record of daily food intake was kept for each bitch, any refusals being weighed back. Blood was withdrawn at the first and eighth week of pregnancy for estimation of haematocrit and of haemoglobin by the cyanomethaemoglobin method (Drabkin & Austin, 1932). Plasma glucose levels were measured (Schmidt, 1961) after a 16-hour fast at mating and at three days pre- and post-whelp. Milk was expressed manually (without hormonal stimulation) at week four of lactation for the estimation of lactose levels (Munson & Walker, 1970). Data were subjected to Student's paired t-test (Snedecor & Cochran, 1968).

Results

Reproductive performance
Table 2 summarises the findings on the reproductive performance of the 12 bitches studied. No effect of diet was found on length of gestation (taken from first day of mating) or in the various parameters of litter size,

Table 2. *Reproductive performance*

	Beagles (n = 5)		Labradors (n = 7)	
	LC	CC	LC	CC
Gestation length (days)	65.6 ± 1.1[1]	66.6 ± 1.0	64.4 ± 1.3	63.9 ± 0.9
Total pups born	29	25	60	52
Total pups born alive	25	24	59	52
Mean live litter size	5.0 ± 0.7	4.8 ± 1.1.	8.4 ± 0.7	7.4 ± 0.6
Mean live litter weight at birth (kg)	1.65 ± 0.23	1.64 ± 0.32	3.39 ± 0.19	3.03 ± 0.16
Mean pup birth weight (g)	330 ± 7	340 ± 9	402 ± 6	408 ± 8
Pups alive at three days (losses)	24(1)	23(1)	58(1)	50(2)

Note:
[1] Mean ± SEM.

weight and viability. There was an apparent tendency for bitches fed the LC diet to produce more puppies (Beagles total 29, Labradors 60) than those fed the CC diet (Beagles 25, Labradors 52) and this was reflected in the slightly greater live litter size (Beagles 5.0 and 4.8 pups; Labradors 8.4 and 7.4 pups). However, none of these differences achieved statistical significance. The mean live weights of the litters at birth were almost identical on the two regimes for the Beagles, but Labradors fed the LC diet produced slightly (but not significantly) heavier litters than those fed the CC diet, corresponding to the greater numbers of puppies born alive. Individual birth weights fell within very narrow ranges, for both breeds, and did not differ according to diet.

Survival to three days was virtually identical in all groups, irrespective of breed or diet. Two puppies were lost in the Labrador CC group, and one in each of the three other groups. These represent survival rates in excess of 96%.

Food intake and bodyweight

Figure 1 shows the mean weekly bodyweights of Beagle and Labrador bitches through gestation and early lactation.

In both breeds mean bodyweight during pregnancy was similar on both dietary regimes and, although the CC diet gave slightly higher mean bodyweights through lactation in Beagles, the intra-individual variability was high and the differences not therefore statistically significant.

Energy intakes through gestation and lactation are shown in Figure 2. Dietary regime did not influence intake during pregnancy in either breed, although in the first week of lactation the intake of Labradors fed the CC diet was significantly less than that of animals fed the LC diet ($p < 0.05$).

Figure 3 shows mean puppy bodyweights from birth to three weeks of age. Labrador puppies gained 970 g (LC) and 980 g (CC); Beagles 900 g (LC) and 890 g (CC). These weight gains are within the range normally seen in this colony in puppies from bitches on a wide variety of dietary regimes.

Milk lactose content

Table 3 shows the lactose content of bitch milk assayed at four weeks after whelping. These figures are complete for Beagles but based on only five Labradors fed each diet as two samples were lost. There was considerable variability in individual levels of milk and no consistent response to dietary regimes.

Bitch plasma glucose

A complete set of plasma glucose values was not obtained. Results are presented in Table 3 where paired values are available for each bitch fed both diets, allowing analysis by dietary regimes for each sampling time. However, the three sampling periods are not strictly comparable with each other as they may refer to different groups of bitches.

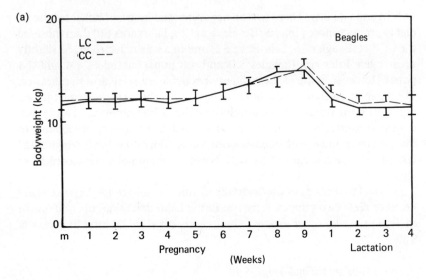

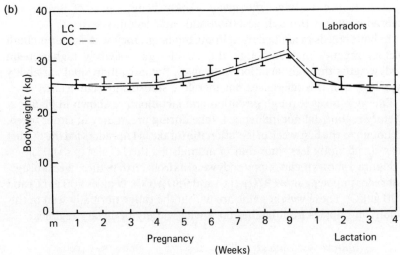

Fig. 1. Mean weekly bodyweight ± SEM of Beagle (1a) and Labrador (1b) bitches through pregnancy and lactation fed the test (LC) or control (CC) diets. m (mating)

Initial samples were taken within the first week after mating with the bitches already being fed the test foods. Pre-whelp samples are those which were taken at an estimated three days before whelping.

At both sampling points in gestation, the plasma glucose levels tended to be higher in bitches fed the diet containing starch. Nevertheless none of these differences achieved statistical significance. Mean post-whelp plasma glucose values are identical in Beagles fed both diets; in the Labradors they tended to be higher for the LC diet, but the differences were not significant.

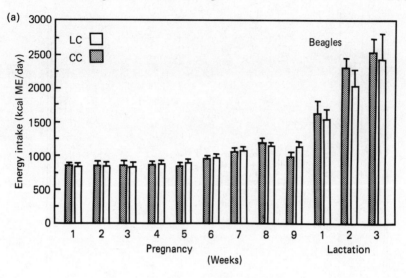

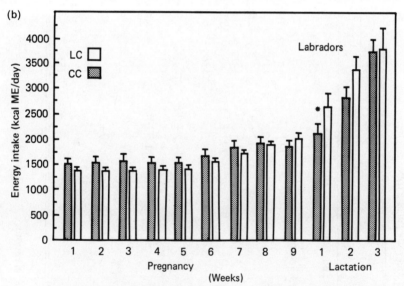

Fig. 2. Mean weekly energy intakes ± SEM of Beagle (2a) and Labrador (2b) bitches through pregnancy and lactation fed test (LC) or control (CC) diets. *$p < 0.05$.

Haematocrit and haemoglobin

Table 3 shows the values for haemoglobin and haematocrit during the first and last weeks of gestation. Levels at the end of gestation were lower than those at the beginning, irrespective of breed or dietary regime. All individual values were within the normal ranges found in this colony for these stages of pregnancy.

Table 3. *Bitch blood and milk parameters*

Parameter	Beagles		Labradors	
	LC	CC	LC	CC
Lactose content bitch milk g/dl	4.7 ± 0.9	5.8 ± 1.3	5.0 ± 0.9	4.2 ± 1.0
(n)		(5)		(5)
Bitch plasma glucose levels mg/dl				
Post-mating	112 ± 3	119 ± 20	101 ± 10	122 ± 11
(n)		(3)		(4)
3 days pre-whelp	117 ± 9	128 ± 12	117 ± 8	124 ± 16
(n)		(5)		(4)
3 days post-whelp	121 ± 13	121 ± 9	120 ± 5	112 ± 8
(n)		(5)		(7)
Bitch haemoglobin level g%				
Week 1	16.2 ± 0.8	15.9 ± 0.7	16.8 ± 1.2	15.7 ± 1.4
(n)		(4)		(7)
Week 9	11.8 ± 0.5	12.8 ± 0.8	12.7 ± 0.8	13.7 ± 1.0
(n)		(5)		(7)
Bitch haematocrit level %				
Week 1	49 ± 2	47 ± 2	50 ± 3	53 ± 2
(n)		(4)		(7)
Week 9	40 ± 2	39 ± 1	41 ± 2	41 ± 3
(n)		(5)		(7)

Notes:
[1] mean ± SEM of *n* observations

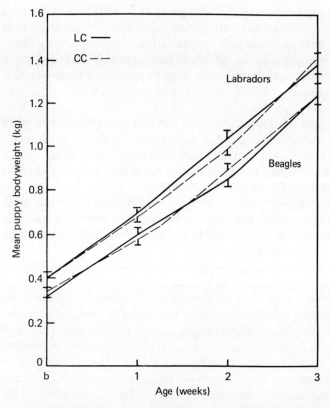

Fig. 3. Mean weekly bodyweights ± SEM of puppies born to Beagle and Labrador bitches fed on test (LC) and control (CC) diets. b (birth)

Discussion

In this study, no differences were found in the reproductive performance of bitches fed either a diet in which there was no available carbohydrate (LC) or one where food grade starch supplied 11% of the estimated metabolisable energy (CC). Litter sizes were within the ranges expected for these breeds (as reviewed by Meyer, 1984), the means being between 12 and 14% of maternal bodyweight at conception. Live litter weight did not differ significantly with dietary regime. Similarly, individual puppy birthweights were in the normal range of this colony and did not vary with maternal diet.

The most striking feature of these data is the complete absence of the elevated puppy mortality observed by Romsos & colleagues (1981) when bitches were fed a carbohydrate-free diet throughout pregnancy: only 35% of puppies from bitches fed the carbohydrate-free diet survived to three days of age. This contrasts with the results of our study where bitches whelped healthy viable puppies on both dietary regimes, survival to three days being in excess of 96% for all combinations of breed and diet. In Romsos's study such survival rates were observed only for the carbohydrate-containing diet.

The similar energy intakes on the two dietary regimes (Figs. 2A, 2B) were reflected by similar mean bodyweight gains in pregnancy (Labradors: LC 24%, CC 26%; Beagles: LC 26%, CC 25%) and maintenance of bodyweight through lactation (Figs. 1A, 1B).

Puppies showed normal weight gains for this colony, independent of maternal diet. The lactose content of bitches' milk varied widely although the mean levels were close to means published elsewhere (Oftedal, 1984). No effect of dietary regime was evident, suggesting that lactose synthesis is independent of a dietary supply of carbohydrate.

Bitches and puppies remained in good health throughout the trial. Routine estimations of haemoglobin and haematocrit were within normal ranges and showed the expected fall in late pregnancy, but there were no signs of any diet-related changes.

Some plasma glucose samples were lost, but those which were obtained allowed comparison between the two dietary regimes. Unlike the dogs studied by Romsos *et al.* (1981), there was no extreme hypoglycaemia observed immediately before whelping in bitches fed the carbohydrate-free diet. It appears that, on this dietary regime, in the present study there was sufficient gluconeogenesis to meet the needs of the bitch and developing puppies.

There are several key differences in the design of the current study and that of the Michigan group (Romsos *et al.*, 1981) which may account for the differences in the results. In the latter study, the bitches were transported from one institute to another three-and-a-half to four weeks after mating, and not assigned to test diets until then, whereas the dogs in this study lived within one building for the entire experiment and were fed the test foods from the day after the first mating. There is some evidence that the composition of the previous diet affects glucose turnover during starvation, the minimum rate being reached more quickly in dogs previously fed a high carbohydrate diet than those fed a high fat or high protein diet (Cowan, Vranic & Wrenshall, 1969) and it may be that this influenced the ability of the bitches to adapt to the carbohydrate-free diet. Variation in previous diet or the stress of transport may also explain why in the study of Romsos *et al.* (1981) the puppies of only four of the eight bitches fed the carbohydrate-free diet accounted for all of the mortality between birth and three days of age. It may be of considerable importance that these bitches all whelped early relative to the predicted date – premature birth has a very strong influence on neonatal mortality.

The Romsos study took no account of bitch variability. All were reported to have demonstrated good mothering characteristics in their previous environment, but there was no control for performance in the experiment. In the study reported here, each bitch acted as her own control, going through pregnancy and lactation on both dietary regimes.

Probably the most important difference between the two studies was in the composition of the diets. In the Romsos study 26% of the energy was derived from protein, a level slightly higher than the NRC recommended level of 22%. In the present study the LC diet supplied approximately 51% of the energy (estimated ME) as protein compared with 45% on the CC diet. This higher protein content may have supplied sufficient gluconeogenic amino acids to allow the maintenance of plasma glucose levels, despite heavy demand.

Amino acids traditionally described as gluconeogenic are those whose degradation leads to pyruvate or intermediates of the tricarboxylic cycle, allowing the formation of carbohydrate. In man, alanine has been shown to be the major gluconeogenic amino acid, accounting for 10% of the total gluconeogenesis in starvation (Felig *et al.*, 1969).

Work which showed that children with fasting hypoglycaemia had low plasma alanine levels (Pagliara *et al.*, 1972) was interpreted as evidence that alanine was being taken up from the blood for gluconeogenesis. Brady *et al.* (1977) observed depressed levels of alanine, glycine and serine in the plasma of fasted dogs, suggesting that these were the principal gluconeogenic amino acids in the dog.

The plasma alanine level was consistently significantly low in dogs fed the carbohydrate-free diet during pregnancy and lactation in the study reported by Romsos *et al.* (1981) indicating possible clearance by gluconeogenesis. The amino acid content of their diet is not given, but it is possible to estimate the concentrations of alanine, glycine and serine using food tables for the amino acid content of kidneys (Paul & Southgate, 1978) and our own data (unpublished) for that of isolated soya protein. This calculation gives a total for the three amino acids of 3.25 g per 400 kcal gross energy which we estimate is about 4.2 g per 400 kcal ME. The LC diet used in the present study contained 10.3 g per 400 kcal ME. Even allowing for small errors of estimation, there remains at least a two-fold difference which may account for much of the difference in outcome between the two studies. Furthermore, the unusually low protein digestibilities in Romsos's study (76 and 69%) may have exacerbated this problem. The hypoglycaemia found immediately after whelping (by Romsos) is then likely to be explained by a lack of gluconeogenic precursors rather than an inability to synthesise sufficient glucose. The observation by Romsos that his dogs were hypoalanaemic before whelping supports this theory.

There is no doubt that the dog has a marked capacity for gluconeogenesis. Kronfeld *et al.* (1977) fed a carbohydrate-free diet to working sledge dogs during the racing season and found no change in plasma glucose levels. This was true even when the dogs were consuming 4,000–5,000 kcal/day (three to four times estimated maintenance requirement) in order to maintain bodyweight.

Romsos *et al.* (1976) fed a range of foods to groups of eight female Beagle puppies (two months old at the start of test) for periods of eight months. There were no differences reported in the plasma glucose levels of dogs fed carbohydrate-free diets and those where carbohydrate made up 20–60% of the diet (on a dry matter basis), indicating that gluconeogenesis supplied the total glucose need of these puppies. The results were in contrast to those of Resnick (1978) who fed puppies a carbohydrate-free diet beginning at seven and nineteen weeks of age and had survival rates of only 46% in puppies fed the carbohydrate-free diet before 17 weeks of age. However, post-mortem examination showed fatty livers in all the dead pups and it may be that the carbohydrate-free diet was not satisfactory for other reasons (for example, it may have contained inadequate levels of choline which was not included in the vitamin supplement). Certainly other studies have shown that the capacity for gluconeogenesis from glycerol is present from a very early age in the dog (Hall *et al.*, 1976).

de Bruijne *et al.* (1981) have shown that, even in prolonged total starvation, dogs maintain plasma glucose levels for long periods within a normal range, and are also resistant to ketosis. In this they are similar to pigs (Wangsness *et al.*, 1981) and guinea pigs (Freminet & Leclerc, 1980) but different from man (Saudek & Felig, 1978; Owen *et al.*, 1979), rats (Goodman *et al.*, 1980; Triscari, Bryce & Sullivan, 1980) and mice (Kleiber, 1975). However, there is little evidence available on how these species react to a low carbohydrate diet in pregnancy. It has been shown that on a low carbohydrate diet pregnant rats become rapidly hypoglycaemic and ketotic with total embryo loss by day 12, half of which was attributable to a depression of energy intake and half to the metabolic consequences of the diet (Taylor *et al.*, 1983). Kertiles *et al.*, (1979) showed that pigs were capable of sustaining normal embryonic development despite total starvation (arguably more severe than carbohydrate deprivation although resulting in a similar fuel supply) through pregnancy. Similarly, pregnant bitches starved for three days pre-whelp produced healthy viable puppies (Kliegman, Miettinen & Adam, 1980).

To sum up, the ability of a bitch to produce viable offspring when fed a carbohydrate-free diet probably depends on two factors: first the capacity for gluconeogenesis which the dog does appear to possess and, secondly, the provision of sufficient appropriate gluconeogenic precursors. In the present study, a high protein, carbohydrate-free diet supported normal pregnancy and lactation in Beagle and Labrador bitches. The apparent requirement for carbohydrate during pregnancy in the dog (Romsos *et al.*, 1981) appears to be due to an inadequate protein intake, or at least an inadequate supply of suitable gluconeogenic amino acids. We therefore conclude that, while carbohydrate is *physiologically* essential, it is not an indispensable component of the diet.

Acknowledgements

The authors would like to thank Mrs L. Lambert, Mrs D. Orson, Mr T. Grant and Mr M. Wilson for technical assistance.

REFERENCES

Brady, L.J., Armstrong, M.K., Muiruri, K.L., Romsos, D.R., Bergen, W.G. & Leveille, G.A. (1977). Influence of prolonged fasting in the dog on glucose turnover and blood metabolites. *Journal of Nutrition*, **107**, 1053–61.

Cowan, J.S., Vranic, M. & Wrenshall, G.A. (1969). Effects of preceding diet and fasting on glucose turnover in normal dogs. *Metabolism*, **18**, 319–30.

de Bruijne, J.J., Altszuler, N., Hampshire, J., Visser, T.J. & Hackeng, W.H.L. (1981). Fat mobilisation and plasma hormone levels in fasted dogs. *Metabolism*, **30**, 190–4.

Drabkin, D.L. & Austin, J.H. (1932). Spectrophotometric constants for common haemoglobin derivatives in human, dog and rabbit blood. *Journal of Biological Chemistry*, **98**, 719.

Felig, P., Owen, O.E., Wahren, J. & Cahill, Jr, G.F. (1969). Amino acid metabolism during prolonged starvation. *Journal of Clinical Investigation*, **48**, 584–94.

Freminet, A. & Leclerc, L. (1980). Effect of fasting on glucose, lactate and alanine turnover in rats and guinea pigs. *Comparative Biochemistry and Physiology*, **65B**, 363–7.

Goodman, M.N., Larsen, P.R., Kaplan, M.M., Aoki, T.T., Young, V.R. & Ruderman, N.B. (1980). Starvation in the rat. II Effect of age and obesity on protein sparing and fuel metabolism. *American Journal of Physiology*, **239**, E277–86.

Hall, S.E.H., Hall, A.J., Layberry, R.A., Berman, M. & Hetenyi, G. (1976). Effects of age and fasting on gluconeogenesis from glycerol in dogs. *American Journal of Physiology*, **230**, 362–7.

Kertiles, L.P., Anderson, L.L., Parker, R.O. & Hard, D.L. (1979). Maternal serum metabolites during prolonged starvation in pregnant pigs. *Metabolism*, **28**, 100–4.

Kleiber, M. (1975). *The fire of life; an introduction to animal energetics*. Huntington, New York: Robert E. Krieger Publishing Co. Inc.

Kliegman, R.M., Miettinen, E.L. & Adam, P.A.J. (1980). Substrate-turnover interrelationships in fasting neonatal dogs. *American Journal of Physiology*, **239**, E287–93.

Kronfeld, D.S., Hammel, E.P., Ramberg, C.F. & Dunlap, H.L. (1977). Haematological and metabolic responses to training in racing sled dogs fed diets containing medium, low or zero carbohydrate. *American Journal of Clinical Nutrition*, **30**, 419–30.

Leibetseder, J. (1984). Fibre in the dog's diet. In *Nutrition and Behaviour in Dogs and Cats*, ed. R.S. Anderson, pp. 71–7. Oxford: Pergamon Press.

Meyer, H. (1984). Mineral metabolism and requirements in bitches and suckling pups. In *Nutrition and Behaviour in Dogs and Cats*, ed. R.S. Anderson, pp. 13–24. Oxford: Pergamon Press.

Munson & Walker (1970). Method for determination of lactose. In *Chemical Analysis of Foods*, 6th edn, ed. D. Pearson. London: Churchill.

National Research Council (1985). *Nutrient Requirements of Dogs*. Washington, DC: National Academy of Sciences.

Oftedal, O.T. (1984). Lactation in the dog: milk composition and intake by puppies. *Journal of Nutrition*, **114**, 803–12.

Owen, O.E., Reichard, G.A., Patel, M.S. & Boden, G. (1979). Energy metabolism in feasting and fasting. *Advances in Experimental Biology and Medicine*, **111**, 169–88.

Pagliara, A.S., Karl, I.E., de Vivo, D.C., Feigin, R.D. & Kipnis, D.M. (1972). Hypoalaninemia: a concomitant of ketotic hypoglycaemia. *Journal of Clinical Investigation*, **51**, 1440–9.

Paul, A.A. & Southgate, D.A.T. (1978). *McCance and Widdowson's The Composition of Foods*, 4th edn, p. 281. London: Her Majesty's Stationery Office.

Resnick, S. (1978). Effect of age on survivability of pups eating a carbohydrate-free diet. *Journal of the American Veterinary Medical Association*, **172**, 145–8.

Romsos, D.R., Belo, P.S., Bennink, M.R., Bergen, W.C. & Leveille, G.A. (1976). Effects of dietary carbohydrate, fat and protein on growth, body composition and blood metabolite levels in the dog. *Journal of Nutrition*, **106**, 1452–64.

Romsos, D.R., Palmer, H.J., Muiruri, K.L. & Bennink, M.R. (1981). Influence of a low carbohydrate diet on performance of pregnant and lactating dogs. *Journal of Nutrition*, **111**, 678–89.

Saudek, C.D. & Felig, P. (1978). The metabolic events of starvation. *American Journal of Medicine*, **60**, 117–26.

Schmidt, F.M. (1961). Determination of plasma glucose by the hexokinase method with deproteinization. *Klinische Wochenschrift*, **39**, 1244.

Snedecor, G.W. & Cochran, W.G. (1968). *Statistical Methods*. Ames, USA: Iowa State University Press.

Taylor, S.A., Shrader, R.E., Koski, K.G. & Zeman, F.J. (1983). Maternal and embryonic response to a 'carbohydrate-free' diet fed to rats. *Journal of Nutrition*, **113**, 253–67.

Triscari, J., Bryce, G.F. & Sullivan, A.C. (1980). Metabolic consequences of fasting in old lean and obese Zucker rats. *Metabolism*, **29**, 377–85.

Wangsness, P.J., Acker, W.A., Burdette, J.H., Krabill, L.F. & Vasilatos, R. (1981). Effects of fasting on hormones and metabolites in plasma of fast-growing lean and slow-growing obese pigs. *Journal of Animal Science*, **52**, 69–74.

13

The effects of carbohydrate-free diets containing different levels of protein on reproduction in the bitch

E. KIENZLE AND H. MEYER

Introduction

In defining the nutrient requirements of the neonate it is often necessary to employ an 'argument by design' making the assumption that the mother's milk represents an optimal diet for the suckling. One remarkable thing about milks is that they and only they contain lactose raising the possibility that carbohydrate may be of special importance in the nutrition of the neonate. That being so it is not unreasonable to suggest that there may be in consequence a particular importance of carbohydrate in the physiological economy of the pregnant or lactating mother.

In pet animal nutrition such ideas received some support when Romsos *et al.* (1981) demonstrated that carbohydrate-free diets fed to pregnant bitches were associated with severe disturbance in the metabolism of the bitch and the development of her puppies.

Even in the maintaining animal carbohydrate is of course physiologically essential, but no dietary requirement appears to exist because gluconeogenesis from amino acids is adequate to meet physiological demand. In our work therefore we have also been concerned to determine the extent to which such a substitution can modify the dietary needs of the bitch, by studying whether the effects of a carbohydrate-free diet in reproduction could be ameliorated by feeding a high protein diet to the bitch.

For this study we used pregnant and lactating bitches (beagles and setters) weighing 8–25 kg with four or more pups. As Table 1 shows these were divided into three dietary groups. Group I were a control group fed a carbohydrate containing medium protein diet (CCMP). Group II were fed a carbohydrate free diet that had a high level of protein (CFHP), to test the hypothesis that dietary protein could ameliorate any effects of carbohydrate deprivation. Group III were fed a carbohydrate-free medium protein diet (CFMP) of closely similar protein content to that used by Romsos et alia in their studies. In all groups the protein intake exceeded NRC (1985) recommendations.

243

Table 1. *Experimental conditions*

Group	I CCMP carbohydrate containing medium protein	II CFHP carbohydrate-free high protein	III CFMP carbohydrate-free medium protein
g digestible crude protein per MJ digestible energy	12	20	10
Percentage of digestible energy from protein	29	48	26
Percentage of digestible energy from carbohydrate	30	0	0
Number of pregnant bitches	5	6	5
Number of lactating bitches	3	3	4
Minimum test period			
days ante-partum	27	27	28
days post-partum	0–35	0–35	5–14

Table 2. *Mean daily intake of digestible energy and digestible crude protein during the last four weeks of pregnancy and during lactation (per $W^{0.75}$)*

Group	I CCMP	II CFHP	III CFMP
Pregnancy			
energy (MJ)	0.91 ± 0.1	0.64 ± 0.1	0.78 ± 09.2
protein (g)	10.6 ± 1.5	12.7 ± 2.8	7.9 ± 1.9
lactation			
energy (MJ)	1.80 ± 0.4	1.60 ± 0.6	1.38 ± 0.2
protein (g)	21.7 ± 5.2	34.1 ± 13.5	12.9 ± 1.2

Pregnant bitches received the diet for at least four weeks before term, lactating bitches were placed on the appropriate diet before the end of the colostral phase, five days after birth.

Energy intake was measured as digestible energy (DE). Intakes and faecal losses were determined directly, with the following energy values being assumed for digestible nutrients: digestible crude protein 24 kJ/g, digestible crude fat 38 kJ/g and digestible nitrogen-free extract 17 kJ/g. Blood of bitches was examined after an overnight fast for glucose (enzymatically), β-hydroxybutyrate (enzymatically), total protein (Biuret method), urea (Bethelot reaction) and free amino acids (ion exchange chromatography). Plasma insulin and cortisol were determined by radioimmunoassay. Liver glycogen reserves were also determined enzymatically. The intravenous glucose tolerance test was performed three weeks before whelping after an overnight fast with a glucose dose of 1 g per kg. bodyweight.

The results, summarised below, showed clearly that although the absence of carbohydrate in a diet of normal protein content (CFMP diet) induced a variety of abnormalities during reproduction essentially similar to those described by Romsos *et al.*, the effects were reduced in intensity or eliminated altogether when a high level of protein was included in the diet (CFHP diet).

Food intakes, mortality and growth

There were differences in the mean daily energy intake between the three groups as Table 2 shows. In part these reflected overall differences in palatability of the three diets, but, in part, they were a reflection of different patterns of food consumption between the groups. There was a marked decrease in intake of the bitches consuming both of the carbohydrate-free diets in the week before whelping, while bitches fed the CFMP diet had very high intakes during the first week of lactation, with very low intakes in the second week, some bitches refusing food altogether.

As Table 3 makes clear, feeding the CFMP diet during pregnancy had a marked effect on the young born. Table 3 shows that the viability of the puppies born to Group III bitches, who received the CFMP diet, was

Table 3. *Effect of nutrition during pregnancy on vitality of puppies*

Group	I CCMP	II CFHP	II CFMP
Number of puppies	47	40	38
stillborn (%)	0	0	18
perinatal mortality (%)	2.3	7.9	75.0
Birth weight			
in g	355 ± 50	320 ± 74	210 ± 43
in % of bitch weight	2.4 ± 0.7	2.0 ± 0.5	2.0 ± 0.5
expected[1]	2.1	2.1	2.8
Changes in birth weight (%)			
in the first 24 h of life	+ 2.2 ± 5.5	+ 2.3 ± 11.6	− 11.9 ± 30.3

Notes:

[1] Calculated after regression: $y = 4.27 - 0.075\ a - 0.097\ b$

y = expected birth weight in % of bitch weight

a = bitch weight in kg b = number of puppies

markedly reduced with 18% stillborn and 75% dying soon after birth. It is also clear from the Table that the viability of puppies born to Group II bitches fed the CFHP diet was not significantly different to that of the control group.

The Group III puppies also had low birth weight, only 71% of that which would be expected from the bitches' weight and the number in the litter (Meyer, Dammers & Kienzle, 1985), and the placental weight was reduced in proportion. By contrast the litters of the Group II bitches which had been fed the CFHP diet were not significantly different from controls in these regards. Group III puppies, as Table 3 also shows, tended to lose weight in the first 24 hours after birth, again in contrast to the pattern in other groups.

Puppies born to Group III bitches were not only small and weak, they also showed typical signs of the runting phenomenon observed by Fox (1966) after prenatal malnutrition. A typical puppy is shown in Figure 1 where, in particular, the disproportionate size of the head relative to the body can be seen, something which gives, consequently, a premature appearance.

Carbohydrate homeostasis

The impression of prematurity was confirmed by the size of the liver which was smaller in puppies to bitches fed both the carbohydrate-free diets (Fig. 2). These puppies had a lower liver glycogen content as well, though the reduction was less in the puppies born to Group II bitches fed the CFHP diet, than those born to Group III bitches (Fig. 2). These two factors, a smaller liver and lower glycogen content in the liver, meant that the total glycogen per unit weight of the pups was markedly reduced (Fig. 2). Within Group III there was a marked difference in liver glycogen between the weak and the viable puppies. Pups born to the Group III bitches had normal blood

Fig. 1. Puppy from Group III.

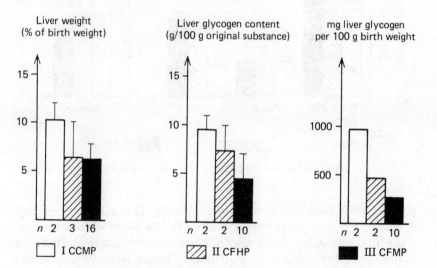

Fig. 2. Liver glycogen reserves of puppies in relation to carbohydrate and protein intake of bitch.

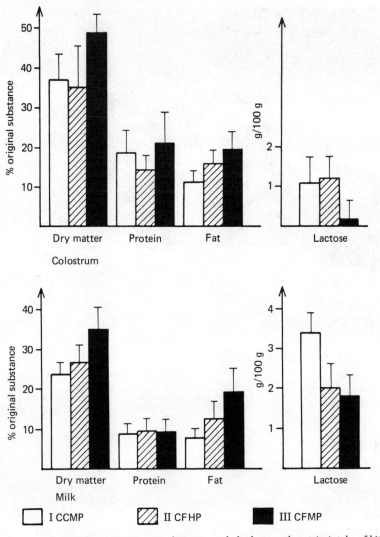

Fig. 3. Milk composition in relation to carbohydrate and protein intake of bitch.

glucose levels immediately after birth, but by 12–72 hours they had
developed severe hypoglycaemia; the blood glucose of the other two groups
was within what Allan, Kornhauser & Schwartz (1966) have found to be the
normal physiological range. In one Group III puppy, no blood glucose was
actually detectable, and, like its litter mates, this puppy showed spasms
which disappeared after a subcutaneous injection of glucose. Blood glucose
and weight changes on day 1 of life were both good predictors of mortality. It
was possible to reduce mortality considerably by repeated application of 2 ml

per 100 g of bodyweight of a 5% glucose solution. Surviving puppies developed normally and reached normal weight by four weeks of age.

Diet and lactation

As Figure 3 shows, the composition of bitches' milk depended strongly on diet with dams fed the carbohydrate-free diets producing a milk with higher dry matter, and fat content, than Group I animals receiving the carbohydrate-containing diet. Protein content of the milk was not, however, changed by feeding carbohydrate-free diets. Bitches fed the carbohydrate-free diets also had lower liver lactose contents.

There were no significant differences in the milk composition of animals that received the carbohydrate-free diets throughout pregnancy and those that only received it after birth. This was due to the fact that milk composition changed very rapidly after beginning a carbohydrate-free diet. Twenty four hours after introducing a carbohydrate-free diet for example, the dry matter content of milk amounted to 30%.

There was a uniform correlation between lactose and the dry matter content of milk, as Figure 4 shows. This was independent of the dietary group or the stage of lactation, and it made no difference whether the dry matter increase was due to a higher protein content (as in colostral milk) or to a rise in fat as in the diet-induced changes described above. This probably reflects a strong relationship between osmolarity and lactose content of the intra-alveolar fluid in the mammary glands.

So the dry matter content of the milk changes in the following manner: from 35% on the CFMP diet (providing 10 g of protein per MJ) to 28% on the carbohydrate free diet of Romsos et alia (1981), which provided 15 g protein per MJ, to 25% in our CFHP diet which provided 20 g protein per MJ, falling to 23% on bitches fed the diet containing carbohydrate (CCMP). This observation raises the question of which composition of bitches' milk is biologically normal. Dogs may scavenge but, in so far as wild dogs prey upon animals, their diet is of course best approximated by the CFHP diet.

The poor growth of the Group III pups whose mothers were fed the CFMP suggests the possibility of impaired food intake in these animals.

The high viscosity of the milk, expecially the colostral milk of the bitches fed the CFMP diets could be expected to make it difficult for all puppies, and especially weak neonates, to suckle, and this may well have contributed to the neonatal weight loss and high mortality. The bitches potential for milk production could not be measured although actual intakes of pups could be estimated from the weight gain of puppies during feeding. On this basis there was no difference in milk intake between Groups I and II, though milk intake of Group III was poor.

During the final days of the study the group III puppies either did not gain weight or showed signs of milk incompatibility or illness.

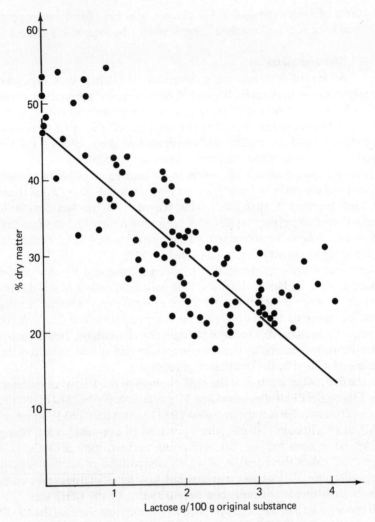

Fig. 4. Relation between lactose and dry matter content in milk
($y = 47.0 - 7.6x$; $r = -0.82$. (From Kienzle, Meyer & Lohrie, 1983.)

Plasma metabolites

The drastic changes in milk composition and foetal development were accompanied by marked changes in plasma metabolites which were indicative of a high gluconeogenic activity in bitches fed a carbohydrate-free diet. These changes began three weeks before term and reached their maximum immediately before whelping. The patterns observed in pregnant and lactating bitches were essentially similar and, for brevity only, the changes observed in pregnant animals are described here.

Table 4. *Blood glucose concentration of puppies in relation to liver glycogen reserves*

Group	I CCMP	II CFHP	III CFMP
Liver glycogen			
mg/100 g birth weight	981	489	277
Blood glucose mg/dl[1]			
Immediately after birth	108.2	No value	80.2
12–72 h after birth	95.7	89.6	37.7

Note:
[1] Mean coefficient of variation amounted to 38.6%

Glucose appears to be a most important metabolite for foetal development, the energy metabolism of foetus and placenta depending, in good part, on glucose supply. As Figure 5 shows, the plasma glucose concentration in the week before whelping of Group III bitches fed the CFMP diet was related to litter weight, a strong negative correlation being evident. In bitches who were not pregnant or had very small litters, plasma glucose levels remained

Fig. 5. Plasma glucose concentration of bitches from group III CFMP in the last week before birth in relation to litter weight $y = 124.9 - 5.03x$; $r = -0.91$. (From Kienzle *et al.*, 1985)

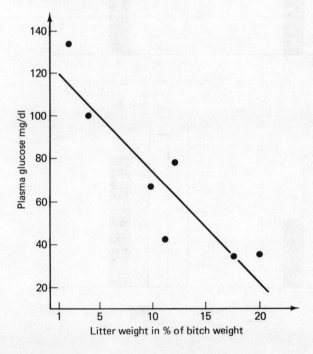

normal and corresponding plasma metabolites only changed to a small degree.

However, bitches with large litters became extremely hypoglycaemic in the last days before whelping, with the lowest value determined being 342 mg/l. Romsos *et al.* (1981) reported similar effects. As Figure 6 makes clear, although plasma glucose concentration of the CFHP-fed bitches also decreased in the last week before birth, it was to a much smaller extent, though again the extent of the fall was correlated with litter weight.

Between the groups the blood β-hydroxybutyrate levels were broadly

Fig. 6. Plasma composition (glucose, β-hydroxibutyrate, free amino acids) of pregnant bitches in relation to carbohydrate and protein intake.

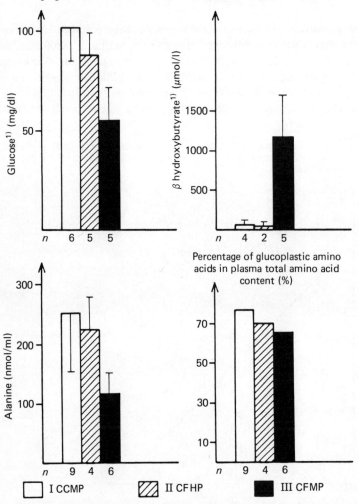

negatively correlated with plasma glucose. There were no significant differences in β-hydroxybutyrate levels between bitches in Groups I and II, but in Group III there was a 20-fold increase in blood β-hydroxybutyrate in the week before birth.

Plasma alanine content appeared to be decreased in both the groups fed carbohydrate-free diets as Figure 6 shows, with a small, non-significant fall being evident in the CFHP animals, but a marked statistically significant decline being evident in CFMP bitches. In the CFMP bitches a positive correlation could be demonstrated between the plasma alanine and plasma glucose concentrations, but no correlation existed for the other two groups. The fraction of the total free amino acids which were glucogenic also declined on the carbohydrate-free diets with the greater decline in the animals fed the CFMP diet.

As might be expected, plasma urea concentration was higher in animals fed the high protein CFHP diet. It was normal in the animals on the CCMP diet but showed a non-significant tendency to rise in animals fed the CFMP diet. Finally, in the week before whelping, even plasma protein levels of the CFMP animals declined.

The regulation of carbohydrate metabolism

Predictably, the most marked changes in plasma insulin and cortisol levels occurred in bitches from Group III fed the CFMP diet. Figure 7 shows the changes which occurred in these animals during pregnancy: insulin levels were extremely low, while the changes in plasma cortisol were the inverse of those occurring for glucose. Similar, but much less marked changes occurred in the Group II animals fed the CFHP diet, and no comparable changes were observed in Group I.

The shift in plasma cortisol levels in lactating bitches from Group III were even more pronounced than those for the pregnant bitches as Figure 8 illustrates, with a negative correlation between cortisol and blood glucose still being evident.

The poor prenatal development of puppies of bitches fed the CFMP diet might not be merely the result of lower blood glucose levels, but also of alterations to the regulation of glucose metabolism, particularly the glucose turnover rate. This possibility is suggested by the results for bitches with small litters. Although these animals had normal glucose levels on carbohydrate-free diets, the viability of their puppies was not much higher than that of puppies of bitches having large litters and low blood glucose levels.

A glucose tolerance test can be used to provide a measure of glucose turnover and Romsos *et al.* (1981) showed a diminished glucose turnover rate in bitches fed carbohydrate-free diets. We have confirmed this observation in our animals.

Pregnant and lactating dogs fed the CCMP diet showed no alterations in

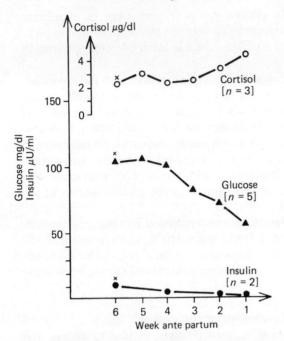

× Values before starting carbohydrate-free feeding

Fig. 7. Changes in plasma glucose, insulin and cortisol concentration in pregnant bitches from Group III CFMP.

glucose tolerance, with the half-time for glucose elimination being 24.2 ± 4.0 minutes.

In Group II animals fed the CFHP diet, there was only a small alteration in glucose tolerance, with the average half-time for glucose elimination being 28.5 ± 8.5 minutes. Figure 9(a) shows a result for a typical animal in this group.

By contrast, glucose tolerance of bitches in Group III, fed the CCMP diet, was poor and average half-times for glucose elimination amounted to 44.0 ± 1.8 minutes for pregnant bitches and 84.4 ± 45.1 minutes for lactating bitches. These results are illustrated in Figure 9(b), which shows the glucose tolerance curve of a pregnant bitch on the CCMP diet three weeks before birth, and Figure 9(c) which shows the curve for a lactating CCMP bitch which was hypoglycaemic before glucose injection.

Conclusions
We have been able to confirm that carbohydrate-free diets fed to pregnant and lactating bitches can cause adverse effects, but have been able to show that performance is not markedly impaired provided that the dietary protein level is sufficiently high.

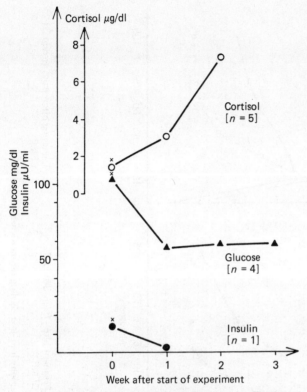

Fig. 8. Changes in plasma glucose, insulin and cortisol concentration in lactating bitches from Group III CFMP.

In diets which provide carbohydrate, 7 g digestible crude protein per unit of metabolic bodyweight ($W^{0.75}$) appear to be adequate for the normal birth and development of puppies, but, if fed without carbohydrate, pregnant bitches require at least 12 g $DCP/W^{0.75}$.

Lactating bitches require 13–18 g $DCP/W^{0.75}$ on carbohydrate-containing diets, and 30 g on carbohydrate-free diets.

Additionally, on carbohydrate-free diets, our view is that the protein level should not fall below 20 g DCP per MJ of digestible energy, (equivalent to a protein:energy ratio of 0.33, calculated on a notional metabolisable energy basis).

It is of note that feeding carbohydrate-free diets with inadequate levels of protein produces a runting phenomenon in the whole litter similar to that observed in single puppies of some litters born to normal bitches. The significance of this observation has yet to be explored.

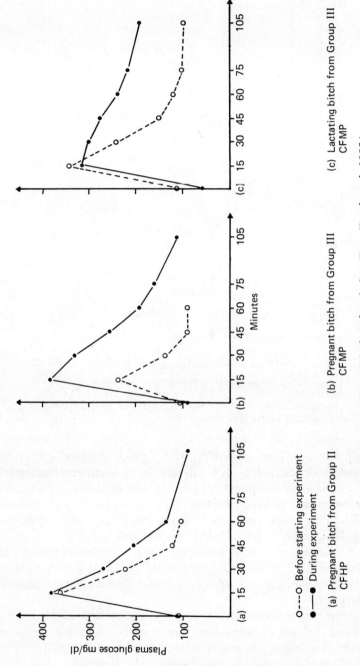

O----O Before starting experiment
●——● During experiment

(a) Pregnant bitch from Group II
    CFHP

(b) Pregnant bitch from Group III
    CFMP

(c) Lactating bitch from Group III
    CFMP

Fig. 9. Changes in glucose tolerance of bitches induced by carbohydrate-free feeding (From Kienzle *et al.*, 1985.)

REFERENCES

Allen, D.T., Kornhauser, K. & Schwartz, R. (1966). Glucose homeostasis in the new born puppy. *American Journal of Diseases of Children,* **112,** 343-50.

Fox, M.W. (1969). *Canine Pediatrics.* Springfield, IL: Charles C. Thomas.

Kienzle, E., Meyer, H. & Lohrie, H. (1985). Influence of carbohydrate free rations with various protein/energy relationships on foetal development, viability of newborn puppies and milk composition. *Advances in Animal Physiology and Animal Nutrition,* **16,** 73–99.

Meyer, H., Dammers, C. & Kienzle, E. (1985). Body composition of newborn puppies and the nutrient requirement of pregnant bitches. *Advances in Animal Physiology and Animal Nutrition,* **16,** 7–25.

NRC (1985). *Nutrient Requirements of Dogs.* National Research Council. Washington DC: National Academy of Sciences.

Romsos, D.R., Palmer, H.J., Muiruri, K.L. & Bensink, M.R. (1981). Influence of a low carbohydrate diet on performance of pregnant and lactating dogs. *Journal of Nutrition,* **111,** 678–89.

14

The use of different sources of raw and heated starch in the ration of weaned kittens

R.O. DE WILDE AND T. JANSEN

Introduction

The natural diet of the cat contains only 1–2% carbohydrates mainly consisting of glycogen and the vegetable carbohydrates in the digestive tract of the herbivorous prey.

In commercial feeds the carbohydrate content is much higher and varies, when expressed on a dry matter (dm) basis, between 0 and 30% for canned feeds and between 30–60% for dry feeds (De Wilde & D'Heer, 1982). More than 50% of this carbohydrate is starch.

The main sources of starch are from cereals or from tubers (potatoes, manioc). The digestibility of cereal starch is influenced by treatment. Heating cereal starch improved the digestibility by 4 to 9 percentage points (Morris *et al.*, 1977) to the value of oligosaccharides. Finely ground cereal starch is more digestible than coarsely ground and wheat starch is better digested than maize starch (Morris *et al.*, 1977).

Among the tuberous starch sources, only potato starch has been examined. At low levels of inclusion (5% dm) the digestibility of partially hydrolysed potato starch is nearly 100% but this declines to 90% as the level of inclusion is raised to 15% dm (De Wilde & Huysentruyt, 1983).

All the results already mentioned concern experiments in adult cats. There are special reasons to examine starch sources in young weaned kittens:

- they lack any adaptation to digest starch, because they have not yet received any source of carbohydrates other than lactose.
- they have a quick accommodation to different feed regimes, because they have not yet developed fixed feeding habits.
- the growth rate itself is an easily measurable parameter of the degree of utilisation of the starch source.

The aim of the present experiment was to examine how young kittens react to a diet in which 25% of the dm of a commercial canned feed is

259

replaced by raw or heated starches of different origins. The level of substitution was such that, relative to the NRC (1978) nutrient requirements for young kittens no deficiency of protein occurred.

Material and methods

Two-month old kittens from six litters and with a weight range from 420 to 1230 g were divided randomly into six pairs and placed in six cages in which faeces and urine could be collected separately. They were fed individually twice daily at a rate of dm intake of 3% of live weight. Water was available *ad libitum* from nipple-flasks. The animals were weighed weekly before feeding.

Before the start of the feeding experiments the animals were treated against endoparasites and vaccinated against panleucopenia and feline rhinotracheitis.

The basal ration was a canned feed (Whiskas, Effem, Brussels) and contained 83.3% moisture, 11.0% crude protein, 4.5% fat, 1.2% ash and 0% starch and crude fibre. The metabolisable energy (ME) content was 21 MJ per kg dm (5 kcal/g dm). Two sources of starch (potato and maize) and two treatments (raw and heated) were combined with the control (basal) ration in two Latin square designs: one for the raw starches and one for the heated starches. The heated starch was treated for 1 hour at 132°C. Starch powder replaced 25% of the dm of the basal ration.

The experimental design is presented in the following scheme.

Cage Number	Period 1	Period 2	Period 3
1	C	RP	RM
2	RP	RM	C
3	RM	C	RP
4	C	HP	HM
5	HP	HM	C
6	HM	C	HP

Note: C = control; R = raw; H = heated; P = potato; M = maize.

Each period lasted 3 weeks: week 1 was considered as an adaptation period for the new feed combination, week 2 and 3 were the experimental weeks, in which Celite 545 was included at 1.5% of dm as a marker for digestibility studies. Week 3 was the collection period of faeces and urine. The following data were collected:
- weight of the animals each week
- measurement of water intake each week
- measurement of urine production
- analysis of feed and faeces, stored at −20°C for dm, N and starch
- determination of faecal pH and composition of the faecal carbohydrates, according to Mayes & Ørskov, 1974

Table 1. *Starch content of the sources*

Method	Potato	Maize
Polarimetric	85.6%	88.9%
Enzymatic on raw material	57.8%	22.0%
Enzymatic on heated material	86.3%	88.2%

– starch analysis according to an acid hydrolysis to glucose (polarimetric method) which corresponds to the 'total' starch content and according to an enzymatic procedure (amyloglucosidase), corresponding to the 'available' starch content.

Results

The starch content of the different sources are given in Table 1. Raw maize starch seems to be less prone to enzymatic *in vitro* attack than raw potato starch. This is not in accordance with the *in vivo* digestibility (see Table 4) where raw maize starch is more digested than raw potato starch. After heating of the starch, the amyloglucosidase test is able to hydrolyse the starch to the same extent as an acid hydrolysis.

All food offered was eaten. The mean weekly weight gain was 103 g with the highest figure (124 g) for the control group and the lowest figure (78 g) for the group on the raw potato starch. The growth rate during the three periods of three weeks each is shown in Figure 1, expressed as a percentage of the liveweight at the beginning of each period. In each of the three periods

Fig. 1. Relative growth rate (during the three-week period) as per cent increase in the liveweight at the beginning of each period. (The figures of the heated groups are based on only two measurements, which make an SD less meaningful.)

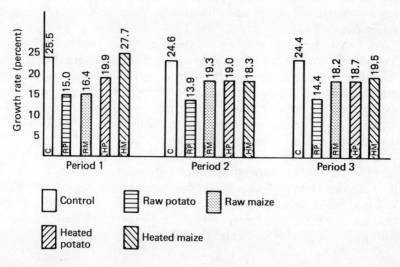

Table 2. *Water intake, faecal water content and urine production on three consecutive diets (mean ± standard deviation of four kittens)*

| Diet | Water consumption (ml/kg body weight) | | Water % of faeces | Urine production (ml/kg BW) |
	From feed[1]	As free water		
Starch (RM + HM)	110a	13 ± 1.4a	75.3 ± 3.0a	81 ± 12.7a
Control	147b	5 ± 0.9b	68.9 ± 7.1a	130 ± 2.8b
Starch (RP + HP)	110a	6 ± 1.5b	69.9 ± 6.7a	75 ± 9.2a

Note:
[1] Fixed amount/kg BW
a-b: significantly ($P < 0.05$) different in the same column

(with different animals per period) the relative growth rate was lowest for the raw potato group (14.4%), is substantially increased in animals fed heated potato starch (19.0%) but still remained less than the control (24.8%). Even for the maize groups, the relative growth rate (18.0% for the raw and 21.8% for the heated maize) is lower than for the control in two of the three periods. Part of the lower growth rate on the starch substitution diets may be due to the lower energy and protein intakes on these diets. Animals were fed constant dry matter intakes but the nutrient content of the dm fell when starch (0% protein and 17 MJ ME/kg dm) replaced the dm of canned feed (27% protein and 21 MJ/kg dm). Part of this difference and the differences between starches may also be due to differences in digestibilities of energy and protein.

The dm intake of 3% of live weight was an attainable percentage for all treatments because there were never any refusals (even with the more unpalatable starch diets) and the growth rate for the controls was acceptable.

The total water intake is shown in Table 2 and varied between 12% of the bodyweight (BW) on the starch diets and 15% on the control diets. The difference was due to the fact that the more concentrated starch diets provided less water, (starch contains only *c.* 10% moisture) and did not induce a compensatory increase in voluntary water consumption.

The faecal water content was not influenced by the water intake but the urine production was significantly higher on the control diet.

The pH values of the faeces are shown in Table 3. The faecal pH on the control diets was 6.4 with a narrow standard deviation of ± 0.1. The pH on the starch diets was in all cases lower than 6.0 with a much wider variation. This may reflect an acid production from carbohydrate fermentation in the hind gut. There is only a slightly lower pH on the raw starch diets, compared with the heated starch diets, and the difference is not significant.

Table 3. *pH of faeces*

Control	6.5 ± 0.1	RP	5.3 ± 0.3	HP	5.8 ± 0.2
Control	6.3 ± 0.1	RM	5.7 ± 0.2	HM	5.6 ± 0.2
Control	6.4 ± 0.1	Raw	5.5 ± 0.3	Heated	5.7 ± 0.2

Table 4. *Digestibility of nutrients (mean ± SD)*

	Starch	Protein
Control	—	85.5 ± 3.0 a
Raw potato	60.4 ± 0.1 a	80.8 ± 1.6 a
Heated potato	96.9 ± 2.1 c	80.2 ± 4.6 a
Raw maize	84.3 ± 0.1 b	79.0 ± 2.1 a
Heated maize	97.0 ± 1.0 c	79.1 ± 1.7 a

Figures in the same column with the same letter are not significantly different.

The digestibility studies are summarised in Table 4. The two heated starch sources have nearly the same very high (97%) digestibility in these young carnivorous animals. When raw starch was fed, the digestibility values dropped to 85% for maize and 60% for potato. These values are significantly lower than those of the heated starch sources and are also significantly different from each other ($P < 0.05$).

The protein ($N \times 6.25$) digestibility was not significantly influenced by adding starch, although the figure for the control diet was about 3% higher than for the starch diets. There was no influence of the starch treatment on protein digestibility.

The composition of the faecal carbohydrates are presented in Table 5. The glucans are defined as the long-chain alpha-glucoside polymers and the dextrins as the shorter chains alpha-glucoside polymers. It can be seen that the heat treatment markedly affected the carbohydrate composition of the faeces. In the raw starch diets most of the faecal carbohydrates are long-chain polymers. In the heated starch diets the faecal carbohydrates are composed equally of dextrins and glucans. This does not mean, however, that dextrins are much more poorly digested in heated diets. The total carbohydrate level in the faeces from the heated starch diets is much lower (i.e. digestibility higher) as Table 4 shows. Thus the absolute content of dextrins in faeces does not appear to have changed substantially.

Otherwise it is very likely (it was not determined in our experiments) that the dextrin content of the starch diets is increased by heating and that the faecal composition is a reflexion of the diet composition.

Table 5. *Composition of the faecal carbohydrates (as % of total)* *(mean ± SD)*

	Glucans[1]	Dextrins[2]
Raw potato	98 ± 0.6	2 ± 0.6
Heated potato	56 ± 4.8	44 ± 4.7
Raw maize	96 ± 0.8	4 ± 0.8
Heated maize	43 ± 5.4	57 ± 5.4

Notes:
[1] Insoluble in 60% alcohol.
[2] Soluble in 60% alcohol.

Discussion

Under our experimental conditions feeding raw starch sources to weaned kittens did not cause severe digestive disturbances, even at an inclusion level of 25% of the dry matter. The starch however was given as a fine powder, which enhances the digestibility (Morris *et al.*, 1977) in comparison to starch, incorporated in a cereal or a tuber.

Heating the starch improves the digestibility and is reflected in a better growth response, especially on the potato starch diets. The fact that there is some relationship between the relative growth rate on the different diets (Fig. 1) and the digestibility (Table 4) suggests that the digestion of the starch was, to some extent, a real one, which occurred in the small intestine and provided increased nutrients to the animal and was not completely due to microbial breakdown of the starch in the hind gut, as was suggested by Drochner (1977). Nevertheless, in our experiments, there was some fermentation in the hind gut, which is reflected in the fall of the faecal pH in the starch diets.

By heating the starch at 132°C during 60 minutes, the nutritive value of the two starch sources became similar. The question arises if, in practice, this thermal treatment can be guaranteed or if a less intensive heating is sufficient. Experiments of Jørgensen & Hansen (1975) with mink demonstrated that cooking potato starch during one hour gave a lower digestibility than three hours at 110°C.

Is there any induction of starch digestion by introducing a source of starch in the diet? This question can be answered in these young kittens, who received starch in two consecutive periods, separated by a period on a starch-free diet. There was not any tendency to observe a better digestibility in the second period than in the first one. This confirms the observations of Jørgensen & Hansen (1975) in mink, who could not demonstrate an improvement of the digestibility of a raw starch during a period of four weeks, in which the digestibility was measured weekly.

When starch is added to the diet of weaned kittens, the risk of diarrhoea is higher especially on raw potato starch, but the incidence depends more on the individual variability than on the source or the treatment of the starch incorporated into the diet. There was no consistent difference in moisture content of the faeces between the treatments.

The apparent digestibility of the crude protein is lower in starch supplemented diets. This is in agreement with Morris *et al.* (1977) and is interpreted as a consequence of undigested starch reaching the large intestine, enhancing microbial growth with the production of more microbial nitrogen. In their experiments with cats, supplemented with oligosaccharides, Drochner & Müller-Schlösser (1978) did not find a consistent decrease in N-digestibility in the supplemented diets. However, they used a control diet based on cereals and soybean meal, which could have been subject to fermentation in the hind gut.

Conclusions

- Replacement of 25% of the dry matter of canned feed by starch reduces the growth rate of young kittens in inverse relation to the digestibility of the starch sources.
- Raw potato starch is less digestible than raw maize starch.
- Heating the starch for 1 hour at 132°C, increases the digestibility to 97% irrespective of the starch source.
- Faecal pH is lower in starch supplemented diets, with a small tendency to the most acid faeces on the raw starch diets (correlation of 0.7 between pH and digestibility for $n = 12$).
- Increasing the dry matter of the diet (by adding starch) drops the urine production because the lower moisture intake is not entirely compensated by an increased voluntary water consumption.
- The consistency of the faeces was more variable on the starch diets but none of the treatments had to be discontinued because of profuse or persistent diarrhoea.

Acknowledgement

The EFFEM Company in Brussels is gratefully acknowledged for providing the starch-free canned feed.

The technical assistance of E. Maes and H. Derycke is fully appreciated.

REFERENCES

De Wilde, R. & F. D'Heer, (1982). Nutritionele evaluatie van commerciële honden- en kattevoeders. *Vlaams Diergeneeskunde Tijdschrift*, **51**, 341–54.

De Wilde, R. & P. Huysentruyt, (1983). De vertering van koolhydraten bij de kat. *Tijdschrift Voor Diergeneeskunde*, **108**, 187–90.

Drochner, W., (1977). Zur Konzentration einiger wichtiger

Stoffwechselabbauprodukte in den Faeces des Hundes nach Zufuhr hoher Mengen nativer Stärke mit dem Futter. *Kleintier-Praxis*, 22, 191–200.

Drochner, W. & S. Müller-Schlösser, (1978). Verdaulichkeit und Verträglichkeitverschiedener Zucker bei Katzen. *Symposium Hanover Veterinary School*, 80–90.

Jørgensen, G. & N.G. Hansen, (1975). Fordøjelighedsforsøg med rene stivelsearter. 422. *Beretning fra Forsøgslaboratoriet København*, 28–36.

Mayes, R.W. & E.R. Ørskov, (1974). The utilisation of gelled maize starch in the small intestine of sheep. *British Journal of Nutrition*, 32, 143–53.

Morris, J.G., J. Trudell & T. Pencovic, (1977). Carbohydrate digestion by the domestic cat (*Felis catus*). *British Journal of Nutrition*, 37, 365–73.

National Research Council, (1978). *Nutrient requirements of cats.* Washington DC: National Academy of Sciences.

15

Pathogenesis of lactose-induced diarrhoea and its prevention by enzymatic splitting of lactose

H.-C. MUNDT, and H. MEYER

Introduction

It is well known that lactose leads to indigestion in adult dogs, depending on the quantity ingested. Large amounts of lactose clinically cause severe diarrhoea with watery faeces. The assumed reasons are an insufficient enzymatic splitting of lactose in the small intestine as well as a subsequent bacterial fermentation in the large intestine. Yet detailed information and quantitative analysis respectively are not available.

The objective of the present investigation was
a) to estimate the lactose flow from the small to the large intestine depending on the amount of lactose ingested
b) to determine the metabolites produced in the large intestine and excreted with the faeces respectively.
c) to investigate the effect of hydrolytic splitting of lactose on the compatibility of rations based on milk.

Material and methods

Pathogenesis of lactose induced diarrhoea

animals: Three adult dogs (7 to 9 years; 14 to 17.5 kg bodyweight; mean 15.4 kg) were housed individually in floor pens (1.7×1.7 m²) and stainless steel metabolism cages ($110 \times 70 \times 100$ cm³) during the collecting periods of colonic chyme and faeces respectively (restricted to a few hours per day); the room was air conditioned (mean 21 °C);

feeding: Dogs were maintained on a balanced canned diet and meal fed once a day (8:00 a.m.); crude nutrients (%; dry matter base): protein 39.7, fat 18.4, fibre 1.45, ash 12.7; 0.5% Cr_2O_3 (dry matter base) was homogeneously mixed to the diet prior to feeding;

lactose: Increasing amounts (0.3 to 4.5 g/kg bodyweight/day) were administered with the feed; each period (dosage) consisted of

267

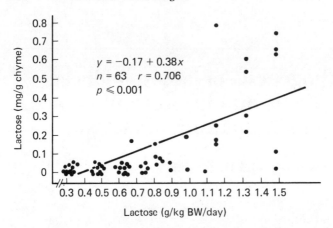

Fig. 1. Lactose concentration in the chyme (proximal colon) of adult dogs following the supply of lactose (minute to moderate amounts).

an adaptation (one week) and experimental period (one week each) respectively;

parameters: feed: intake; apparent digestibility of crude nutrients, minerals, and trace elements (results not presented)

chyme: dry matter, pH, concentration of lactose and lactic acid

faeces: frequency, apparence, dry matter, pH, lactic acid, osmolarity.

Prevention of lactose-induced diarrhoea by enzymatic splitting of lactose

To test the relevance of the results obtained under the conditions described above, practical experience was collected in dogs under experimental (laboratory hydrolysis of skimmed milk by β-galactosidase) and practical conditions (balanced commercial diet based on hydrolysed skimmed milk). The detailed experimental conditions and results are reported elsewhere (Hannes, 1983; Meyer *et al.*, 1984).

Results

Lactose concentration in the colonic chyme (proximal) following lactose intake (Fig. 1)

No or only minute amounts of lactose were detectable in the chyme 5 to 7 hours after feeding up to 1 g lactose/kg BW/day. Further increase led to considerable amounts as demonstrated by the strong correlation up to 1.5 g lactose/kg BW/day, when a maximum level of lactose could be observed in the chyme (~ 0.8 mg lactose/g chyme). The lactose concentration in the chyme rarely exceeded these values, which was even true after very high dosages of lactose (4.5 g/kg BW/day).

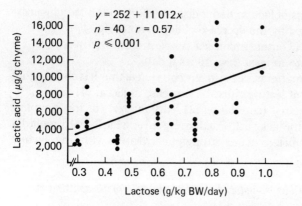

Fig. 2. Concentration of lactic acid in the chyme (proximal colon) of adult dogs following the supply of lactose (minute to moderate amounts).

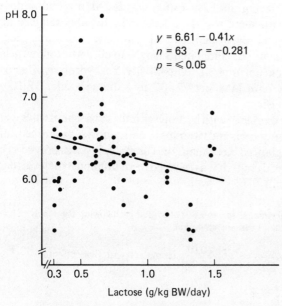

Fig. 3. pH in the chyme (proximal colon) of adult dogs following the supply of lactose (minute to moderate amounts).

Effects of lactose on chyme and faeces

The lactic acid concentration of the chyme was positively correlated with the lactose ingested. High values (15,000 µg lactic acid/g chyme) were already reached after the supply of 1 g lactose/kg BW/day (Fig. 2). Again they were not exceeded following extreme supply of lactose (data not presented). The pH tended to decrease with increasing lactose intake. However, there was a large variation (Fig. 3).

Increasing amounts of lactose had a distinct effect on the faecal consistency as demonstrated by the decrease of dry matter (Fig. 4).
The frequency of faecal output increased from less than twice a day ($\leqslant$ 1 g lactose/kg BW/day) to at least three times a day.

Faecal pH and dry matter were positively correlated; thus it is obvious that increasing amounts of lactose caused a decrease of faecal pH (Fig. 5).

The base of the findings (correlation pH – dry matter – lactose amount) could be specified for the lactic acid fraction (Fig. 6) and the faecal osmolarity (Fig. 7); both parameters were strongly correlated with the lactose consumption.

Prevention of lactose-induced diarrhoea by enzymatic splitting of lactose

Laboratory experience

A milk and flaked grain basic ration was tested in six adult dogs. 50 g skimmed milk equivalent to 2.46 g lactose/kg BW/day reduced dry matter of faeces in all dogs and caused diarrhoea in three dogs. Supplementation with various components (soybean oil, cattle tallow, gelatine, citric acid, agar) did not improve compatibility. Yet hydrolysis of lactose by β-galactosidase (1.7 ml Maxilact®/1000 ml skimmed milk; 37 °C, 4 hours) had a distinct effect.

Based on the good experience with hydrolysed milk, subsequent trials were conducted adding fat, minerals and vitamins to the hydrolysed skimmed milk in order to obtain a balanced diet. Again the compatibility was proved to be very good and lactic acid content was reduced to an unimportant level (data not presented in detail).

Fig. 4. Faecal dry matter (mean ± SD) of adult dogs following the supply of lactose (moderate to extreme amounts).

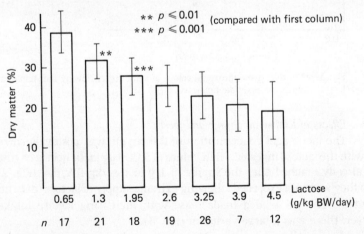

Table 1. *Faecal parameters following test meals with milk and hydrolysed milk in six adult dogs*

ration		number of obser-vations	faecal parameters dry matter %	pH	lactic acid (μg/g)
skimmed milk (50 g/kg BW/day)		8	20.3 ± 5.32	6.3 ± 0.35	683 ± 392
skimmed milk (100 g/kg BW/day)		7	17.5 ± 3.43	6.1 ± 0.44	849 ± 483
skimmed milk (50 g/kg BW/day)	+ flaked grain (once daily)	7	24.5 ± 2.02	6.3 ± 0.40	1.248 ± 974
	+ flaked grain (twice daily)	5	21.3 ± 4.13	6.1 ± 0.59	1,437 ± 889
skimmed milk hydrolysed (50 g/kg BW/day)	+ flaked grain	5	25.1 ± 4.13	6.5 ± 0.44	81.2 ± 65.3
skimmed milk hydrolysed (50 g/kg BW/day)	+ flaked grain	4	27.2 ± 5.49	6.7 ± 0.33	86.0 ± 30.6

Values expressed as mean ± SD.

Practical experience

Experience was collected with a commercial canned balanced diet for adult maintenance based on hydrolysed milk. Acceptability, compatibility and digestibility were very good, even in suckling and weaned pups (apparent digestibility (%): crude protein 89.8 ± 0.01, crude fat 98.4 ± 0.40; $n = 17$).

Discussion and conclusion

The negative effects of increasing amounts of lactose were demonstrated in the experimental series. A marked decrease of faecal consistency is clinically obvious after increasing amounts of lactose. All dogs metabolised up to 1 g lactose/kg BW/day by assumed endogenic hydrolysis (enzymes) which is in accordance with published experience. Yet an influence upon the chyme was already observed after small amounts of lactose, but without clinical relevance. Further increase of lactose led to considerable amounts of lactose and lactic acid in the chyme, yet already reaching a maximum level after moderate amounts (~ 1.5 g/kg BW/day). This indicates a steady state between flow, absorption, and distal transport. The milieu seemed to be affected towards superior conditions of replication for the lactobacilli compared with proteolytic bacteria. Any further metabolic increase of lactobacilli obviously caused severe osmotic diarrhoea. The negative influ-

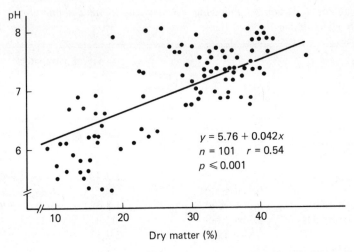

Fig. 5. Correlation between pH and dry matter in the faeces of adult dogs following the supply of lactose (moderate to extreme amounts).

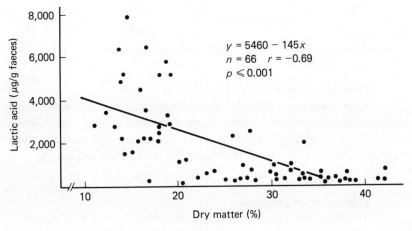

Fig. 6. Correlation between lactic acid and dry matter in the faeces of adult dogs following the supply of lactose (minute to moderate amounts).

ence on the milieu was also demonstrated by the apparent protein digestibility (data not presented).

The practical relevance led to the development of rations based on hydrolysed milk which satisfied in all respects (Biological value, acceptability, compatibility, digestibility).

Thus the following conclusions can be drawn.

1. The enzymatic splitting of lactose is limited in the small intestine of adult dogs which is in accordance with published data about the lactase activity.

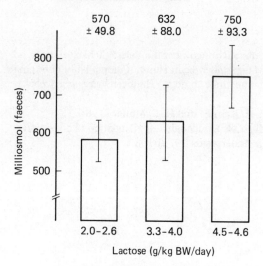

Fig. 7. Faecal osmolarity following the supply of lactose (moderate to extreme amounts). $n = 7$ for each column; correlation equation: $y = 402 + 72.3x$; $n = 21$; $r = 0.67$.

2. Subsequent bacterial fermentation of lactose in the large intestine leads to lactic acid production, increased osmolarity, decreased pH, and decreased absorption of water.
3. It is possible to increase the compatibility of milk by enzymatic hydrolysis of lactose.

Summary

The effect of increasing amounts of lactose (0.3 to 4.5 g/kg BW/day) was tested in three adult dogs (7 to 9 years; mean 15.4 kg BW). Dogs were maintained on a balanced canned diet. Lactose was supplied parallel to feeding (once a day).

Lactose had a dose-related effect on faecal consistency (decrease of dry matter), pH (decrease), osmolarity (increase), and lactic acid concentration (increase). Similar changes were true for the colonic chyme (pH and lactic acid). The colonic findings (including lactose concentration) already reached a maximum level following moderate amounts of lactose (~ 1.5 g/kg BW/day). Any further increase led to severe osmotic diarrhoea. The negative effect of lactose on the milieu of the chyme was assumed to alter the replication conditions for proteolytic bacteria.

The practical consequence of the limited enzymatic splitting of lactose in the small intestine of adult dogs as well as its subsequent bacterial fermentation in the large intestine led to the development of a ration based on hydrolysed milk which satisfied in all respects.

REFERENCES

Hannes, M. (1983) Untersuchunen über die Verträglichkeit nativer
und hydrolysierter Magermilch beim Hund. (Compatibility of original
and hydrolysed skimmed milk in dogs.) Hanover *Veterinary School*
Ph.D. thesis.
Meyer, H., E. Kienzle, M. Hannes und H.-C. Mundt (1984)
Aufgeschlossene Milch als Hundenahrung. (Nutrition of dogs with
hydrolysed milk.) Kleintierpraxis **29**, 301–8.

16

Salt intake, animal health and hypertension: should sleeping dogs lie?

A.R. MICHELL

Introduction

The nutrition of dogs and cats has been intensively studied over many years for commercial reasons and dogs have consistently been used as important experimental animals in research on electrolyte metabolism. Despite this activity, the dearth of data on the nutritional requirement for sodium in either species remains a recurrent theme in the scientific literature.

> 'The dog definitely needs calcium, phosphorus, sodium and chloride . . . Relatively little is known about the exact requirements for any of the minerals.'
>
> (Michaud & Elvehjem, 1944)
>
> 'Very little has been done on the requirement for Na and K in the dog'.
>
> (Shaw & Phillips, 1953)
>
> 'Inadequate data are available to set a minimal requirement for sodium'.
>
> (NRC, 1974)
>
> 'Salt in the dog's diet is one of the most variable, and in some ways troublesome of nutrients'.
>
> (Corbin, 1984)
>
> 'There is a paucity of data on the cat's qualitative and minimal quantitative requirements for minerals . . . Cats undoubtedly require sources of sodium and chloride but quantitative requirements of these elements have not been experimentally defined.'
>
> (MacDonald, Rogers & Morris, 1984).
>
> 'Specific recommendations on appropriate levels of dietary sodium are difficult to make at present.'
>
> (Gaskell, 1985).

This extraordinary lack of information about a major nutrient of the greatest physiological importance is not confined to the needs of carnivores but extends to herbivores (Michell, 1985a), pigs (Michell, 1985b) and humans, in fact to mammals in general (Michell, 1984a). Inadequate information, in itself, is not necessarily a justification for research; it could be

275

argued that, provided water is available, salt is generally sufficiently well tolerated by the healthy animal that it seems safe to provide a liberal excess without worrying about requirement. Unfortunately it is probable that salt causes, and extremely likely that it exacerbates many cases of hypertension, a widespread and serious clinical condition in humans (Michell, 1984a). Concepts of adequate and excessive sodium intake in mammals therefore become important in assessing human salt consumption. Information about dogs is particularly important since they too can develop clinical hypertension.

Physiological basis of requirement

Sodium is the osmotic skeleton of the extracellular fluid and the main effect of changes in body sodium is to alter the volume of extracellular fluid and, therefore, plasma (Michell, 1985a). Sodium is ejected from cells by the energy consuming 'sodium pump' (Na–K ATPase) which maintains concentration gradients which are used, directly or indirectly, to facilitate the transport of other ions and essential substances, as well as establishing the electrical potentials which underlie excitability. The concentration of extracellular sodium is stabilised by the appropriate retention and excretion of water, i.e. by thirst and the effect of antidiuretic hormone on renal function, both of which react sensitively to small changes in sodium concentration. Sodium depletion leads to changes in plasma sodium only indirectly, once contraction of extracellular volume becomes sufficiently severe to intensify thirst and renal water retention and so override the defence of sodium concentration.

In an adult animal the main determinant of the maintenance requirement for sodium is the obligatory loss, which in most species is remarkably low; as Von Bunge (1902) perceived, 'the constant supply of salt in considerable quantities is not a necessity for the adult'. The major route of sodium excretion in dogs is urine, with only 10% or less of intake lost in faeces (though higher values can occur, particularly on high fibre diets) (Davis et al., 1959; Berger, 1960; Smith et al., 1964; Meyer, 1975). Even in ruminants, with their extensive faecal sodium excretion, dietary sodium is almost 100% available for absorption (Michell, 1985c). It is vital that a maintenance requirement is adequately defined otherwise the superimposed requirements for growth, pregnancy and lactation must themselves be suspect. This is the case for the published sodium requirements for farm animals, which tend to be inflated by over-generous estimates of maintenance requirement (Michell, 1985a,b).

The requirement is the minimum intake to provide for health and production (growth, pregnancy, lactation) in typical individuals. Feeding recommendations may be more liberal to allow for aberrant individuals, unforeseen stresses, or increased growth or production through pharmacological effects or improved palatability (Michell, 1985b)[1]. The quantitative basis of a requirement arises from estimates of all available calls on ingested sodium (factorial experiments) or assessment of criteria of health and growth on different levels of intake (empirical experiments). Seemingly straightforward to perform, neither type of experiment is simple to interpret (Michell, 1984a).

Obligatory losses can only be established by allowing sufficient time for adaptation to low levels of sodium intake; the transfer from a high to a low sodium diet is followed by some days of negative sodium balance, due to the gradual excretion of excess sodium as the body pool contracts from that maintained at higher levels of intake (Strauss *et al.*, 1958). Criteria of health (seldom easy to define) are often taken to include the absence of adrenally mediated salt conservation, i.e. the presence of plentiful sodium in urine, faeces and saliva. This reflects an entirely arbitrary assumption that adaptation to high salt is 'normal' and adaptation to low salt is 'abnormal' (Michell, 1984a). The same fallacy underlies the use of criteria, such as extracellular volume, which change with sodium intake; the fact that extracellular volume falls with a reduced salt intake is not necessarily any more sinister than the fact that it rises with an increase. Thus in sheep adaptive changes in salivary, urinary and faecal Na/K ratios can occur without evidence of harmful effects (Sinclair & Jones, 1968). Indeed during lactation both ewes and lambs are apparently unharmed by sodium intakes less than a quarter of those needed to maintain a high salivary Na/K ratio (Morris & Petersen, 1975).

The generosity of published sodium requirements seems to derive, therefore, from a philosophy that it is 'better to have a little in reserve', which is no basis for establishing a nutrient requirement: there has to be objective evidence that lower levels are harmful, either normally or under stress. It is not sufficient, for example, to assume that high salt intakes protect against losses in diarrhoea; such losses are potentially very severe and there needs to be evidence that high intakes significantly increase resistance to their effects. Such caveats are necessary because throughout cultural and scientific history the general belief has been that salt is beneficial and even now adherence to this belief is sometimes more akin to religion than science (Michell, 1984a).

While it might appear that the kidney does more reabsorptive work in responding to a low salt intake, this ignores the fundamental determinant of renal sodium reabsorption, which is not external sodium balance but the need to retrieve the huge amounts of sodium filtered at the glomerulus. An increased glomerular filtration rate thus increases the amount of sodium reabsorption which constitutes the main energy expenditure of the kidney (Pitts, 1974; Katz, 1982; de Wardener, 1985; Mandel, 1986). Despite the tight regulation of glomerular filtration rate (GFR), one of the few factors which can potentially increase it, and hence the reabsorptive work of the kidney, is expansion of extracellular fluid volume (ECF), a likely effect of increased salt intake (Katz & Genant, 1971; Pitts, 1974; Bell & Navar, 1982; Roos et al., 1985). Nevertheless in the short term, at least, dogs seem able to accommodate changes in sodium intake and ECF volume without changes in GFR (Reinhardt & Behrenbeck, 1967).

Pure carnivores (cats) and predominant carnivores (dogs) inevitably have substantial amounts of sodium in the animal tissues of their natural diets. This is an incidental effect, however, and does not necessarily indicate a need, any more than 'salt appetite' in normal herbivores defines an actual need (Michell, 1985b). Salt appetite is not prominent in carnivores (Denton, 1982), indeed some authors claim it is not even stimulated by artificial sodium depletion (Fregly, 1980) though others claim that it is (Ramsay & Reid, 1979). Nevertheless the acceptability of sodium in milk can be influenced by gastric sodium receptors in dogs with high intragastric concentrations leading to rejection of milk with increased sodium content (Chernigovsky, 1964).

Is 'low' sodium intake normal?

Hollenberg (1980, 1982) has argued that the theoretical set point around which human sodium balance fluctuates is the amount of sodium in the body when in balance on zero intake, i.e. with sodium being neither ingested nor excreted. In the absence of extrarenal losses most people are usually in a state of surfeit resulting from chronic excess of dietary salt. While rejecting the idea of a single set point, Bonventre & Leaf (1982) concede that the steady-state characteristics of sodium regulation depend on the underlying level of intake. Similarly, Walser (1985) emphasises that both body content and therefore excretion of sodium are crucially determined by the level of intake. A rigorous definition of requirement should thus incorporate the lowest steady state (and intake) which is safely compatible with good health – there is no such thing as a 'normal' body sodium, per se, since it depends on dietary intake.

In this context, recent work by Brensilver et al. (1985) in rats is extremely pertinent. They found that weight gain was slowed (though growth and positive sodium balance continued) if sodium intake was lower than 250 μmol/day. Grunert et al. (1950) had shown that normal rats required 22 μmol/g diet (0.05% Na) for the first 5 weeks of life, 13 μmol/g subsequently; with food intake at 10g/d in the first 5 weeks, 20 g/d subsequently, these estimates are very similar (220, 260 μmol/d) and close to Brensilver's minimum value. Again, Toal & Leenen (1983) showed that dietary sodium for rats could be reduced to 0.05% without affecting growth essentially confirming the estimate by Louis, Tabei & Spector (1971) of 0.1% NaCl (0.04% Na). The data of Ganguli, Smith & Hanson (1970) and the NRC (1978) suggest that even the demands of pregnancy and lactation can be accommodated by 0.03–0.06% dietary sodium, despite the fact that rats, unlike most species, tend to grow rapidly throughout the period during which they are likely to be studied thus accumulating sodium and remaining in positive sodium balance (Mohring & Mohring, 1972). All these estimates of dietary sodium fall close together but the perspective added by Brensilver *et al.* (1985) is that they only slightly exceed the intake at which urine becomes virtually free of sodium. There is, therefore, nothing inherently 'harmful' about intakes which lead to very low sodium excretion, indeed they may be close to the requirement.

Is 'low' sodium intake harmful?

It is notoriously difficult to induce clinical signs of sodium deficiency by dietary means alone. Nevertheless, with the increasing pressure to reduce human salt intake, fears have been expressed about more subtle adverse effects (Swales, 1980; Laragh & Pecker, 1983; Brown *et al.*, 1984), though some are fairly implausible (Michell, 1980; 1984b; de Wardener, 1984).

Perhaps the most worrying data appear to be those showing that salt restriction reduced the resistance of rats to haemorrhage (Gothberg *et al.*, 1983). It must be emphasised, however, that this experiment concerned intakes below requirement (30–50% of maintenance) not reductions of *excess* intake *above* requirement. Even greater reductions (to 20–30% of requirement) can induce hypertension in rats though only those with one kidney removed (Seymour *et al.*, 1980), an extremely artificial situation. Similarly, rats with malignant hypertension artificially induced by renal artery stenosis showed worse hypertension on a low salt diet (Mohring *et al.*, 1976) which is not particularly surprising since this unusual form of hypertension is sustained by renin secretion which in turn increases with low sodium intake. Others have failed to confirm this finding (Rojo-Ortega *et al.*, 1979) and clinical management of humans with renal hypertension currently favours mild salt restriction (Ledingham, 1985). High salt intake protects against acute renal failure (Bidani, Churchill & Fleischmann, 1979) but insurance against such a rare catastrophe is scarcely a justification for a constant high intake (Michell, 1980).

An interesting series of experiments by Barger and colleagues used dogs to examine the effect of dietary salt on resistance to blood loss. Dogs on low sodium intakes (10–15 mmol/d; 0.4 mmol/kg/d) showed greater protective rise of arterial pressure after constriction of the renal artery (Fray, Johnson & Barger, 1977) and a greater release of renin/angiotensin after haemorrhage. The renin/angiotensin rise, in particular, provided an important defence against the effects of haemorrhage and this response was greater in 'low salt' dogs even though their fall in blood pressure immediately after haemorrhage was less than normal dogs (Kopelman *et al.*, 1983). On the other hand low sodium intake did reduce the response to carotid occlusion (Rocchini, Cast & Barger, 1977) something which could contribute to the prominence of postural hypotension as a sign of sodium depletion in humans. Nevertheless the overall implication was that although dogs on low sodium intake were more dependent on the renin/angiotensin system for protection against haemorrhage their defence was no less effective.

It is, of course, possible for the adverse effects of excessively low sodium intake to be mediated by interactions with ions other than sodium itself (e.g. through effects on their absorption and excretion) or by effects on the gut flora (Michell, 1984). It is also arguable that since sodium transport accounts for as much as 20% of resting energy expenditure (Milligan & McBride, 1985) the enhanced sodium transport associated with adrenally mediated salt conservation is an unfavourable aspect of adaptation to low sodium intake. Nevertheless, both humans and sheep can manage on less than 10 mmol/d, even during the increased demands associated with pregnancy, growth or (in humans) sweating (Frohlich, Messerli, 1982; Hunt, 1983; Michell et al., 1988). This represents intakes below 0.2 mmol/kg/d and in considering the sodium requirement of carnivores, a supposition that their maintenance requirement is any higher implies that their ability to reduce urinary losses is inferior or that they have a major source of obligatory loss other than faeces.

Sodium requirement in dogs and cats

The maintenance requirement of most species has been exaggerated by a number of factors, notably confusion of obligatory losses with those due to prior excesses of salt and scaling up of obligatory loss with bodyweight (Michell, 1985b). Thus a 70 kg sheep is probably just as able to reduce its sodium excretion to 1 mmol/d as a 20 kg sheep; the crucial factors are minimum concentration achieved in urine and faeces, the minimum net output of both, and the accuracy with which these can be measured at such low levels. Dogs, however, are unusual in the wide variety of sizes which have developed within the species as a result of selective breeding and over such a large range obligatory losses will clearly vary with bodyweight; those of a Chihuahua must be less than those of a Great Dane. How obligatory loss should be extrapolated between extremes of bodyweight, i.e. to what power of weight they should be related, is a matter of conjecture and part of the interesting general problem of allometry in dogs (see Chapter 6).

Cats

The 1978 edition of Nutrient Requirements of Cats (NRC, 1978b) avoids the question of sodium requirement by substituting the observation that recommended levels of sodium range from 40–60 mg/100 g of dry diet and should provide 1500 mg (26 mmol) of salt per day for growing kittens. Subsequent publications provide little further information (MacDonald et al., 1984) with Kronfeld (1983) describing feline mineral requirements as 'orthodox'. They may equally be 'catholic' since the suggestion by Brewer

(1983) that cats require 0.88% salt in the diet has a remarkable pedigree, namely a chain of citations originating in a 1941 paper on poultry. The revised NRC publication on nutrient requirements of cats (NRC, 1986) shows little change[2]. Whatever the minimal requirement in cats, however, there may be good reason to provide them with additional salt in order to maintain their water turnover on dry diets (Burger, 1979).

Dogs

With dogs, too, the tendency has been to recommend a dietary salt content of around 1% in the dry matter (NRC, 1974) rather than define the actual requirement. Meyer (1984) in summarising a substantial body of data from his group suggests that the sodium requirement of dogs is 4.6 mmol/kg/d during pregnancy, 5.4 mmol/kg/d during lactation but this includes 2.6 mmol/kg/d against endogenous losses (maintenance) and an assumption that only 75% of dietary sodium is utilised. The actual availability of sodium is probably well above 90% since only 10% of ingested sodium appears in faeces. The figure for endogenous loss, if it is interpreted as obligatory loss, is remarkably high compared with other mammals. Liebetseder (1980) similarly suggests a requirement of 3.8 mmol/kg/d, 0.4% sodium in the dry matter (1% salt).

The requirement probably exceeds 0.16 mmol/kg/d since Drochner, Kersten & Meyer (1976) saw restlessness, polydipsia, polyuria and a rise in PCV at this intake. Morris, Patton & Teeter (1976), however, saw no adverse effects at 0.2 mmol/kg/d. As early as 1933, Loeb *et al.* observed that 14 kg dogs remained in positive sodium balance on 10 mmol daily, i.e. 0.7 mmol/kg/d. On the other hand Gupta *et al.* (1981) and Linden *et al* (1982) found that 0.55 mmol/kg/d produced lower plasma volumes than intakes of 2.6 mmol/kg/d or more. This does not necessarily imply an adverse effect, however, nor do the increases in plasma renin noted in three studies on intakes below 1 mmol/kg/d (NRC, 1985). The likelihood is that the requirement lies in the range 0.2–0.7 mmol/kg/d equal to about 0.03 to 0.09% sodium in diet dry matter, i.e. well below the traditional '1% salt'.

Data on obligatory loss in dogs are scarce. Field *et al.* (1954) found an obligatory faecal loss of 0.2–0.5 mmol/d. The weights, unfortunately, are unstated, but since the dogs received 1250 calories daily they were probably around 20 kg, suggesting obligatory faecal losses of 0.01–0.03 mmol/kg/d. Meyer (1975), however, suggests that faecal losses are of the order of 0.65 mmol/kg/d and arrives at a total obligatory loss of 1.1–1.3 mmol/kg/d including a substantial allowance for dermal loss (0.2–0.4 mmol/kg/d). On the other hand the data of Drochner *et al.* (1976) indicate that on low sodium intakes faecal losses fell to 0.07 mmol/kg/d and urinary losses fell to 0.09–0.13 mmol/kg/d (a figure which probably includes some residual excretion of excess salt ingested during the control period on 3 mmol/kg/d). If we assume that the dogs came to equilibrium after excreting one day's excess intake (Strauss *et al.*, 1958) and that requirement is about 0.5 mmol/kg/d (NRC, 1985), the excess intake is 2.5 mmol/kg/d which, averaged over the 62 days of measurement, is 0.04 mmol/kg/d. The obligatory urinary loss may thus be closer to 0.07 mmol/kg/d and the combined obligatory loss in urine and faeces would be 0.14 mmol/kg/d. Similar considerations apply to the combined loss in urine and faeces of 0.14 mmol/kg/d during two weeks on an intake of 0.37 mmol/kg/d observed by Hamlin *et al.* (1964). Since the control intakes were around 1.8 mmol/kg/d, the excretion of one day's excess would amount to 1.3 mmol/kg/d which, averaged over 14 days of measurement, would be 0.09 mmol/kg/d. This leaves an obligatory loss of 0.05 mmol/kg/d – very close to values observed in sheep (Michell *et al.*, 1988). The obligatory loss in urine and faeces is thus unlikely to exceed 0.14 mmol/kg/d and may be below 0.1 mmol/kg/d. Dermal losses are negligible both at rest and after exercise (Meyer, 1978), as expected. An earlier suggestion (Meyer, 1975) that they were substantial essentially equated them with unmeasured losses, many of which are likely to arise from experimental error in balance studies. Limited data suggest that there could be additional losses in hair growth, up to 0.2 mmol/kg/d (Meyer, 1978).

The most recent *Nutrient Requirements of Dogs* (NRC, 1985) suggests an adult maintenance requirement of of 0.4–0.5 mmol/kg/d (equal to 0.06% sodium in diet dry matter) with 1.3 mmol/kg/d recommended for growth. This is a welcome reduction; the true maintenance requirement remains uncertain, however, and might well be lower.

Is excess salt harmful: hypertension?

The immediate effect of excess salt is to increase water consumption and, provided that there is no impairment of drinking, elevation of plasma sodium concentration is avoided and thereby the signs of toxicity. The most important risk attributed to excess salt is a long-term effect on human blood pressure. The association between salt and hypertension certainly remains controversial (Scribner, 1983; Brown *et al.*, 1984*b*; Nicholls, 1984) but even if the verdict is 'not proven' the evidence is strong enough to call into question casual acceptance of dietary salt levels grossly exceeding require-ment (Michell, 1984*a*). It is likely that intake in Western populations exceeds requirement 10 to 100-fold, which in some species would span the range between requirement and toxicity for nutrients such as vitamin A, vitamin D, iron, zinc, copper and manganese (Suttle, 1983; Howell, 1983; Michell, 1984*a*). Our paleolithic ancestors probably managed on 30 mmol/d (Eaton & Kenner, 1985), well below the level likely to influence hypertension (Houston, 1986).

Perhaps the strongest reasons for moderating human salt consumption are that differences in sodium intake have small but definite effects on blood pressure in infants even during their first six months (Hofman, Hazebroek & Valkenburg, 1983) and that among 'low-salt' cultures there is none with a significant incidence of hypertension or a marked age-related rise in blood pressure (which high-salt populations accept as normal). Among high-salt cultures there are, supposedly, four exceptions to the greater incidence of hypertension. Whyte (1958) found that natives of New Guinea whose salt appetite was satisfied had no higher blood pressure than those who were salt hungry, i.e. had a lower salt intake; however this evidence is flimsy. The former group was assumed to have the higher salt intake but intakes were not measured. Malhotra (1970) observed that Northern Indians on 12.5 g of salt per day had less hypertension than Southerners on 8 g/day. Neither intake is low, however, and the data show that both groups have an age-related rise in arterial pressure (which, in the case of diastolic pressure, is faster in Northerners.) Henry & Cassel (1969) however include a report of Thai farmers without an age-related rise in blood pressure and with a low incidence of hypertension despite sodium intakes of 340 mmol/d (twice the Western average). There are similar data in a study of a Russian community (Denton, 1982).

Whatever the evidence in humans, however, the crucial issue for the present discussion is whether hypertension is common in dogs or cats and whether salt is a contributory factor?

Hypertension in dogs and cats; comparative aspects

The majority of human hypertension is 'essential' or primary with only about 5% attributable to renal disease. Nevertheless renal disease, whether acute or chronic, is frequently associated with hypertension and is usually the cause rather than the result (Alfrey, 1976; McDonald, 1976; Sullivan & Johnson, 1983; de Wardener, 1985). The main mechanisms underlying hypertension in renal disease are salt retention or excessive renin release. It is sometimes said that glomerular disease rather than interstitial disease or pyelonephritis tends to cause hypertension (MacGregor, 1977; Ledingham, 1985) but others would dispute this (Brod, 1978), and more recently it has been appreciated that medullary damage could reduce the production of antihypertensive lipids including prostaglandins (Muirhead, 1983). Both acute and chronic hypertension produce characteristic retinopathies though vision is not necessarily impaired (de Wardener, 1985).

There have been few studies of the incidence of hypertension in dogs and none in cats (Hahn & Garner, 1977; Kittleson & Olivier, 1983; Bovee, 1984). Nevertheless it seems clear that hypertension is much less common in dogs than humans. This is not just because blood pressure is seldom measured in practice; if hypertension were common the ocular effects (Rubin, 1976; Magrane, 1977; Gelatt, 1981) should have been detected more frequently. Indeed, the most common presenting sign of canine hypertension is blindness. Spangler, Gribble & Weiser (1977) reported a less than 1% occurrence of hypertension among 1,000 young dogs; of nine affected animals six appeared to have essential hypertension. Blanchard *et al.* (1979) reported essential hypertension in a 5-year old Husky. Where hypertension does occur it appears most likely to accompany renal disease (Anderson & Fisher, 1968; Valtonen & Oksanen, 1972; Weiser, Spangler & Gribble, 1977) though it has also been associated with diabetes mellitus (secondary renal effects) and hyperadrenocorticalism, as in man. Familial glomerular disease causes hypertension in young Cocker Spaniels (Steward & Macdougall, 1984). In a recent study, dogs with renal failure had higher diastolic but not systolic pressures (Coulter & Keith, 1984). Oedema is more definitely associated with salt retention but it is also an unusual consequence of canine renal disease (Michell, 1983).

Dogs are perfectly capable of developing hypertension as a result of experimental renal or renovascular damage or induced stress and they have been repeatedly used as hypertensive models (Selkurt, Abel, Edwards & Nunn, 1973; Freeman, Davis & Watkins, 1977; Yinchang, Guiyun & Zhong, 1980). Nevertheless the clinical view suggests that they are either resistant to salt retention, hypertension or both and other evidence lends support. Swales (1981) notes that hypertension is harder to induce experimentally in dogs than rabbits and Ladd & Raisz (1949) found the dogs could resist very high levels of salt intake without weight gain or oedema and their renal function adjusted much more rapidly than in humans. Similarly Wilhelmj, Waldmann & McGuire (1951) observed that dogs on very high sodium intake (40 mmol/kg/d; 20 times even Western consumption in man) only developed transient hypertension and recovered beyond 3 weeks. Vogel (1966) produced hypertension with massive salt loading in dogs (34 mmol/kg/d) but his experiments did not continue beyond 26 days so it is impossible to say whether pressures would have returned to normal subsequently. Salt loading certainly heightens vascular sensitivity to the hypertensive effects of catecholamines but only at extreme intakes, i.e. 6–12 mmol/kg/d (Cowley & Lohmeir, 1979).

In summary, the fact that chronic renal disease is fairly common in dogs (Bovee, 1984) while hypertension appears to be rather unusual suggests that hypertension is less of a risk than in humans. Moreover the dog appears to be more resistant to salt loading and salt-induced hypertension. There seems little reason, therefore, for salt restriction in canine renal disease particularly since sodium depletion can cause a rapid decline in renal function. There is, however, one caveat. Brenner (1985) documents recent evidence that compensatory hyperfiltration ultimately becomes a cause of renal damage in chronic renal failure. Data from Polzin *et al.* (1984) show that on 1.6 mmol/kg/d there was some recovery of renal function in experimentally-induced canine chronic renal failure whereas on high salt intakes (8–9 mmol/kg/d) GFR did not recover.

Conclusion

Although the usual levels of salt fed to dogs and cats are probably well above requirement, it is unlikely that any harm results, provided their kidneys are healthy and water is accessible. Nevertheless it is important to establish what the requirements are, for three reasons.

(a) In physiology: renal function is greatly affected by salt intake yet dogs in most experiments are maintained on a chronic excess or, worse still, their intake is unspecified thus making the experiments unrepeatable.

(b) In veterinary medicine: we can scarcely 'restrict' salt, e.g. in cardiac disease without knowing what constitutes an adequate or excessive intake.

(c) In comparative medicine: it is important that over-generous estimates of mammalian sodium requirement are modified in order to appreciate the real excess in many 'customary' intakes. This is particularly important in human hypertension but the very fact that it seems less so in canine hypertension may offer important opportunities to understand the basic causes of this condition.

Acknowledgements

I thank Mrs E. Michell for translation from German publications and Mrs R. Forster for preparing the text.

Notes

[1] A more satisfactory approach to the expression of nutrient requirement, where feasible, is to define the statistical likelihood that individual intake will be adequate within particular ranges of dietary intake (Beaton, 1988).

[2] Although levels of sodium in the range from 400–600 mg/kg diet have been fed successfully, little evidence exists for determination of a minimum requirement.

Note added in proof

The literature review was completed in mid-1985 and, with one or two exceptions, subsequent revisions have been purely editorial.

REFERENCES

Alfrey, A.C. (1976). Chronic renal failure. In: *Renal and Electrolyte Disorders*. Ed. R.W. Schrier. pp. 319–48. Boston: Little Brown.

Anderson, L.J. & Fisher, E.W. (1968). The blood pressure in canine interstitial nephritis. *Research in Veterinary Science*, **9**, 304–13.

Beaton, G.H. (1988). Nutrient requirements and population data. *Proceedings of the Nutrition Society*, **47**, 63–78.

Bell, P.D. & Navar, L.G. (1982). Intrarenal feedback control of GFR. *Seminars in Nephrology*, **2**, 289–301.

Berger, E.Y. (1960). Intestinal Absorption and Excretion. Ch. 8 in *Mineral Metabolism*, Vol. 1, Part A; Ed. C.L. Comar & Bronner. pp. 263-6. New York: Academic.

Bidani, A., Churchill, P. & Fleischmann, L. (1979). Sodium chloride-induced protection in nephrotoxic acute renal failure: independence from renin. *Kidney International*, 16, 481–90.

Blanchard, G.L., Eyster, G.E., Carrig, C.B., Rouner, D.R., Bailie, M.D. & Padgett, G.A. (1979). Essential hypertension in a Siberian Husky Dog. *Federation Proceedings*, 38, 1350.

Bonventre, J.V. & Leaf, A. (1982). Sodium homeostasis: steady states without a set point. *Kidney International*, 21, 880–5.

Bovee, K.C. (1984). *Canine Nephrology*. pp. 463, 517, 597, 632. USA: Harvard.

Brenner, B.M. (1985). Nephron adaptation to renal injury or ablation. *American Journal of Physiology*, 249F, 324–37.

Brensilver, J.M., Daniels, F.H., Lefavour, G.S., Malseak, R.M., Lorch, J.A., Ponte, M.L. & Cortell, S. (1985). Effect of variations in dietary sodium intake on sodium excretion in mature dogs. *Kidney International*, 27, 497–502.

Brewer, N.R. (1983). Nutrition of the cat. *Journal of the American Veterinary Medical Association*, 180, 1179–82.

Brod, J. (1978). Hypertension and renal parenchymal disease: mechanisms and management. In: *Hypertension: Mechanisms, Diagnosis and Treatment*. Eds. G. Onesti & A.N. Brest. pp. 137–64. Philadelphia: E.A. Davis.

Brown, J.J., Lever, A.F., Robertson, J.I.S., Semple, P.F., Bing, F.R., Heagerty, A.M., Swales, J.D., Thurston, H., Ledingham, J.G.G., Laragh, J.H., Hansson, L., Nicholls, M.G. & Espiner, A.E. (1984*a*). Salt and hypertension. *Lancet*, 2, 456.

Brown, J.J., Lever, A.F., Robertson, J.I.S. & Semple, P.F. (1984*b*). Should dietary sodium be reduced; the sceptics' position. *Quarterly Journal of Medicine*, 53, 427–38.

Burger, I. (1979). Water balance in the dog and cat. *Pedigree Digest*, 6, 10–11.

Chernigovsky, V.N. (1964). Experimental study of role of interoreceptive signals in the alimentary behaviour of animals. In: *Brain and Behaviour*, Vol. 2, Ed. M.A.B. Brazier. Washington DC: American Institute of Biological Sciences.

Corbin, J. (1984). Feeding of Dogs and Cats. In: *Livestock Feeds and Feeding*, Ed. D.C. Church. pp. 446–59. Corvallis, Oregon: O. & B. Books.

Coulter, D.B. & Keith, J.C. (1984). Blood pressure obtained by indirect measurement in conscious dogs. *Journal of the American Veterinary Medical Association*, 184, 1375–8.

Cowley, A.W. & Lohmeir, T.E. (1979). Changes in renal vascular sensitivity and arterial pressure associated with sodium intake during long-term intrarenal norepinephrine infusion in dogs. *Hypertension*, 1, 549–58.

Davis, J.O., Ball, W.C., Bahn, R.C. & Goodkind, M.J. (1959). Relationship of adrenocortical and anterior pituitary function to fecal excretion of sodium and potassium. *American Journal of Physiology*, 196, 149–52.

Denton, D.A. (1982). *The Hunger for Salt.* pp. 54, 571, Berlin: Springer-Verlag.

de Wardener, H.E. (1984). Salt and hypertension. *Lancet,* **2**, 688.

de Wardener, H.E. (1985). *The Kidney* (5th edn.) pp. 74–5; 208–10. Edinburgh: Churchill Livingstone.

Drochner, W., Kersten, U. & Meyer, H. (1976). Auswirkungen einer Na-Depletion und anschliessenden Repletion auf den Stoffwechsel von Beaglehunden. *Zentralblatt fur Veterinarmedizin,* **23A**, 739–53.

Eaton, S.B. and Konner, M. (1985). Paleolithic nutrition. *New England Journal of Medicine,* **312**, 283–9.

Field, H., Dailey, R.E., Boyd, R.S. & Swell, L. (1954). Effect of dietary sodium on electrolyte composition of the contents of the terminal ileum. *American Journal of Physiology,* **179**, 477–81.

Fray, J.C.S., Johnson, M.D. & Barger, C. (1977). Renin release and pressor response to renal arterial hypotension: effect of dietary sodium. *American Journal of Physiology,* **233**, H191–5.

Freeman, R.H., Davis, J.O. & Watkins, B.E. (1977). Development of chronic perinephritic hypertension in dogs without volume expansion. *American Journal of Physiology,* **233**, F278–81.

Fregly, M.J. (1980). On the spontaneous NaCl intake by dogs. In: *Biological and Behavioural Aspects of NaCl Intake,* ed. M. Kare, R. Bernard & M.J. Fregly. New York: Academic.

Frolich, E.D. & Messerli, F.H. (1982). Sodium and hypertension. In: *Sodium: Its Biological Significance.* Ed. S. Papper. Florida: C.R.C. Press.

Ganguli, M.C., Smith, J.D. & Hanson, L.E. (1970). Sodium metabolism and its requirements during reproduction in female rats. *Journal of Nutrition,* **99**, 225–34.

Gaskell, C.J. (1985). Nutrition in diseases of the urinary tract in the dog and cat. *Veterinary Annual,* **25**, 383–91. Bristol: Wright Scientechnica.

Gelatt, K.N. (1981). *Textbook of Veterinary Ophthalmology.* pp. 507, 716. Philadelphia: Lea & Febiger.

Gothberg, G., Lundin, S., Aurell, M. & Folkow, B. (1983). Response to slow, graded bleeding in salt-depleted rats. *Journal of Hypertension,* **1**, 24–6.

Grunert, R.R., Meyer, J.H. & Phillips, P.H. (1950). The sodium and potassium requirements of the rat for growth. *Journal of Nutrition,* **42**, 609–18.

Gupta, B.N., Linden, R.J., Mary, D.A.S.G. & Weatherill, D. (1981). The influence of high and low sodium intake on blood volume in the dog. *Quarterly Journal of Experimental Physiology,* **66**, 117–28.

Hahn, A.W. and Garner, H.E. (1977). Indirect measurement of blood pressure in animals. *Advances in Veterinary Science and Comparative Medicine,* **21**, 1–16.

Hamlin, R.L., Smith, R.C., Smith, C.R. & Powers, T.E. (1964). Effects of a controlled electrolyte diet, low in sodium, on healthy dogs. *Veterinary Medicine/Small Animal Clinician,* **59**, 748–51.

Henry, J.P. & Cassel, J.C. (1969). Psychosocial factors in essential hypertension: recent epidemiological and animal experimental evidence. *American Journal of Epidemiology*, **90**, 171–200.

Hofman, A., Hazebroek, A. & Valkenburg, H.A. (1983). A randomized trial of sodium intake and blood pressure in newborn infants. *Journal of the American Medical Association*, **250**, 370–3.

Hollenberg, N.K. (1980). Set point for sodium homeostasis: surfeit, deficit and their implications. *Kidney International*, **17**, 423–9.

Hollenberg, N.K. (1982). Surfeit, deficit and the set point for sodium homeostasis. *Kidney International*, **21**, 883–4.

Houston, M.C. (1986). Sodium and hypertension. *Archives of Internal Medicine*, **146**, 179–85.

Howell, J. McC. (1983). Toxicity problems associated with trace elements in domestic animals. In: *Trace Elements in Animal Production and Veterinary Practice*, ed. N.F. Suttle, R.G. Gunn, W.M. Allen, K.A. Linklater and G. Weiner, pp. 107–17. Occasional Publications No. 7 – British Society of Animal Production.

Hunt, J.C. (1983). Sodium intake and hypertension: a cause for concern. *Annals of Internal Medicine*, **98(2)**, 724–8.

Katz, A.I. (1982). Renal Na-K-ATPase: its role in tubular sodium and potassium transport. *American Journal of Physiology*, **242**, F207–19.

Katz, A.I. & Genant, H.K. (1971). Effect of extracellular volume expansion on renal cortical and medullary Na-K-ATPase. *Pflugers Archives*, **330**, 136–48.

Kittleson, M.D. & Olivier, N.B. (1983). Measurement of systemic arterial blood pressure. *Veterinary Clinics of North America: Small Animal Practice*, **13**, 321–36.

Kopelman, R.I., Dzau, V.J., Shimabukuro, S. & Barger, C.A. (1983). Compensatory response to hemorrhage in conscious dogs on normal and low salt intake. *American Journal of Physiology*, **244**, H351–6.

Kronfeld, D.S. (1983). Feeding cats and feline malnutrition. *Compendium of Continuing Education for Practising Veterinarians*, **5**, 419–20.

Ladd, M. & Raisz, L.G. (1949). Response of the normal dog to dietary sodium chloride. *American Journal of Physiology*, **159**, 149–52.

Laragh, J.H. & Pecker, M.S. (1983). Dietary salt and hypertension: some myths, hopes and truths. *Annals of Internal Medicine*, **89**, 735–42.

Ledingham, J.M. (1985). Hypertension and the kidney. In: *Postgraduate Nephrology*. Ed. F. Marsh. pp. 325–53. London: Heinemann Medical.

Liebetseder, J. (1980). *The Nutrition of the Dog*. pp. 11–14. Roche Information Service.

Linden, R.J., Mary, D.A.S.G. & Nickalls, R. (1982). The relationship between blood volume and dietary salt in the dog. *Journal of Physiology*, **330**, 66P.

Loeb, R.F., Atchley, D.W., Benedict, E.M. & Celod, J. (1933). Electrolyte balance studies in adrenalectomised dogs with particular reference to the excretion of sodium. *Journal of Experimental Medicine*, **57**, 775–92.

Louis, W.J., Tabei, R. & Spector, S. (1971). Effects of sodium intake on inherited hypertension in the rat. *Lancet,* **2**, 1283–6.

McDonald, K.M. (1976). The kidney in hypertension. In: *Renal and Electrolyte Disorders.* Ed. R.W. Schrier. pp. 263–88. Boston: Little Brown.

Macdonald, M.L., Rogers, Q.R. & Morris, J.G. (1984). Nutrition of the domestic cat, a mammalian carnivore. *Annual Review of Nutrition,* **4**, 521–62.

MacGregor, G.A. (1977). High blood pressure and renal disease. *British Medical Journal,* **2**, 624.

Magrane, W.G. (1977). *Canine Ophthalmology.* p. 261. Philadelphia: Lea & Febiger.

Malhotra, S.L. (1970). Dietary factors causing hypertension in India. *American Journal of Nutrition,* **23**, 1353–63.

Mandel, L.J. (1986). Primary active sodium transport, oxygen consumption and ATP. *Kidney International,* **29**, 3–9.

Meyer, H. (1975). Kalzium- , Phosphor- und Natrium-versargung bei gesunden und kranken Hunden. *Praktische Tierarzt,* **56**, 42–5.

Meyer, H. (1978). Kalzium und Phosphor bedarf des Hundes. *Ubersichten zur Tiernahrung,* **6**, 31–54.

Meyer, H. (1984). Mineral metabolism and requirements in bitches and suckling pups. In: *Nutrition and Behaviour in Dogs and Cats.* Ed. R.S. Anderson. pp. 13–24. Oxford: Pergamon.

Michaud, L. & Elvehjem, C.A. (1944). The nutritional requirements of the dog. *The North American Veterinarian,* **25**, 657–66.

Michell, A.R. (1980). Salt and hypertension. *Lancet,* **1**, 1358.

Michell, A.R. (1983). Abnormalities of renal function. In: *Veterinary Nephrology.* Ed. L.W. Hall. p. 196. London: Heinemann.

Michell, A.R. (1984a). Sums and assumptions about salt. *Perspectives in Biology and Medicine,* **27**, 221–33.

Michell, A.R. (1984b). Salt and hypertension. *Lancet,* **2**, 634.

Michell, A.R. (1985a). Sodium in health and disease: a comparative review with emphasis on herbivores. *Veterinary Record,* **116**, 653–7.

Michell, A.R. (1985b). Sodium and research in farm animals: problems of requirement, deficit and excess. *Outlook on Agriculture,* **14**, 179–82.

Michell, A.R. (1985c). The gut: the unobtrusive regulator of sodium balance. *Perspectives in Biology and Medicine,* **29**, 203–13.

Michell, A.R., Moss, P., Hill, R., Vincent, I. & Noakes, D.E. (1988). Effect of pregnancy and sodium intake on water and electrolyte balance in sheep. *British Veterinary Journal,* **144**, 147–57.

Milligan, L.P. & McBride, B.W. (1985). Energy costs of ion-pumping by animal tissues. *Journal of Nutrition,* **115**, 1374–82.

Mohring, J. & Mohring, B. (1972). Evaluation of sodium and potassium balance in rats. *Journal of Applied Physiology,* **33**, 688–92.

Mohring, J., Petri, M., Szokol, M., Haack, D. & Mohring, B. (1976). Effects of saline drinking on malignant course of renal hypertension in rats. *American Journal of Physiology,* **230**, 849–57.

Morris, J.G. & Petersen, R.C. (1975). Sodium requirements of lactating ewes. *Journal of Nutrition,* **105,** 595–8.

Morris, M.L., Patton, R.L. & Teeter, S.M. (1976). Low sodium diet in heart disease: how low is low? *Veterinary Medicine/Small Animal Clinician,* **71,** 1225–7.

Muirhead, E.E. (1983). Depressor functions of the kidney. *Seminars in Nephrology,* **3,** 14–29.

NRC (1974). Nutrient Requirements of Domestic Animals No. 8; *Nutrient Requirements of Dogs,* pp. 14, 35, 36. Washington DC: National Academy of Sciences.

NRC (1978a). *Nutrient Requirements of Domestic Animals No. 10; Nutrient Requirements of Laboratory Animals,* p. 19 (The Laboratory Rat). Washington DC: National Academy of Sciences.

NRC (1978b). *Nutrient Requirements of Cats.* p. 10. Washington DC: National Academy of Sciences.

NRC (1985). *Nutrient Requirements of Dogs.* pp. 17, 44–5. Washington DC: National Academy Press.

NRC (1986). *Nutrient Requirements of Cats* p. 17. Washington DC: National Academy Press.

Nicholls, M.G.N. (1984). Reduction of dietary sodium in Western society. *Hypertension,* **6,** 795–801.

Pitts, R.F. (1974). *Physiology of the Kidney and Body Fluids,* 3rd edn. pp. 109, 249–50. Chicago: Year Book.

Polzin, D.J., Osborne, C.A., Hayden, D.W. & Stevens, J.B. (1984). Influence of reduced protein diets on morbidity, mortality and renal function in dogs with induced chronic renal failure. *American Journal of Veterinary Research,* **45,** 506–17.

Ramsay, D.J. & Reid, I.A. (1979). Salt appetite in dogs. *Federation Proceedings,* **38,** 971.

Reinhardt, H.W. & Behrenbeck, D.W. (1967). Untersuchungen an wachen Hunden über die Einstellung der Natriumbilanz. *Pflugers Archiv,* **295,** 266–92.

Rocchini, A.P., Cant, J.R. & Barger, A.C. (1977). Carotid sinus reflex in dogs with low to high sodium intake. *American Journal of Physiology,* **233,** H196–202.

Rojo-Ortega, J.M., Querioz, F.P. & Genest, T. (1979). Effects of sodium chloride on early and chronic phases of malignant hypertension in rats. *American Journal of Physiology,* **236,** H665–71.

Roos, J.C., Koomans, H.A., Mees, E.J.D. & Delawi, I.M.K. (1985). Renal sodium handling in normal humans subjected to low, normal and extremely high sodium supplies. *American Journal of Physiology,* **249F,** 941–7.

Rubin, L.F. (1976). *Atlas of Veterinary Ophthalmoscopy.* p. 132. Philadelphia: Lea & Febiger.

Scribner, B.H. (1983). Salt and hypertension. *Journal of the American Medical Association,* **250,** 388–9.

Selkurt, E., Abel, F.L., Edwards, J.L. & Num, M.N. (1973). Renal function in dogs with hypertension induced by immunologic nephritis. *Proceedings of the Society for Experimental Biology and Medicine,* **144,** 295–303.

Seymour, A.A., Davis, J.O., Freeman, R.H., de Forrest, J.M., Rowe, B.P., Stevens, G.A. & Williams, G.M. (1980). Hypertension produced by sodium depletion and unilateral nephrectomy: a new experimental model. *Hypertension*, 2, 125–9.

Shaw, R.K. & Phillips, P.H. (1953). The potassium and sodium requirements of certain mammals. *Lancet*, 73, 176–80.

Sinclair, K.B. & Jones, D.I.H. (1968). Comparison of the weight gain and composition of blood and saliva in sheep grazing timothy and ryegrass swards. *British Journal of Nutrition*, 22, 661–6.

Smith, R.C., Haschem, T., Hamlin, R.L. & Smith, C.M. (1964). Water and electrolyte intake and output and quantity of faeces in the healthy dog. *Veterinary Medicine/Small Animal Clinician*, 59, 743–8.

Spangler, W.L., Gribble, D.H. & Weiser, M.G. (1977). Canine hypertension; a review. *Journal of the American Veterinary Medical Association*, 170, 995–8.

Steward, A.P. & Macdougall, D.F. (1984). Familial nephropathy in the Cocker Spaniel. *Journal of Small Animal Practice*, 25, 15–24.

Strauss, M.B., Lamdin, E., Pierce Smith, W. & Bleifer, D.J. (1958). Surfeit and deficit of sodium. *Archives of Internal Medicine*, 102, 527–36.

Swales, J.D. (1980). Dietary salt and hypertension. *Lancet*, 1, 1177–9.

Swales, J.D. (1981). Blood pressure and the kidney. *Journal of Clinical Pathology*, 34, 1233–40.

Sullivan, J.M. & Johnson, J.G. (1983). The management of hypertension in patients with renal insufficiency. *Seminars in Nephrology*, 3, 40–51.

Suttle, N.F. (1983). The nutritional basis for trace element deficiencies in ruminant livestock. In: *Trace Elements in Animal Production and Veterinary Practice*. Eds. N.F. Suttle, R.G. Gunn, W.M. Allen, K.A. Linklater and G. Weiner, pp. 19–25. Occasional Publication No. 7 – British Society of Animal Production.

Toal, C.B. & Leenen, F.H.H. (1983). Dietary sodium restriction and development of hypertension in spontaneously hypertensive rats. *American Journal of Physiology*, 245, H1081–4.

Valtonen, M.H. & Oksanen, A. (1972). Cardiovascular disease and nephritis in dogs. *Journal of Small Animal Practice*, 13, 687–97.

Vogel, J. (1966). Salt-induced hypertension in the dog. *American Journal of Physiology*, 210, 186–90.

Von Bunge, G. (1902). *Textbook of Physiological and Pathological Chemistry*. Eds. F.A. Starling and E.H. Starling. 2nd English edn. from 4th German edn. Lecture 7. Philadelphia: Blakiston.

Walser, M. (1985). Phenomenological analysis of renal regulation of sodium and potassium balance. *Kidney International*, 27, 837–41.

Weiser, M.G., Spangler, W.L. & Gribble, D.H. (1977). Blood pressure management in the dog. *Journal of the American Veterinary Medical Association*, 171, 364–9.

Whyte, H.M. (1958). Body fat and blood pressure of natives in New Guinea. *New Zealand Journal of Medicine*, 7, 36–46.

Wilhelmj, C.M., Waldmann, E.B. & McGuire, T.F. (1951). Effect of prolonged high sodium chloride ingestion and withdrawal upon blood pressure of dogs. *Proceedings of the Society for Experimental Biology and Medicine,* **77**, 379–82.

Yinchang, J., Guiyun, Z. & Zhong, W. (1980). Development of hypertension in dogs under intensified tension of higher nervous activities. *Scientia Sinica,* **23**, 665–72.

17

Calcium metabolism and skeletal development in dogs

H.A.W. HAZEWINKEL

Introduction

Besides being an essential structural component of the skeleton, calcium is involved in many biological processes including muscle contraction, blood coagulation and hormonal release. Circulating calcium levels must therefore be closely regulated.

Plasma calcium concentration varies within narrow limits, and is normally between 2.2–3.0 mmol/l in adult dogs (Schärer, 1970) and 2.4–2.9 mmol/l in the young growing dog (Hazewinkel *et al.*, 1985*b*).

Approximately 50% of the circulating calcium is in the ionised (i.e. unbound) form, which is the biologically active portion. Since calcium is bound primarily to albumin, the concentration of albumin in the blood has to be taken into account in the dog when judging calcium concentration (Meuten *et al.*, 1982).

Regulatory systems

Three systems are involved in the regulation of plasma calcium concentrations: the intestine, the kidney and the bone. After ingestion some of the calcium crosses the intestinal epithelium via a carrier mediated, active transport system, or by passive diffusion (Allen, 1982). The latter is the predominant pathway for different species at an early age (Allen, 1982). The endogenous faecal fraction is added to the non-absorbed calcium portion. The endogenous faecal fraction of calcium originates from within the blood and is secreted or diffuses into the intestinal lumen, and is later not reabsorbed. This fraction is of a lesser magnitude in the young animal than that found in the adult (Levine, Walling & Coburn, 1982).

It is generally recognised that, although the kidneys contribute to the overall regulatory process of calcium metabolism, they play a minor role (Massry, 1982). In dogs, only 60% of circulating calcium is ultrafiltrable at the level of the glomeruli (Edwards, Sutton & Dirks, 1973) and more than 99% of this filtered calcium is reabsorbed (Poulos, 1957).

There is a close link between calcium deposition into the skeleton (i.e.

293

accretion) and calcium loss from the bone (i.e. resorption) (Parfitt, 1976; Raisz, 1984). In the adult dog, under balanced conditions, both accretion and resorption values are 0.1–0.2 mmol per kg bodyweight per day (Goldman, 1970; Meyer, 1978). However, both values are increased to at least 100 times higher in the young growing dog (Hedhammar et al., 1980; Hazewinkel, 1985a).

In addition to the above described regulatory systems of plasma calcium concentration, three specific hormonal regulators are known: parathyroid hormone (PTH), calcitonin (CT) and 1,25 dihydroxycholecalciferol (1,25 vit D), the most active vitamin D metabolite. These calciotropic hormones have in common that their release is under the influence of plasma calcium concentration. They, in turn, modify calcium metabolism through their respective target tissues.

Calciotropic hormones

PTH is synthesised in the parathyroid glands. Low calcium concentration stimulates PTH synthesis and secretion whereas high calcium concentration stimulates fragmentation of PTH into its biologically inactive components (Mayer et al., 1979). PTH has no known effects upon the intestine (Walling, 1982). However, it does effect the kidney: an increased circulating PTH concentration results in decreased calcium and increased phosphorus excretion (Sutton, Wong & Dirks, 1976). When PTH acts on bone, osteoblasts (i.e. bone-matrix producing cells) shrink and change their shape. This allows the brush border of the osteoclasts (i.e. bone resorbing cells) to contact the bone-matrix surface (Chambers & Moore, 1983). Enzymatic dissociation of mineral from the matrix, as well as breakdown of collagen fibres, takes place in the contact area between the brush border and osteoid surface.

Calcitonin (CT) is produced in C-cells which are mainly located within the thyroid and internal parathyroid glands (Copp et al., 1967). In dogs, as in other mammals, concentration of circulating CT increases after intravenous as well as after an oral calcium load (Hazewinkel et al., 1985a). Sustained calcium excess with hypercalcemia may cause hyperplasia of C-cells in dogs (Kameda, 1970). CT has no known effect on the kidney (Clark et al., 1969; Puschett et al., 1974) nor intestine (Cramer, Parkes & Copp, 1969; Barbezat & Reasbeck, 1983) in this species. CT may also influence the satiety centre, so as to decrease food intake (Freed, Perlow & Wyatt, 1979; Levine & Morley, 1981). When CT acts on bone, the size of the brush border as well as the motility and number of osteoclasts decreases and, therefore, also bone resorption (Chambers & Moore, 1983; Cramer, et al., 1969).

1,25 vit D is synthesised in the mitochondria of the renal tubular cells. Its synthesis in the dog is determined by the concentration of calcium in the plasma (Halloran et al., 1982) and it stimulates the synthesis of transport and enzyme proteins, which increase the transport of calcium and phosphate through the intestinal wall via independent mechanisms (Kanis, Guillard-Cumming & Russell, 1982). In the kidney, 1,25 vit D decreases the excretion of calcium and phosphorus (Kanis *et al.*, 1982), whereas in bone it has two effects: increased mineralisation as well as increased bone resorption (Kanis *et al.*, 1982; Ornoy & Zusman, 1983). From studies in growing Great Dane dogs, raised on diets meeting the US NRC Nutrient Requirements of Dogs (1974), it is known that whereas the plasma 1,25 vit D concentration stays within narrow limits, those of PTH and CT decrease during the first half year of life (Hazewinkel, 1985a).

Therefore, age has to be taken into consideration when interpreting the concentrations of calciotropic hormones in the young dog.

Calcium metabolism in the dog

In studies involving *adult* dogs, the percentage of calcium which is absorbed from the food varies between 0 and 90% (Hoff-Jorgensen, 1946; Gershof, Legg & Hegstedt, 1958; Jenkins & Phillips, 1960; Schmidt, 1977; Hedhammar *et al.*, 1980). The percentage absorbed depends upon several factors, including the composition of the food and the calcium content (Meyer & Mundt, 1983). The calcium content in these studies ranged from 0.11–2.0%.

In *young* individuals food composition may noticeably influence several physiologic mechanisms which bear consequences in later life (Zoppi, Mantovanelli & Cecchettin, 1982). In order to test the influence of one single food constituent, all other factors must be kept constant. It is only in this situation, then, that the produced effects can, either directly or indirectly, be attributed to the food constituent in question. In a study of the influence of different calcium contents in a food, which otherwise met the US-NRC requirements (1974), the physical, biochemical and calcium metabolical changes in growing Great Dane dogs were recorded (Hazewinkel, 1985a; Hazewinkel *et al.*, 1985b). Five dogs fed completely according to the US-NRC-requirement 1974 (1.1% Ca on dry matter base) served as controls (NC-dogs); five dogs were fed 0.55% Ca (LC-dogs); and six dogs were fed 3.3% Ca (HC-dogs) (Fig. 1).

The dogs, originating from several litters, were studied from 6 to 26 weeks of age whereas the LC-dogs had to be euthanised after 5–11 weeks due to severe locomotor disturbances.

Studies in Great Danes

LC-dogs had a higher food intake and grew more rapidly in height and weight than the NC-dogs. The HC-dogs remained behind both groups. At the age of 12 weeks (i.e. 6 weeks after the start of the study), both the LC and HC-dogs developed signs of disturbed locomotion. The LC-dogs developed valgus deviation of the front legs, lordosis and painful locomotion. HC-dogs developed signs of radius curvus syndrome (Carrig, 1977) and

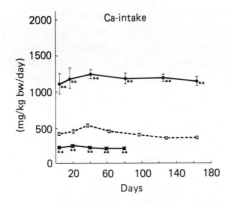

Fig. 1. Calcium intake (mean ± SEM, $n \geq 10$) of NC ($\bigcirc$), HC ($\bullet$), and LC ($\blacksquare$) dogs. Differences from the NC-dogs are indicated with* when $0.01 \leq P \leq 0.05$ and with** when $P < 0.01$.

posterior paresis with the neurologic signs of the Canine Wobbler Syndrome (Chrisman, 1982). On dermatological and ophthalmological examination these dogs did not reveal any abnormalities.

Plasma calcium and alkaline phosphatase concentrations were higher in both the LC- and HC-dogs as compared with the NC-dogs. Plasma phosphate concentrations were higher in the LC-dogs and lower in the HC-dogs than in the NC-dogs. The glomerular filtration rate (as reflected by the circulating urea and creatinine concentrations) was inversely related to the calcium intake (Fig. 2).

Studies of calcium metabolism conducted with the aid of [^{45}Ca] revealed a calcium absorption coefficient of 45% for NC, 80% for LC, and 45% for HC-dogs. Calcium losses in the endogenous faecal fraction (max 1.5 mmol/kg BW/day) and kidneys (max 0.4 mmol/kg BW/day) were of minor magnitude, especially in the NC- and HC-dogs. In both the LC- and HC-dogs the calcium accretion was higher than in the NC-dogs. The calcium resorption was lower in the HC and higher in the LC-dogs than that found in the NC-dogs. As a result calcium balance was more positive in HC-dogs and less positive in the LC-dogs than in NC-dogs (Fig. 3).

In summary, in LC-dogs calcium metabolism was characterised by a high absorption coefficient and a high bone turnover rate together with a less positive balance. In the HC-dogs, calcium absorption and balance were both high and almost proportional to calcium intake, and further associated with a low bone turnover.

The differences in bone turnover were reflected in pathologic appearance. The mineral content of cortical and cancellous bone was higher in the HC than in the NC-dogs. Decreased bone turnover with decreased modelling of the skeleton probably led to delayed expansion of the spinal canal proportion to the growth of the spinal cord. Stenosis of the vertebral canal leading to degeneration of the spinal cord was seen more often in HC than in NC-dogs and was totally absent in LC-dogs. However, the LC-dogs had generalised osteoporosis with greenstick and compression fractures appearing 6 weeks after the start of the study. In addition, severe disturbances of enchondral ossification (i.e. the process of cartilage turning into bone) were noticed in the HC-dogs whereas in the NC-dogs only mild disturbances were seen and in the LC-dogs, no disturbances were found in this respect (Fig. 4). These disturbances, known as osteochondrosis (Olsson, 1981), included retained cartilage cones in all growth plates, especially at the distal ulna, and at the predilection sites for osteochondritis dissecans, particularly the proximal and distal humerus.

Discussion

From these findings it can be concluded that these young dogs do not have a mechanism to protect themselves against excessive calcium feeding. Under the influence of the calciotropic hormones the calcium excess

Fig. 2. Plasma concentrations of total calcium, inorganic phosphate, urea and creatinine for NC (○), HC (●) and LC (■) dogs. For further explanation see legend for Figure 1.

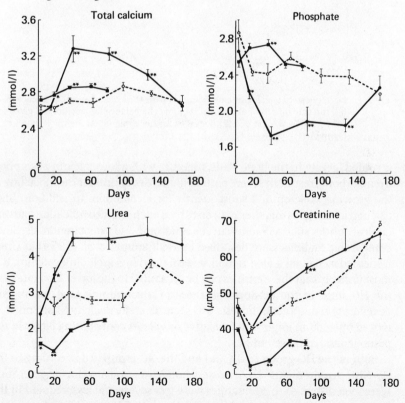

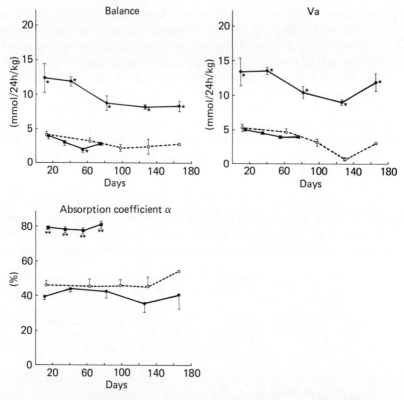

Fig. 3. Results of calcium kinetic studies with ⁴⁵Ca. The absorption coefficient (a), the true intestinal absorption (V_a) and the balance for calcium are given for NC (○), HC (●) and LC (■) dogs. For further explanation see the legend for Figure 1.

is routed largely to the bone. This, together with a low calcium absorption from the bone, results in severe pathologic consequences for the modelling of the growing skeleton and subsequently for locomotion. In addition, high calcium intake goes together with disturbed enchondral ossification causing clinical entities such as radius curvus syndrome and osteochondrosis. Since comparable findings were described by Hedhammar et al. (1974) in Great Danes fed ad libitum a food rich in several nutrients including calcium, it is most unlikely that these entities can be attributed to the lower food intake in the HC-dogs. In the LC-dogs (on 0.55% Ca containing food) no myelin degeneration or osteochondrosis was seen. Therefore calcium itself must be looked upon as an important causative or at least contributing factor in the pathogenesis of these entities.

Both in the HC-group (3.3% Ca) and the NC-group (fed according to the US-NRC requirements 1974) osteochondrosis and myelin degeneration were seen although these changes were less severe and less frequent in the latter group and were absent in the LC-group (0.55% Ca). Therefore it may

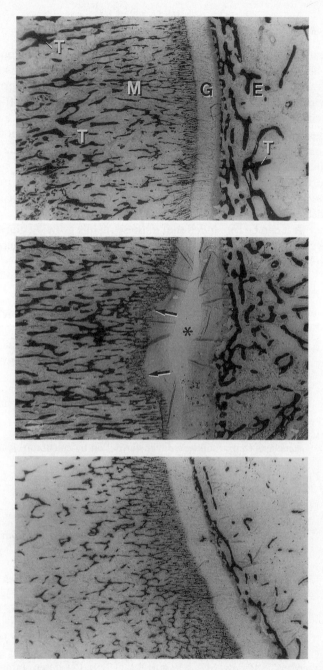

Fig. 4. Sections made through the epiphysis (E), growth plate (G) and
metaphyseal area (M) of the distal tibia of 3-month-old Great Danes fed a
normal calcium (NC, top), high calcium (HC, middle) or low calcium (LC,
bottom) diet. More and larger trabeculae are present in the E and M in the HC-
dog than in the NC-dog, due to diminished osteoclasia. In the HC-dog disturbed
enchondral ossification is the cause of irregular widenings (←) of the G (* is an
artefact). In the LC-dog the trabeculae in both the E and the M are small due to
generalised osteoporosis. No disturbances in enchondral ossification are present
in the LC-dog.

be questioned whether 1.1% Ca on dry matter base is the correct content of
the food for these growing Great Dane dogs. However, 0.55% Ca intake
(which is almost the NRC 1985 requirement) appears to be also incorrect, as
can be concluded from the incidence of pathological fractures in the axial
and abaxial skeleton, due to generalised osteoporosis.

However, as the phosphate content of all diets was maintained at 0.9% P
(on dry matter), the Ca:P ratio of diet LC was low (0.61) and may have
contributed to the adverse effects observed in this group. It is open to
question, therefore, whether a LC diet with a *correct* ratio (around 1.0)
would have given optimum bone development. Further studies are needed
to establish this.

Meanwhile it can be concluded that in these rapidly growing giant breed
dogs, the calcium content of the food is very critical for normal development
of the skeleton.

REFERENCES

Allen, L.H., (1982). Calcium bioavailability and absorption: a review.
American Journal of Clinical Nutrition, **35,** 783–808.
Barbezat, G.O. & Reasbeck, P.G. (1983). Effect of Bombesin, Calcitonin
and Enkephalin on canine jejunal water and electrolyte transport.
Digestive Diseases and Sciences, **28,** 273–7.
Carrig, C.B. (1977). Comparative radiology: dysplasia of the canine
forelimb. In *Current concept in radiology,* vol. III, ed. E.J. Potchem, pp.
217–60. Saint Louis: Mosby Company.
Chambers, T.J. & Moore, A. (1983). The sensitivity of isolated
osteoclasts to morphological transformation by calcitonin. *Journal of
Clinical Endocrinology and Metabolism,* **57,** 819–24.
Chrisman, C.L. (1982). *Problems in small animal Neurology,*
Philadelphia: Lea & Febiger.
Clark, M.B., Byfield, P.G.H., Boyd, C.W. & Foster, G.V. (1969). A
radioimmunoassay for human calcitonin. *Lancet,* **2,** 74–7.
Copp, D.H., Cockcroft, D.W. & Kueh, Y. (1967). Calcitonin from
ultimobranchial glands of dog, fishes and chickens. *Science,* **158,**
924–6.
Cramer, C.F., Parkes, C.O. & Copp, D.H. (1969). The effect of chicken
and hog calcitonin on some parameters of Ca, P, and Mg metabolism
in dogs. *Canadian Journal of Physiology and Pharmacology,* **47,** 181–4.
Edwards, B.R., Sutton, R.A.L. & Dirks, J.H. (1973). Effect of calcium
infusion on renal tubular reabsorption in the dog. *American Journal
of Physiology,* **227,** 13–18.
Freed, W.J., Perlow, M.J. & Wyatt, R.J. (1979). Calcitonin: Inhibitory
effect on eating in rats. *Science,* **206,** 850–2.
Gershoff, S.N., Legg, M.A. & Hegstedt, D.M. (1958). Adaptation to
different calcium intakes in dogs. *Journal of Nutrition,* **64,** 303–12.
Goldman, M. (1970). Skeletal mineralization. In *The Beagle as an
experimental dog,* ed. A.C. Anderson, pp. 216–25. Ames, Iowa: Iowa
State University Press.

Halloran, B.P., Hulter, H.N., Toto, R.D. & Levens, M. (1982). Regulation of the plasma concentration of 1,25 (OH)$_2$D by parathyroid-hormone, calcium and phosphate. In *Vitamin D, chemical, biochemical and clinical endocrinology of calcium metabolism*, ed. A.W. Norman, K. Schaefer, D. von Herrath & H.G. Grigoleit, pp. 467–9. Berlin: H.G. de Gruyter.

Hazewinkel, H.A.W. (1985*a*). Influences of different calcium intakes on calcium metabolism and skeletal development in young Great Danes, PhD Thesis Utrecht State University.

Hazewinkel, H.A.W., Goedegebuure, S.A., Poulos, P.W. & Wolvekamp, W.Th.C. (1985*b*). Influences of chronic calcium excess on the skeletal development of growing Great Danes. *Journal of the American Animal Hospital Association*, **21**, 377–91.

Hedhammar, A., Wu, F.M., Krook, L., Schrijver, H.F., de Lahunta, A., Wahlen, J.P., Kallfelz, F.A., Nunez, E.A., Hintz, H.F., Sheffy, B.E. & Ryan, G.D. (1974). Overnutrition and skeletal disease. An experimental study in growing Great Dane dogs. *Cornell Veterinarian*, **64**, suppl. 5, 11–160.

Hedhammar, A., Krook, L., Schrijver, H. Kallfelz, F. (1980). Calcium balance in the dog. In *Nutrition of the dog and cat*, ed. R.S. Anderson, pp. 119–27. Oxford: Pergamon Press.

Hoff-Jorgensen, E. (1946). The influence of phytic acid on the absorption of calcium and phosphorus I. In dogs. *Biochemical Journal*, **40**, 189–92.

Jenkins, K.J. & Phillips, P.H. (1960). The mineral requirements of the dog II. The relation of calcium, phosphorus and fat levels to minimum calcium and phosphorus requirements. *Journal of Nutrition*, **70**, 241–6.

Kameda, Y. (1970). Increased mitotic activity of the parafollicular cells of the dog thyroid in experimentally induced hypercalcemia. *Archivum Histologicum Japonicum*, **32**, 179–92.

Kanis, J.A., Guillard-Cumming, D.F. & Russell, R.G.G. (1982). Comparative physiology and pharmacology of the metabolites and analogues of vitamin D. In *Endocrinology of calcium metabolism*, ed. J.A. Parsons, pp. 321–62. New York: Raven Press.

Levine, B.S. & Morley, J.E. (1981). Reduction of feeding in rats by calcitonin. *Brain Research*, **222**, 187–91.

Levine, B.S., Walling, M.W. & Coburn, J.W. (1982). Intestinal absorption of calcium: its assessment, normal physiology and alterations in various disease states. In *Disorders of Mineral Metabolism*, vol. II, ed. F. Bronner & J.W. Coburn, pp. 103–188. New York: Academic Press.

Mayer, G.P., Keaton, J.A., Hurst, J.G. & Habener, J.F. (1979). Effects of plasma calcium concentration on the relative proportion of hormone and carboxyl fragment in parathyroid venous blood. *Endocrinology*, **104**, 1778–4.

Massry, S.G. (1982). Renal handling of calcium. In *Disorders of Mineral Metabolism*, vol. II, ed. F. Bronner & J.W. Coburn, pp. 189-235. New York: Academic Press.

Meyer, H. (1978). Calcium and phosphorus requirement of the dog. *Animal Nutrition*, **6**, 31–54.

Meyer, H. & Mundt, H.C. (1983). Untersuchung zum Einsatz von Knochenschrot in Futterrationen für Hunde. *Deutsche Tierärztliche Wochenschrift*, **90**, 81–6.

Meuten, D.J., Chew, D.J., Capen, C.C. & Kociba, G.J. (1982). Relationship of serum total calcium to albumin and total protein in dogs. *Journal of the American Veterinary Medical Association*, **180**, 63–7.

National Research Council (1974 & 1985). *Nutrient Requirements of Dogs*, Washington DC: National Academy of Sciences.

Olsson, S.E. (1981). Pathophysiology, morphology and clinical signs of osteochondrosis (chondrosis) in the dog. In *Pathophysiology in small animal surgery*, ed. M.J. Bojrab, pp. 604–17. Philadelphia: Lea & Febiger.

Ornoy, A. & Zusman, J. (1983). Vitamin and cartilage, chapt. 8. In *Cartilage, development, differentiation and growth*, vol. 2, ed. B.K. Hall, pp. 297–326. New York: Academic Press.

Parfitt, A.M. (1976). The actions of parathyroid hormone on bone: relation to bone remodeling and turnover, calcium homeostasis and metabolic bone disease. *Metabolism*, **25**, 809–44.

Poulos, P.P. (1957). The renal tubular reabsorption and urinary excretion of calcium by the dog. *Journal of Laboratory and Clinical Medicine*, **49**, 253–7.

Puschett, J.B., Beck, W.S., Jelonek, A. & Fernandez, P.C. (1974). Study of the renal tubular interactions of thyrocalcitonin, cyclic adenosine 3′, 5′-monophosphate, 25-hydroxycholecalciferol, and calcium ion. *Journal of Clinical Investigation*, **53**, 756–67.

Raisz, L.G. (1984). Studies on bone formation and resorption *in vitro*. *Hormone Research*, **20**, 22–7.

Schärer, V. (1970). *Die Bestimmung der Normalwerte von Calcium und Phosphor im Serum beim Hund*, Bern (Switzerland): Hell & Co.

Schmidt, M. (1977). Einfluss überhöhter Eiweissgaben auf die Verdauungsvorgänge sowie den intermediären Stoffwechsel beim Hund. Hanover Veterinary School Dissertation in Veterinary Medicine.

Sutton, R.A.L., Wong, N.L.M. & Dirks, J.H. (1976). Effects of metabolic acidosis and alkalosis on sodium and calcium transport in the dog kidney. *Kidney International*, **15**, 520–33.

Walling, M.W. (1982). Regulation of intestinal calcium and inorganic phosphate absorption. In *Endocrinology of calcium metabolism*, ed. J.A. Parsons, pp. 87–102. New York: Raven Press.

Zoppi, G., Mantovanelli, F. & Cecchettin, M. (1982). Metabolic-endocrine responses to feeding different formulas during the first months of life, in *Monographs in paediatrics*, vol. 16, pp. 88–95. Basel: S. Karger.

18

The effects of the overfeeding of a balanced complete commercial diet to a group of growing Great Danes

R.B. LAVELLE

Introduction

Orthopaedic problems of large and giant breeds pose a major challenge to the veterinary profession and dog breeders. Along with other breeds, Great Danes have an increased prevalence of some orthopaedic problems which are seen in the developing bone. Amongst these are: Osteochondrosis in many sites (Olsson, 1975; Grondalen, 1982a), hip dysplasia (Lust, Geary & Sheffy, 1973) and metaphyseal osteopathy (Grondalen, 1976). While the clinical behaviour of these problems and others is well recognised and their radiological features have been clearly defined, too little is known of their aetiology. More understanding is called for, not least to aid clinical treatment. For, although in some situations, e.g. osteochondrosis of the shoulder joint, this is very successful, in many others, such as disproportionate growth of the radius and ulna, it is variable. In some conditions the outlook is hopeless – as it is for example, in severe hypertrophic osteodystrophy, or metaphyseal osteopathy as it is also known.

Nutrition may play a part in these orthopaedic problems, though its contribution is still poorly understood. But it is increasingly clear that incorrect nutrition, particularly excessive nutrition, will affect bone development in a number of ways. Nutritional secondary hyperparathyroidism is one such problem that is well recognised and understood (Bennett, 1976). This is caused by feeding a diet with an imbalanced calcium:phosphorus ratio and in dogs this is usually due to a primarily muscle meat diet with its very low calcium content and highly adverse calcium:phosphorus ratio. Hypervitaminosis D is another such problem.

Other problems of over-feeding probably exist as can be inferred from work on other species. For example, the work of Saville & Lieber (1969) with rats which led them to conclude 'maximal growth rate may be incompatible with optimal skeletal characteristics', or the observations of Dluzniewska, Obtulowicz & Koltek (1965) that rapid growth in children is associated with skeletal abnormalities. Over-feeding has also been associated with skeletal

303

Table 1. *Details of the composition of the combination of two parts canned:1 part dry food (% as fed) compared with those of the diet used by Hedhammar et al. and the NRC recommendations*

Constituent	Lavelle		Hedhammar	NRC (1974)
	Wet weight	Dry weight	Dry weight	Dry weight
Dry matter	43.9	—	—	—
Protein	13.0	29.6	36.0	22
Fat	6.3	14.4	13.7	5
Linoleic acid	0.7	1.6	—	1
Calcium	1.0	2.3	2.05	1.1
Phosphorus	0.7	1.6	1.44	0.9
Metabolisable energy (kJ/100 g)	790	1,802	2,094	1,680

problems in poultry (Wise & Jennings, 1972), pigs (Reiland, 1978), cattle (Reiland *et al.*, 1978) and horses (Wyburn, 1977).

In dogs it is also recognised that many orthopaedic problems probably have a hereditary component. For example, Wobbler syndrome in Dobermans and Great Danes (Wright, Rest & Palmer, 1973) and OCD of the elbow in the Labrador Retriever (Mason *et al.*, 1980) and Rottweiler (Grondalen, 1982*b*). But the genetic factors are only one part of the aetiology, as has been quantified for hip dysplasia which in some breeds is currently thought to have a hereditary component of 40%, the rest being due to unspecified 'environmental factors' (Willis, 1986).

Nutrition must be considered as among the most plausible of these 'environmental factors', as it must in the aetiology of hypertrophic osteodystrophy, about the aetiology of which little is certainly known.

Hedhammar and his colleagues recently focussed attention on such nutritional factors when they suggested, on the basis of their experimental studies, that generalised over-nutrition may play an important part in the aetiology of orthopaedic problems in large and giant breeds (Hedhammar *et al.*, 1974).

This is the first such experimental report for the dog, although there are some limitations to Hedhammar's valuable study which both reduce its generality and obscure the nutritional interpretation of the results. Firstly, the study was undertaken using a breed in which the expected orthopaedic problems are naturally of high prevalence. Secondly, the dogs were artificially reared and their clinical behaviour was dissimilar to that of the naturally occuring cases. Finally, although 'over-feeding' was the highlighted term in the title of the study, the diet fed showed a number of departures from the National Research Council's (NRC) recommendations (Table 1). These were high metabolisable energy (ME), high calcium and

Table 2. *Distribution of great danes into the dietary groups with their litter 'pairing'*

	Unrestricted feeding	Restricted feeding
	Group U	Group R
Litter A	Dog 1 (F), Dog 2 (F)	Dog 3 (F)
Litter B	Dog 4 (F)	Dog 5 (M)
Litter C	Dog 6 (M), Dog 7 (F)	Dog 8 (M), Dog 9 (F), Dog 10 (F)

Note:

F = Female   M = Male

phosphorus levels and a high protein content. We decided to set up this study on the effects of the over-feeding of a balanced commercial diet to a group of Great Danes in the hope that it would eliminate some of the problems in the interpretation of Hedhammar's work.

Experimental method

The Great Danes used in the study came from three litters, A, B and C. The dogs in litter A, three blue bitches, were donated by a member of the South Australian Great Dane Club. Litters B and C were bred from donated bitches using a sire belonging to a member of the Great Dane Club of Victoria. Litter B, came from 'Elsa', who produced a brindle bitch and a fawn dog. 'Chim' produced litter C, five fawn puppies, three females and two male.

The puppies were vaccinated at the time of admission to the dog colony and given routine worm prevention treatment. They were housed in outdoor concrete runs which were surrounded by wire mesh. The runs were 6 m by 3 m with a diagonally placed division at the centre. An above-ground kennel approximately 1 m × 1 m × 1 m was available in each half of the run. The dogs from each litter were housed adjacent to their litter mates. At weekly intervals the dogs were weighed before they were given their morning feed.

The ten dogs were divided into matched pairs, one of each assigned to unrestricted feeding (Group U) or restricted feeding (Group R). As far as possible the dogs were paired with a litter mate for comparative purposes (Table 2). The dogs were fed a mixture of a complete balanced commercial canned dog food and a complete commercial balanced dry dog food using two parts by weight of the canned food to one part by weight of dry food. The composition of the two food products is given in Table 1.

The dogs in the Group U were fed twice a day with a weighed quantity of food in excess of their anticipated appetite. The first feed was given in the morning and in the mid afternoon the feed bowls were removed and the remaining food weighed. The second feed was given in the late afternoon and the bowls collected the following morning when the remaining food was weighed. The dogs in Group R were given 60% by weight of the food eaten by the Group U paired member in the previous feeding period. At the end of the feeding period the food remaining was weighed.

The clinical assessment of the dogs was carried out in a variety of ways. The dogs were inspected twice daily at feeding times and casually observed by the dog colony staff on numerous occasions during the day. Particular note was taken of the dogs' appetite, activity, evidence of lameness and nature of faeces. The dogs were assessed more closely each time they were brought to the clinic and hospital for radiographic examination.

Haematological and biochemical analyses were made on blood samples collected whenever that radiography was undertaken. Calcium, phosphorus, magnesium, alkaline phosphatase and total solids were determined and a full blood examination made. Parvovirus serological tests were carried out when the disease affected the dog colony.

The radiographic examination was undertaken with the dogs anaesthetised and the studies were repeated at approximately three-week intervals. The areas examined were the cervical spine, shoulder joint, radius and ulna in both cranio-caudal and lateral projections, pelvis and stifle joint. A variety of measurements of both length and width were made. Potential sites of osteochondrosis were examined for evidence of the development of this problem. The timing of the appearance and development of ossification centres and growth plates were recorded.

Results

General health

The general health of the dogs was excellent throughout the project apart from a brief spell of gastro-intestinal upset caused by the inadvertent introduction of canine parvovirus into the dog colony. One dog, R3, required intravenous fluid therapy for 48 hours but the others needed nothing more than gastro-intestinal sedatives and they continued to have a fair appetite.

The 'U' fed dogs ate the major portion of their meal shortly after being fed but continued to eat throughout the rest of the period it was present. The 'R' fed group ate all their food immediately after it was offered. None of the dogs showed evidence of obesity and the 'R' fed group did look a little thin in the earlier weeks. However, with time they caught up the 'U' fed group in their general body condition although overall they were a little smaller.

Throughout the study all the dogs were very active and ran freely in their runs. The only exception was dog R3 when it was hospitalised for a few days with parvovirus infection. On its return to the colony it continued its former active lifestyle. When the dog colony staff walked among the dogs at feeding time, or if there were visitors to the colony, the dogs all jumped up at the fencing of the runs with great excitement and wagging of tails. The gait of the dogs was examined closely at the time of each radiological examination but at no stage was lameness detected. The only clinical skeletal problem was cranial bowing of the radius and ulna in dog R3. This change was not associated with a valgus deformity.

Weight gain

The average weights of the dogs in the two groups are shown in Fig. 1. The 'U' fed group were heavier than the 'R' fed group throughout the study. The weight of the 'U' group began to plateau towards the end of the study but the 'R' group continued to put on weight each month. Thus, before the 5th month of the study, group 'U' grew faster, after which the 'R' group had the greater rate of weight gain. This is also true when specific weight gain (as grams per $W^{0.75}$ per day) is calculated as shown in Fig. 2. Food intake, of course, increased with increasing size (Fig. 3(a)) and food intake per unit weight (Figure 3(b)) or per unit metabolic body size ($W^{0.75}$) (Fig. 3(c)) correspondingly declined with increasing age and size. But whichever scheme is used to express the results the 'U' dogs had the higher ME intake throughout the study. Initially their intake was 900 kJ/kg/day but this gradually decreased to approximately 600 kJ at 6 months and 450 kJ at 10 months. The corresponding intakes of the 'R' group were 600 kJ, 500 kJ and 300 kJ respectively. When expressed on a metabolic bodysize basis there was no major change except that the difference between the two groups was a little greater and the decline in energy intake was more gradual. The 'U' fed group were eating approximately 1500 kJ/kg$^{0.75}$/day for the first 7 months and the 'R' fed group approximately 1100 kJ.

Fig. 1. The weights of the 'U' group and 'R' group dogs shown as the mean of the weekly weighing.

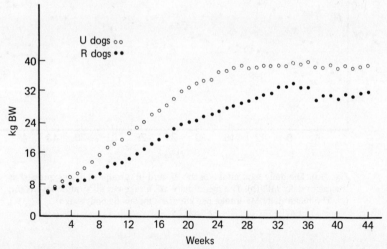

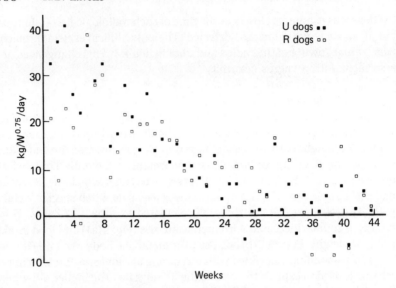

Fig. 2. The daily weight change (grams per kilogram metabolic bodyweight) of the 'U' group and 'R' group dogs.

(a)

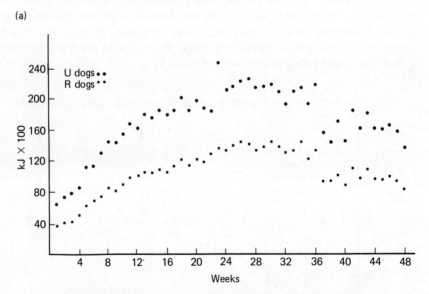

Fig. 3(a). The daily food intake of the 'U' and 'R' groups of dogs expressed as the mean daily ME; (b). The mean daily ME intake per kilogram bodyweight; (c). The mean daily ME intake per kilogram metabolic bodyweight.

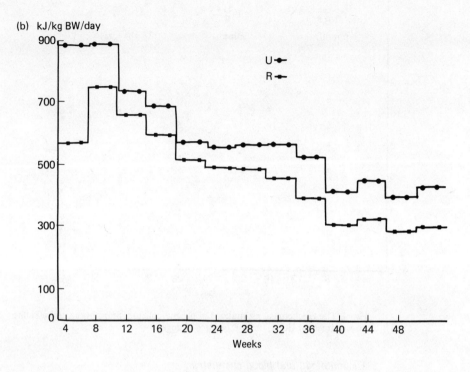

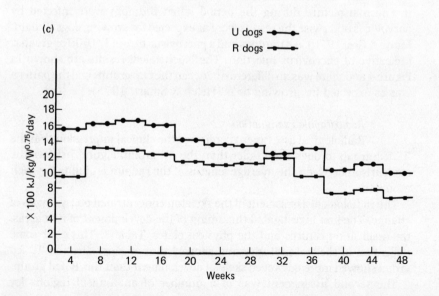

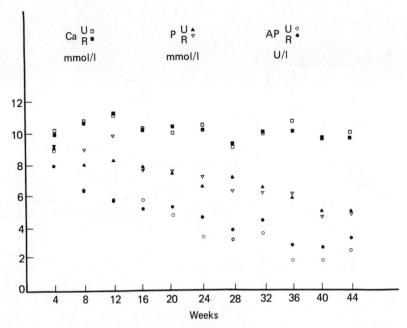

Fig. 4. The biochemical estimations of plasma calcium, phosphorus and alkaline phosphatase for the 'U' and 'R' groups of dogs.

Haematology and blood chemistry

The full blood examination results showed a brief departure from the normal picture during the period when the dogs were infected by parvovirus otherwise the results were as expected for growing dogs (Bulgin, Munn & Gee, 1970). All the dogs had a parvovirus titre of 1/1600 or greater, indicative of parvovirus infection. The biochemical results are shown in Figure 4 and there was no difference between the two groups and the pattern was as expected for growing dogs (Fletch & Smart, 1973).

Radiographic examination

Radiological assessment confirmed the clinical impression that the 'U' fed group to dogs were taller than the 'R' group. Figure 5 shows the comparison between the average lengths of the radium and ulna for each group.

The radiological assessment of the skeleton concentrated on two types of change. The first investigated the timing of the development of two areas, the ossification centres and the physical plates, Table 3. This gave some indication of skeletal maturity and showed that at most sites the 'U' fed group showed more advanced skeletal development than the 'R' fed group.

The second assessment was of a number of anatomical regions for evidence of pathological change associated with bone problems seen in Great Danes (Table 4). There was no consistent difference between the two groups of dogs. The cervical region was examined for evidence of the 'Wobbler Syndrome' but, in only one dog, R5, was there evidence of a problem. The shoulder joint was examined for evidence of osteochondrosis. Three dogs in the 'U' fed group, U4, U6 and U7 showed an alteration to the outline of the humeral head compared with only one R fed dog, R8. However, this radiological change was not evident clinically in any animal.

The distal ulnar cartilage core was retained in four of the five 'U' fed group. However, in each case the core was narrow and central in position and did not result in any disproportionate growth of the radius and ulna. None of the

Fig. 5. The mean lengths of the radius and ulna of the 'U' and 'R' groups of dogs during their growth.

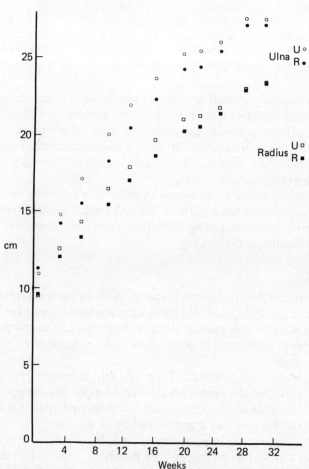

Table 3. *The mean (± standard error of mean) times of the radiological appearance of some important anatomical features which indicate the stage of skeletal development*

		'U' dogs	'R' dogs
Scapula tuberosity	Adult formation	24±0.95	31±2.5
Proximal humeral physis	Closure	52±1.41	50 —
Tubercle humerus	Adult formation	30±1.73	34±2.64
Medial condyle humerus	Adult formation	31±1.44	37±2.90
Lateral condyle humerus	Adult formation	30±2.25	34±2.64
Proximal ulna physis	Closure	37±2.14	37±2.90
Anconeus	Appearance	14±0.49	16.5±0.87
Anconeus	Closure	17.5±1.86	20.5±1.71
Distal ulna physis	Closure	48±2.00	46 —
Distal radial physis	Closure	48±2.00	46 —
Acetabulum	Adult formation	24±0.77	27.5±4.09
Ilium secondary centre	Appearance	27±3.52	34±1.00
Ischium secondary centre	Appearance	12.5±2.1	16±1.15

dogs had an ununited anconeal process but two in each group showed evidence of osteophyte formation on the anconeal process.

Two dogs in each group showed some minor changes in the hip joint that were indicative of mild hip dysplasia. There was minor subluxation of the femoral heads but there were no degenerative changes. There was no clinical evidence of a problem.

All forelimb radiographs were examined for evidence of deformity but the only changes were seen in dog R3. There was radiological confirmation of the clinical finding of cranial bowing of the radius and ulna. There was no indication of the cause of the bowing.

Discussion

It is clear that the present study failed to duplicate the observations of Hedhammar *et al.* (1974) that *ad libitum* fed giant breed dogs have an increased incidence of severe musculo-skeletal problems, and radiological changes associated with endochondral ossification, secondary to their excessive food intake.

There are a variety of possible reasons for this difference between Hedhammar's work and the present study. For example, chance genetic differences could be involved, as might be inferred, for example, from the fact that Hedhammar's animals had a greater stature at the outset.

Undoubtedly, however, a primary reason must lie in the difference in the response of dogs to *ad libitum* feeding. In both studies the *ad libitum* fed dogs over-ate relative to the NRC (1974) standards. The consumption of the

Table 4. *The radiological assessment of subclinical skeletal changes at sites of orthopaedic problems in growing dogs*

	'U' dogs					'R' dogs				
	1	2	4	6	7	3	5	8	9	10
Cervical vertebra										
C5	−	−	−	−	−	−	−	−	−	−
C6	−	−	−	−	−	−	−	−	−	−
C7	−	−	−	−	−	−	+	−	−	−
OCD humeral head	−	−	+	+	+	−	−	+	−	−
Anconeus osteophyte	+	+	−	−	−	+	+	−	−	−
Distal ulnar core	+	+	−	+	+	−	−	−	−	−
Cranial bowing radius	−	−	−	−	−	+	−	−	−	−
OCD femur	−	−	−	−	−	−	−	−	−	−
Hip dysplasia	+	+	−	−	−	+	+	−	−	−

+ Evidence of radiological change during study
− Normal radiological appearance during study

animals in the present study was not excessive if compared to Blaza's (1983) suggestion that giant breeds need 50% more energy than the 1974 NRC allowance. Intakes of Hedhammar's *al libitum* fed group were still high by this standard, being approximately twice NRC levels in the growing animals.

A major difference between the two studies was in activity. The animals in the present study were kept in conditions allowing considerable activity, and clearly availed themselves of this. Hedhammar's animals by contrast had only limited opportunities for exercise, and, within three weeks of starting the experiment, there was a noticeable fall-off in voluntary activity in his *ad libitum* fed animals. As discussed in Chapter 6 of this book, activity costs could be a quite significant mode of energy expenditure particularly for large and giant breeds, so that the activity differences would add to the excessive amount of energy available for growth in Hedhammar's animals.

Differences in voluntary activity may be one factor accounting for the energy intake differences of the *ad libitum* fed animals, animals with restricted opportunity for activity have been reported as showing a tendency to overeat. Another factor may have been the nature of the diet, that used by Hedhammar being very palatable because of its high meat content. Both diets had a high fat content (13.7% dry matter, as opposed to 14.4% in the present study), there being a known tendency in laboratory rodents to overeat high fat diets.

The greater amount of energy available for growth in Hedhammar's animals, was reflected in their greater growth rates, which were above the already high levels reported in the present study.

Interestingly, however, neither of these excessive growth rates seemed to be associated with obesity in the animals observed; even in Hedhammar's high intake/low activity animals "over-nutrition did not result in excessive fat deposition". (Page 32 Hedhammar *et al.*, 1974)

The higher energy consumption of Hedhammar's animals meant that they also had a high intake of other nutrients, and Hedhammar pointed out that the calcium intakes were particularly excessive, as the calcium levels in the diet were more than twice that recommended by NRC (1974).

If a hypothesis of dietary cause is accepted then the particularly excessive intakes of Hedhammar's study are illustrated by the fact that problems even occurred in his restricted animals which, since they were fed two-thirds the intake of the *ad libitum* animals, still had an intake which, relative to NRC, was as high as the *ad libitum* animals in the present study. By contrast neither the U or R fed animals in the present study showed signs of diet-related bone disease. Only one dog, and that from the restricted group, showed a clinical skeletal abnormality. This animal had cranial bowing of the radius of both forelegs. While there was some radiological evidence of the bone problems of growing Great Danes, their degree of development was of no significance, and there were no consistent differences between the U-fed and R-fed groups.

It seems possible, therefore, that while over-consumption may well be a problem in giant and large breeds, it is also possible that, with care given to husbandry conditions, consumption and the degree of energy balance in even *ad libitum* fed animals may be self-limited enough to avoid problems.

Nevertheless breeders and owners of large and giant breeds should take care to ensure that they do not over-supplement their dogs' diets with minerals and that the ME intake is kept to the recommended levels.

Young dogs should be given exercise as this will help to keep them lean and may aid bone development by giving a greater safety margin during the period of rapid growth.

Careful collation of breeding records is necessary to investigate the nature of any suggested genetic basis of bone problems.

REFERENCES

Bennett, D. (1976). Nutrition and bone disease in the dog and cat. *Veterinary Record*, **98**, 313–20.

Blaza, S.E. (1983). The Nutrition of Giant Breeds of Dog. *Pedigree Digest*, **8**, (3), 8–9.

Bulgin, M.S., Munn, B.A. & Gee, W. (1970). Haematological changes to $4\frac{1}{2}$ years of age in clinically normal Beagles. *Journal of the American Veterinary Medical Association*, **157**, 1064–70.

Dluzniewska, K.A., Obtulowicz, A. & Koltek, K. (1965). On the relationship between diet, rate of growth and skeletal deformities in school children (English summary). *Folia Medica Craciviensia*, **7**, 115–26.

Fletch, S.M & Smart, M.E. (1973). Blood chemistry of the giant breeds – bone profile. *Bulletin American Society of Veterinary Clinical Pathology*, 2, 30.

Grondalen, J. (1976). Metaphyseal osteopathy (hypertropic osteodystrophy) in growing dogs. A clinical study. *Journal of small Animal Practice*, 17, 721–35.

Grondalen, J. (1982a). Arthrosis in the elbow joint of young rapidly growing dogs. VI Interrelation between clinical, radiographical and pathoanatomical findings. *Nordisk Veterinaer Medicin*, 34, 65–75.

Grondalen, J. (1982b). Arthrosis in the Elbow Joint of Young Rapidly Growing Dogs. VII Occurrence in the Rottweiler Breed. *Nordisk Veterinaer Medicin*, 34, 76–82.

Hedhammar, A., Wu, F., Krook, L., Schryver, H.F., Delahunta, A., Whalen, J.P., Kallfez, F.A., Numez, E.A., Hintz, H.F., Sheffy, B.E. & Ryan, G.D. (1974). Overnutrition and skeletal disease. An experimental study in growing Great Dane dogs. *The Cornell Veterinarian*, 64, Supplement 5, 1–159.

Lust, G., Geary, J.C. & Sheffy, B.E. (1973). Development of Hip Dysplasia in Dogs. *American Journal of Veterinary Research*, 34, 87–91.

Mason, T.A., Lavelle, R.B., Skipper, S.C. & Wrigley, W.R. (1980). Osteochondrosis of the elbow joint in young dogs. *Journal of Small Animal Practice*, 21, 641–56.

NRC (1974). *Nutrient Requirements of Dogs.* National Research Council Report No. 8, Washington DC: National Academy of Sciences.

Olsson, S.E. (1975). *Lameness in the dog: A review of lesions causing osteoarthrosis of the shoulder, elbow, hip, stifle and hock joints.* American Animal Hospital Association, 42nd Annual Meeting Cincinnati Ohio Proceedings, 1, 363–70.

Reiland, S. (1978). The effect of decreased growth rate on frequency and severity of osteochondrosis in pigs. An experimental investigation. *Acta Radiologica* [Suppl] 358, 107–22.

Reiland, S., Stromberg, B., Olsson, S-E., Dreimanis, I. & Olsson, I.G. (1978). Osteochondrosis in growing bulls. *Acta Radiologica,* [Suppl] 358, 179–96.

Saville, P.D. & Lieber, C.S. (1969). Increases in skeletal calcium and femur thickness produced by undernutrition. *Journal of Nutrition*, 99, 141–4.

Willis, M.B. (1986). Hip Scoring: Review of 1985–86. *Veterinary Record*, 118, 461–2.

Wise, D.R. & Jennings, A.R. (1972). Dyschondroplasia in domestic poultry. *Veterinary Record*, 91, 285–6.

Wright, F., Rest, J.R. & Palmer, A.C. (1973). Ataxia of the Great Dane caused by stenosis of the cervical vertebral canal: Comparison with similar conditions in the Bassett Hound, Doberman Pinscher, Ridgeback and the Thoroughbred Horse. *Veterinary Record*, 92, 1–5.

Wyburn, R.S. (1977). A degenerative joint disease in the horse. *New Zealand Veterinary Journal*, 25, 321–2 & 335.

19

The role of zinc in canine and feline nutrition

CHARLES A. BANTA

Introduction

In terms of basic inorganic chemistry, zinc is a relatively uninteresting element. Its biochemistry, however, is far more fascinating and its critical role in the practical formulation of pet foods is now seen to make it of central concern to the pet animal nutritionist and the pet food manufacturer. In the last twenty years zinc deficiency has been recognised as a practical problem in the nutrition of a wide range of animal species including humans and their domestic animals. In dogs and cats zinc deficiences have been observed and some of these instances have been directly due to inappropriate formulation of commercial diets. It is particularly important, therefore, that the background to the physiological role of this trace element is properly understood.

Zinc is a member of group IIb of the periodic arrangement of elements which means it has a full complement of electrons in the d subshell and an 'inert pair' of electrons in its outer s subshell. As a result, it has a relatively high ionisation potential and is stable in both the free form and in compounds with a $+2$ valence. In organic chemistry zinc is occasionally found as a minor catalyst.

One of the most important characteristics of zinc in biological systems is its ability to form complex ions and that it can accommodate up to six ligands (Riordan, 1976). No evidence exists to indicate that zinc is either oxidised or reduced in biological reactions, nor does it participate in single electron transfers (Sugarman, 1983).

The essentiality of zinc in a biological system was first demonstrated in 1869 when it was reported that zinc was required for growth of *Aspergillus niger*. It was more than a half century before this observation was extended to higher green plants, and an additional decade passed before a mammalian requirement was reported (Prasad, 1976). Since that time, according to Prasad (1976), zinc deficiency has been reported in a wide variety of animal species including rats, mice, swine, calves, lambs, dogs, birds and man. A zinc deficiency has also been reported in cats by Kane *et al.* (1981).

317

Biochemistry of zinc

Zinc is widespread within the biochemical systems of individual cells. There are at least 70 zinc-containing enzymes (Riordan, 1976) and possibly many more (Hanson, Fernandes & Good, 1982; Sugarman, 1983). According to Riordan (1976), there is at least one zinc-containing enzyme in each of six enzyme categories recognised by the Commission on Enzyme Nomenclature of the International Union of Biochemistry. The zinc enzymes may be divided into two general groups on the basis of their affinity and specificity for zinc (Chester, 1978). In zinc–enzyme complexes, the binding between the protein and zinc ion is weak and the specificity for zinc to produce enzyme activation is low. In zinc metalloenzymes, zinc is firmly bound with a definite stoichiometry and a relatively high degree of specificity. While this type of classification is convenient, it may tend to confuse because, in some cases, it is an oversimplification of the relationship between the protein and the metal.

Zinc functions in two ways in enzyme systems. It can, through interaction with various functional groups on the protein, help to maintain the tertiary structure of the enzyme; or it may, through reaction at the active site, participate directly in the catalytic action (Chesters, 1978). The most commonly implicated ligands include cysteinyl groups, histidyl groups, tyrosyl groups and the carboxyl groups of aspartic and glutamic acids (Riordan, 1976).

With the large number of enzymes depending on zinc for structure and function, it might be assumed that the signs of zinc deficiency could be explained on the basis of reduced or absent function of one or more of these enzymes. Efforts to demonstrate this have generally proved unfruitful because the studies reported on the effect of zinc deficiency on enzyme activities have provided equivocal results. Mills *et al.*, (1967) showed that, while the zinc containing pancreatic enzyme carboxypeptidase A was reduced in zinc deficiency in rats, there was no reduction in protein digestion. Robinson & Hurley (1981) examined the accumulation of procarboxypeptidase A in the pancreata of foetuses of zinc deficient as compared to zinc sufficient dams. While there was a quantitative reduction of the proenzyme in foetuses from deficient dams there was no qualitative difference. Foetal chymotrypsinogen was also measured and this zinc-independent proenzyme was found to be similarly reduced in foetuses from deficient dams. It was concluded that the limited factor was protein synthesis and not zinc incorporation into the apoenzyme.

Zinc has been shown to be an integral part of a variety of nucleic acid polymerases (Riordan, 1976; Chesters, 1978; Sugarman, 1983). Reduction in the activities of these enzymes could be translated into reduced protein synthesis and in turn to a reduced rate of growth. While there have been many zinc-containing enzymes identified, efforts to attribute the effects of zinc deficiency to a reduction in activity of any one or a combination of these have been unsuccessful.

Other efforts aimed at the biochemical explanation of the effects of zinc deficiency revolve around the regulatory role of zinc in cell membranes. One of the difficulties associated with the study of this role of zinc is that regulatory zinc represents only a small fraction of the total cellular zinc, and the effect of a severe deficiency is usually seen before there is a measurable change in cell zinc concentration (Chesters, 1978). On the basis of this, and other studies in which marginal deficiencies allowed for minimal growth and tissue zinc depletion, Chesters (1978) concluded that the major zinc complexes do not constitute a reserve of zinc for use in the regulatory function.

According to Sugarman (1983), zinc exerts its regulatory function by binding to proteins on the membrane surface. These may be structural proteins, thereby altering the stability of the membrane, or they may be functional proteins, such as surface enzymes, with a resulting activation or deactivation. In either case, the cell function could be changed. Chvapil (1976) has studied the effect of zinc on specific cells and membranes. In cultured lymphocytes he reported that zinc acts as a nonspecific mitogen with both increased blastogenic transformation and mitosis. While much of the evidence suggests that this is the result of the zinc containing DNA polymerase, cell mitosis was also stimulated by zinc compounds to which the cell membrane is impermeable, leading Chvapil to consider the possibility of a cell membrane effect. Evidence was stronger for a cell membrane effect on the inhibition by zinc of function in mast cells, platelets and macrophages.

Both Chvapil (1976) and Sugarman (1983) have reviewed the literature on the regulatory function of zinc in cells and membranes. Neither was able to conclude with any degree of certainty exactly how zinc exerts its regulatory effect. It is clear, however, that zinc can alter the function of a variety of mammalian cells.

Zinc deficiency

Zinc deficiency has been reported in a wide variety of animal species (Prasad, 1976). The signs of zinc deficiency common to most species studied include growth retardation, skin and hair changes including parakeratosis, hyperkeratosis, achromotrichia and alopecia, and gastrointestinal disturbances including anorexia, emesis and diarrhoea (Burch & Sullivan, 1976; Prasad, 1976; Li & Vallee, 1980; Prasad, 1985). In addition, reproductive abnormalities including delayed sexual maturation (Prasad, 1976), impaired spermatogenesis (Kane *et al.*, 1981), teratogenesis (Hurley & Svenerton, 1966), and dystocia (Apgar, 1968) have been reported in several species as have both cell mediated immune deficiency (Chandra & Au, 1980) and disordered immunoglobulin profiles (Beach *et al.*, 1980).

The growth retardation associated with zinc deficiency appears to be unrelated to the reduced food intake and may be due to an effect on growth hormone or a direct effect on cell division (Prasad, 1985). Disrupted cell division and cell maturation could explain many of the signs of zinc deficiency.

Experimental zinc deficiency has been produced in dogs by a number of workers including Robertson & Burns (1963), Sanecki, Corbin & Forbes (1982) and this author (unpublished). It has also been produced by Kane *et al.* (1981) in cats. Clinical zinc deficiency has also been reported in dogs by Kunkle (1980) and by Sousa *et al.* (1983).

Robertson & Burns (1963) fed a corn-soy basal diet containing 0.3% calcium and 33 mg zinc/kg to control dogs. The experimental groups received the basal diet with either 0.8% added calcium (2% calcium carbonate), or 0.8% added calcium and 104 mg zinc/kg diet (200 mg zinc carbonate). The diets were fed over a period of ten months and while age of dogs was not reported, it appeared that they were young and actively growing. Dogs fed either the basal diet or basal plus calcium and zinc appeared normal throughout the experimental period. Those fed the basal diet supplemented with calcium began to exhibit signs of zinc deficiency within two to three months. Dogs fed the calcium supplemented diet gained less than 4 kg over the experimental period as compared to 11.6 kg for the basal diet group and 12.8 kg for those supplemented with calcium and zinc. Other changes in the high calcium, low zinc group included reduced plasma and foecal amylase activity and reduced plasma and urine zinc in comparison to other groups. The calcium-supplemented dogs also showed a dull and rough hair coat, desquamating skin lesions along the posterior aspect of the abdomen, and on hindlegs, and three of four dogs had an ocular discharge and severe conjunctivitis. These same three dogs exhibited anorexia and emesis prior to death. Post-mortem examination of these dogs showed marked emaciation with no subcutaneous or abdominal fat, fatty change of the liver with distended gall bladder, and evidence of kidney damage with calcium deposits in the renal pelvis. Parakeratosis was not observed.

Sanecki *et al.* (1983) also used a corn-soy diet to study zinc deficiency in English Pointer pups. Pups were weaned on a nutritionally adequate control diet at five weeks of age and assigned to either the treatment or control group one week later. The control diet contained 2.6% calcium and 120 to 130 mg zinc/kg while the zinc deficient diet contained the same level of calcium and 20 to 35 mg zinc kg. Skin lesions began to appear in the zinc-deficient group in the second week on the deficient diet. The development of lesions followed a definite pattern with lesions first appearing in areas of contact or trauma (foot pads), then in areas of stretch (skin over joints), followed by areas of friction (axilla or groin), and finally mild involvement of the mucocutaneous junction, ear canal and toenail bed. Lesions first appeared as small erythematous spots which enlarged and formed nodules or pustules which finally coalesced to form a brown, grainy crust. No lesions were observed in dogs fed the zinc-adequate control diet. After five weeks, the diet groups were reversed. Improvement was noted within one week on the diet adequate in zinc and complete healing of all skin lesions was seen by the fifth week. Dogs initially fed the control diet began to exhibit signs after two weeks and lesions increased in severity through the sixth week on the zinc-deficient diet, at which time the study was concluded. Microscopically, the lesions showed parakeratosis and hyperkeratinisation leading to a build-up of surface debris, inflammatory infiltration by neutrophils, lymphocytes or macrophages, and sloughing of epithelium. Sanecki *et al.* (1983) concluded that zinc deficiency disrupts the normal maturation process of skin cells.

This author (unpublished) fed Labrador Retriever bitches a practical diet containing corn, soybean meal, meat and bone meal, wheat middlings and vitamin-mineral supplement with or without added zinc chloride. The diets contained approximately 1.8% calcium and either 210 mg zinc/kg (DDAZ) or 50 mg zinc/kg (DDAX). Bitches were placed on the diets prior to breeding and carried through gestation and lactation. One bitch was fed DDAZ and two were fed DDAX. No signs of zinc deficiency were noted through gestation or lactation and litter size and birth weights were normal for the colony. All pups were normal through weaning at six weeks of age at which time pups from each litter were selected for either the control or treatment group. Pups fed DDAZ showed no signs of zinc deficiency regardless of the zinc status of the mother. Pups fed DDAX began to exhibit hair coat changes two to three weeks post-weaning. These consisted of dryness and loss of colour which began on the extremities and face and progressed to total body involvement. One to two weeks after the appearance of hair coat changes, erosions were noted on the foot pads and lesions similar to those reported by Sanecki *et al.* (1983) began to appear on the abdomen, medial aspect of the hindlimbs and face. Lesions progressed in severity with time until all dogs were placed on DDAZ. Response to the zinc-adequate diet was dramatic with almost total disappearance of skin lesions within 72 hours. The only remaining evidence of lesions was areas of hyperpigmentation where lesions had been. Normal hair coat pigmentation returned within two to three months.

Subsequent to completion of this study, during the course of routine eye examinations, it was noted that there was a marked change in the colour of the tapetum lucidum of dogs which had been fed the zinc-deficient diet. There were no other gross changes apparent in the eye and there was no noticable affect on vision. The difference in appearance of the tapetum persisted for at least eight months following the first examination in one dog. Tjälve & Brittebo (1982) reported a marked uptake of [^{65}Zn] in the tapetum lucidum of both pigmented and albino ferrets. The significance of these observations is unclear at this time.

Experimental zinc deficiency in the cat has been reported by Kane *et al.* (1981). They conducted several studies using both near-mature cats and weanling kittens. In the first, using four-month old cats, they fed a semi-purified soy protein basal diet and the same diet with either calcium or zinc supplement. Diet 1 (basal) was calculated to contain approximately 0.8% calcium and was reported to contain 15 mg zinc/kg. Diet 2 was the same as Diet 1 with an additional 0.6% calcium (2% dicalcium phosphate). Diet 3 was the same as Diet 1, but supplemented to a level of 67 mg zinc/kg. The diets were fed for eight months with no effect noted on appearance, weight gain or food consumption. Plasma zinc concentrations were reduced in cats fed both low zinc diets as compared to those fed the zinc supplemented diet. Examination of testes from both low zinc groups revealed significant degeneration of semiferous tubules which was not reversed by switching to the zinc supplemented diet for eight weeks.

In a second study, Kane *et al.* (1981) further reduced zinc content of the diets through the use of EDTA washed soy protein. Diet 4 contained EDTA washed soy protein and was reported to contain 0.7 mg zinc/kg while Diet 5 was the same but supplemented to a level of 52 mg zinc/kg. A third diet (Diet 6) was of similar composition except that a crystalline amino acid mixture was used in place of the soy protein and it was reported to have a zinc content of 4.8 mg/kg. Kittens were weaned directly on to the experimental diets at eight weeks of age and fed for fourteen weeks. Kittens fed Diet 5 maintained a linear growth rate through an eight week period of observation while those fed Diet 4 began to lose weight after three weeks on diet. Daily food intake of the Diet 4 kittens was less than a third of that of kittens fed the zinc supplemented diet. Kittens fed the amino acids diet (Diet 6) gained at a reduced rate and had food intakes of approximately 60% of that of the zinc adequate kittens. Two of three kittens fed Diet 4 exhibited perioral skin lesions after four weeks on a diet containing 0.7 mg zinc/kg while one of four kittens fed the diet containing 4.8 mg zinc/kg developed similar lesions after six weeks on the diet. Biopsy specimens from the lesions showed focal crusting, parakeratosis, acanthosis, intracellular edema and mild spongiosis of the epidermis with dermal edema and mild mixed inflammatory infiltrate. Kittens fed a diet containing 52 mg of zinc showed no pathological changes.

Not only has zinc deficiency been produced experimentally but several syndromes in dogs have been described which have been related by presenting signs to zinc deficiency or alleviated by zinc therapy. Sousa *et al.* (1983) described a syndrome which they referred to as the 'generic dry dog food disease' because it was seen only in dogs fed inexpensive, unbranded dry dog foods sold in supermarkets. The disease presented with variable amounts of crusting and scaling which appeared in generally bilaterally symmetric patterns involving the bridge of the nose, the mucocutaneous junctions, pressure points and distal extremities. Lesions were well demarcated with an erythematous border, variable hyperpigmentation and lichenification of older lesions. Occasional findings included alopecia, focal erosions, papules and pustules. The most severe cases were accompanied by fever, lymphadenopathy and dependent pitting oedema. Histopathological findings included subacute to chronic dermatitis with areas of epidermal necrosis and mixed dermal inflammatory infiltrate. Treatment with systemic antibiotics was unsuccessful and systemic corticosteroids were only slightly effective. Improvement was rapid when the sole treatment was changing to a national brand dog food. Data from preliminary studies into the actual cause of this disease were indicative of zinc deficiency, however, the data were not discussed. The disease had no apparent age or breed predilection.

Kunkle (1980) reported on an entity presenting with similar signs but with an apparent breed predilection for Alaskan Malamutes, Siberian Huskies and Doberman Pinschers. The signs of this syndrome included erythema and oedema, scaling, exudation, thick crusts and alopecia with occasional mild to moderate weight loss and a dull, dry hair coat. The signs are totally responsive to treatment with oral zinc sulfate but usually return when therapy is terminated. While not discussed, it must be assumed that diet is not a factor. Due to the breed predilection, a genetic defect affecting zinc absorption or utilisation might be suspected.

Genetic disorders affecting zinc absorption are not common, but have been reported in both man (Michaëlsson, 1974) and dogs (Brown *et al.*, 1978). Michaëlsson (1974) reported on the treatment of a case involving an 18 year old male suffering from the age of five months with acrodermatitis enteropathica (AE). AE is an autosomal, recessive genetic desease characterised by eczematisation around natural body openings and on the hands and feet, particularly in the paronychial regions with achromothicia and alopecia. Enteropathy with diarrhoea is frequent (Weissmann, 1976). Michaëlsson (1974) reported that 18 days after treatment with oral zinc sulphate was begun, the dermatitis had completely healed and within five months, hair growth was normal. Brown *et al.* (1978) studied Alaskan Malamute dogs with chondrodysplastic dwarfism of apparently genetic origin. Affected males exhibited delayed sexual maturation and mature dwarfed males produced spermatozoa with a high percentage of acrosomal defects. Intestinal zinc absorption was compared in normal and affected dogs using [^{65}Zn], and it was found that zinc absorption in affected dogs was only 25% of that of normal dogs. They also found that in normal dogs zinc was initially bound to intestinal cellular protein and then released into the soluble fraction, while in affected dogs, zinc was rapidly bound to the protein fraction, but was not subsequently released. A number of workers, including Duncan & Hurley (1978), have proposed a low molecular weight zinc-binding ligand, found in milk and the intestines of mature rats but not in neonatal rats, as a necessary factor for normal zinc absorption. Efforts to date aimed at characterising the zinc binding ligand have produced only theories. While the identity of the factor (or factors) involved in the facilitation of zinc absorption has not been determined, it appears to be under genetic control with the inherent danger of genetic defects reducing the availability of zinc for general or specific uses in metabolism with resultant signs of zinc deficiency.

One final factor which can alter zinc absorption is that of diet composition. As discussed above, increasing dietary calcium can induce signs of zinc deficiency. Pécoud, Donzel & Shelling (1975) studied the effect of a variety of diets on the absorption of zinc from an oral dose of zinc sulfate in humans. They found that in the fasting state, zinc absorption, as measured by the post ingestinal rise in serum zinc, was directly related to the dose of zinc sulfate. When zinc sulphate was given with black, unsweetened coffee, zinc absorption was reduced by 50%. When given with a meal of brown bread, milk and cheese, a diet high in phytate, calcium and phosphorus, zinc absorption was totally inhibited. Brown bread alone and dairy products alone also substantially reduced zinc absorption. Giving zinc sulphate with purified phytate, phosphate, and/or calcium produced similar results. *In vitro* studies showed the zinc was precipitated by phytate and that this effect was enhanced by calcium.

Van Campen (1969) and Van Campen & Scaife (1967) showed that copper and zinc compete for absorption in the intestine of rats and that an excess of either element reduced the absorption of the other. According to the scheme presented by Cousins (1979), this effect may be due to competition for binding to intestinal metallothionine.

Conclusions

One cannot carry out a review of the literature of zinc biochemistry, metabolism and nutrition without being in awe of its vastness, and most of it has appeared over the past 30 years. While much has been learned about the specific role of zinc in specific biological systems, it appears that the overall regulatory role in basic cellular function has eluded a highly competent group of investigators. The ubiquity of zinc in nature has led Prasad (1976, 1985) to marvel at the incidence of zinc deficiency in humans. This has, in all likelihood, been due to the inclusion of substances in the diet that reduced the availability of zinc rather than to an absolute dietary deficiency. Such has also been the case in dogs in the United States at least twice in the past 20 years.

In the mid 1960s, a syndrome appeared in which young growing dogs fed commercial diets developed grey, lustreless hair coats with scaly skin and patches of dermatitis (personal observation). The condition generally appeared at three to four months of age and disappeared spontaneously within two to three months. At the time, the American petfood industry was in a period of change with increasing competition and increasing ingredient costs. This was met by the substitution of high fibre, high phytate cereal byproducts and vegetable protein sources for whole grain cereals and animal protein sources. The inclusion of increased zinc supplements in these diets eliminated reports of the condition. Hard economic times of the early 1980s brought an increased popularity of generic (unbranded) commercial dog foods and again, reports of a diet related skin condition appeared (Sousa *et al.*, 1983).

Dietary zinc deficiencies, whether absolute or relative resulting from dietary interactions, may be controlled by zinc supplementation of diets. Such problems may be controlled and eliminated by adequate education of pet food manufacturers. Areas which deserve further attention involve genetic-metabolic defects which may alter zinc utilisation such as that described by Brown *et al.* (1978) in Alaskan Malamutes. Other conditions affecting dogs which might be associated with disruptions in zinc metabolism include demodectic mange and idiopathic epilepsy. This is highly speculative and only offered as food for thought.

Demodicosis is an inflammatory disease of the skin of dogs which is characterised by an abnormal increase in the numbers of the mite, *Demodex canis*, a normal inhabitant of dog skin. It has been associated with an immunodeficiency state (Muller, Kirk & Scott, 1983). These authors reported an apparent breed predilection and other predisposing factors including poor nutrition and stress. An unattributed comment in a report prepared by a petfood company for lay readers (*Friskies Research Digest*, 1974, vol. 10, no. 3, p. 7, Carnation Co. Los Angeles) indicated that demodicosis can sometimes be cured or arrested by feeding supplemental zinc and that relapse often occurs on termination of zinc therapy.

The association of zinc and epileptic seizures is based on a highly speculative discussion by Barbeau & Donaldson (1974) who found that the amino-sulphonic acid, taurine, was in deficit in epileptic foci in human and in experimentally produced epileptic rats. They presented evidence to suggest that taurine acts as an 'inhibitory' transmitter in neuron stabilisation. They discussed two mechanisms through which zinc provides a stabilising effect, one in which zinc acts directly and the second in which zinc inhibits the release of taurine. They hypothesised that a metabolic defect in zinc utilisation could lead to hyperexcitability in the central nervous system through either one or both of these mechanisms and thus lead to epileptic seizures. Idiopathic epilepsy may, like demodicosis, have a genetic predisposition. This is based on the author's observation of a large Beagle colony in which the incidence of this disorder was reduced by judicious culling of breeding stock. On the basis of the report by Barbeau & Donaldson (1974), this author initiated oral zinc therapy in several epileptic dogs. While there appeared to be a reduction in both incidence and severity of seizures, traditional therapy was required for complete control of the disease.

It is obvious that we have much to learn about zinc in nutrition and metabolism. It is hoped that this work will stimulate discussion and further research with the aim of improving the nutrition and health of pets.

REFERENCES

Apgar, J. (1968). The effect of zinc deficiency on parturation in the rat. *American Journal of Physiology*, **215**, 160–3.

Barbeau, A. & Donaldson, J. (1974). Zinc, taurine and epilepsy. *Archives of Neurology*, **30**, 52–8.

Beach, R.S., Gershwin, M.E., Makishima, R.K., & Hurley, L.S. (1980). Impaired immunologic ontogeny in postnatal zinc deprivation. *Journal of Nutrition*, **110**, 805–15.

Brown, R.G., Hoag, G.N., Smart, M.E., Mitchell, L.H. (1978). Alaskan Malamute Chondrodysplasia V. Decreased gut zinc absorption. *Growth*, **42**, 1–6.

Burch, R.E. & Sullivan, J.F. (1976). Clinical and nutritional aspects of zinc deficiency and excess. *Medical Clinics of North America*, **60**, 675–85.

Chandra, R.K. & Au, B. (1980). Single nutrient deficiency and cell-mediated immune responses. I Zinc. *American Journal of Clinical Nutrition*, **33**, 736–8.

Chesters, J.K. (1978). Biochemical functions of zinc in animals. *World Review of Nutrition and Dietetics*, **32**, 135–64.

Chvapil, M. (1976). Effect of zinc on cells and biomembranes. *Medical Clinics of North America*, **60**, 799–812.

Cousins, R.J. (1979). Regulation of zinc absorption: role of intracellular ligands. *American Journal of Clinical Nutrition*, **32**, 339–45.

Duncan, J.R. & Hurley, L.S. (1978). Intestinal absorption of zinc: a role for a zinc binding ligand in milk. *American Journal of Physiology*, **235**, E556–9.

Hanson, M.A., Fernandes, G. & Good, R.A. (1982). Nutrition and Immunity: The influence of diet on autoimmunity and the role of zinc in the immune response. In *Annual Review of Nutrition*, vol. 2, ed. W.J. Darby, H.P. Broquist & R.E. Olson, pp. 151–77. Palo Alto, California: Annual Reviews, Inc.

Hurley, L.S. & Swenerton, H. (1966). Congenital malformations resulting from zinc deficiency in rats. *Proceedings of the Society for Experimental Biology and Medicine*, **123**, 692–6.

Kane, E., Morris, J.G., Rogers, Q.R., Ihrke, P.J. & Cupps, P.T. (1981). Zinc deficiency in the cat. *Journal of Nutrition*, **111**, 488–95.

Kunkle, G.A. (1980). The differential diagnosis of facial dermatitis. In *Current Veterinary Therapy VII*, ed. R.W. Kirk, p. 440.

Li, T-K. & Vallee, B.L. (1980). The biochemical and nutritional roles of other trace elements. In *Modern Nutrition in Health and Disease*, 6th edn, ed. R.S. Goodhart & M.E. Shils, pp. 408–41. Philadelphia: Lea & Febiger.

Michaëlsson, G. (1974). Zinc therapy in acrodermatitis enteroppathica. *Acta Dermatovener (Stockholm)*, **54**, 377–81.

Mills, C.F., Quarterman, J., Williams, R.B. & Delgarno, A.C. (1967). The effects of zinc deficiency on pancreatic carboxypeptidase activity on protein digestion and absorption in the rat. *Biochemical Journal,* **102**, 712–18.

Muller, G.H., Kirk, R.W. & Scott, D.W. (1983). *Small Animal Dermatology,* Philadephia: W.B. Saunders Company.

Pécoud, A., Donzel, P. & Schelling, J.L. (1975). Effect of foodstuffs on the absorption of zinc sulfate. *Clinical Pharmacology and Therapeutics,* **17**, 469–74.

Prasad, A.S. (1976). Zinc deficiency in man. *American Journal of Diseases of Children,* **130**, 359–61.

Prasad, A.S. (1985). Clinical manifestations of zinc deficiency. In *Annual Review of Nutrition,* vol. 5, ed. R.E. Olson, E. Beutler & H.P. Broquist, pp. 341–63. Palo Alto, California: Annual Reviews, Inc.

Riordan, J.F. (1976). Biochemistry of zinc. *Medical Clinics of North America,* **60**, 661–74.

Robertson, B.T. & Burns, M.J. (1963). Zinc metabolism and zinc-deficiency syndrome in the dog. *American Journal of Veterinary Research,* **24**, 997–1002.

Robinson, L.K. & Hurley, L.S. (1981). Effect of maternal zinc deficiency or food restriction on rat fetal pancreas. I Procarboxypeptilase A and chymotrypsinogen. *Journal of Nutrition,* III, 858–68.

Sanecki, R.K., Corbin, J.E. & Forbes, R.M. (1982). Tissue changes in dogs fed a zinc-deficient ration. *American Journal of Veterinary Research,* **43**, 1642–6.

Sousa, C., Ihrke, P., Reinke, S., Schmeitzel, L. & Stannard, A.A. (1983). Generic dog food and skin disease. *Journal of the American Veterinary Medical Association,* **182**, 198.

Sugarman, B. (1983). Zinc and infection. *Reviews of Infectious Diseases,* **5**, 137–47.

Tjälve, H. & Brittebo, E.B. (1982). Uptake of [^{65}Zn] in tapetum lucidum. *Medical Biology,* **60**, 112–15.

Van Campen, D.R. (1969). Copper interference with the intestinal absorption of zinc-65 by rats. *Journal of Nutrition,* **97**, 104–8.

Van Campen, D.R. & Scaife, P.U. (1967). Zinc interference with copper absorption in rats. *Journal of Nutrition,* **91**, 473–6.

Weisman, K. (1976). The biological role of zinc in normal and pathological conditions: A survey. *Danish Medical Bulletin,* **23**, 146–52.

20

Factors determining the essential fatty acid requirements of the cat

J.G. McLEAN AND E.A.MONGER

Introduction

It will be evident from earlier sections of this book, and most especially the review by Morris and Rogers in Chapter 5, that a large number of the nutritional idiosyncrasies exhibited by the cat can be regarded as evolutionary adaptations to a carnivorous diet. The concept that the protein or vitamin A requirement can be seen in this way is well accepted. Less clear is the way in which essential fatty acid (EFA) metabolism in cats can be related to the suggestion by Rivers, Sinclair & Crawford (1975) that the cat has lost the normal mammalian capacity for EFA metabolism, and that, as far as its EFA requirements are concerned, it is an obligate carnivore. This hypothesis as originally propounded was over-simplified, though not entirely invalid. The purpose of this chapter is to examine the complexities of EFA metabolism, with particular reference to the way in which the desaturation process relates to nutrient requirements of the cat.

Essential fatty acids in mammalian nutrition

In all mammalian species so far studied there exists a dietary requirement for all *cis*-polyunsaturated fatty acids; this is the EFA requirement. This requirement was first described in the rat by Burr & Burr (1929). The exact chemical nature of the EFA has been confused: the essentiality of linoleic acid (9, 12–18:2; *n*-6)[1] has been clearly established, but alpha-linolenic acid (9, 12, 15–18:3; *n*-3)[1] was at first described as 'partly

[1] In the formula 9,12 – 18:2, the numbers before the hyphen specify double bond positions according to the IUPAC convention of numbering from the carboxyl group. The 18:2 gives the chain length of the fatty acid and its double bond number. Metabolic relationships between fatty acids are not immediately apparent from their chemical formulae listed in this way, and the system exemplified 18:2 *n*-6 is commonly used, at least for fatty acids with methylene interrupted double bonds; in this system *n*-6 refers to the position of the final double bond from the carboxyl moiety, where *n* is the carbon number. Reference to Figure 1 will make it clear that this double bond position is not modified in metabolism, and thus can be used to identify metabolic families of polyunsaturated fatty acids.

essential', because it led to partial but not complete remission of the clinical signs of EFA deficiency. The current consensus is that there are two distinct EFA, or more precisely families of EFA, the *n*-6 and *n*-3 families (Mead, 1986). Classically described, EFA deficiency is in fact an *n*-6 deficiency, due, in most mammals, to a lack of dietary linoleic acid. There is also a separate deficiency of *n*-3 EFA, due to an inadequate intake of alpha-linolenic acid. This deficiency appears to be more difficult to induce and involves different deficiency signs. Additionally, alpha-linolenate will ameliorate, but not cure, some of the signs of linoleic acid deficiency.

The signs of linoleic acid deficiency in most mammals are skin lesions, an increased rate of trans-epidermal water loss, retarded growth and an increased metabolic rate, decreased reproductive capacity and a variety of metabolic changes and pathological effects in various tissues (see Holman, 1971 for a review). The signs of alpha-linolenic acid deficiency and its metabolic role are less well defined (see Tinoco, 1982; Mead 1986).

The dietary 18 Carbon EFA, linoleic acid and alpha-linolenic acid, are metabolised to long chain polyunsaturated fatty acids (LCP), which play a vital role in the structure and function of cell membranes as well as acting as precursors for the biologically active eicosanoids. The most important *n*-6 LCP metabolite of linoleic acid is arachidonic acid (5, 8, 11, 14-20:4; *n*-6), which is usually produced by a pathway involving the Δ6- and Δ5-desaturases, illustrated in Figure 1. There is, however, an alternative pathway shown in Figure 1, via the Δ8- and Δ5-desaturases, which has been identified in the rat testicle (Albert & Coniglio, 1977) and human bladder and colon (Nakazawa, Mead & Yonemoto, 1976). Alpha-linolenic acid is metabolised to *n*-3 LCP by a similar enzymic sequence, and the pathway is also set out in Figure 1.

The methods for assessing EFA status have been reviewed by McLean & Sinclair (1986). In most species an indication of EFA deficiency may be obtained from the detection of 5, 8, 11 − 20:3; *n*-9 in the blood and tissues. This fatty acid is derived from oleic acid by the action of the Δ6-desaturases and Δ5-desaturases (Fig. 1). It is not normally synthesised in significant amounts because both alpha-linolenic acid and linoleic acid are the preferential substrates for the Δ6-desaturases enzymes.

Essential fatty acids in feline nutrition

In 1975, Rivers, Sinclair & Crawford reported observations on cats fed on purified diets containing vegetable oils which provided EFA only as linoleate or as a mixture of linoleate and alpha-linolenate. These cats developed clinical signs compatible with EFA deficiency in other animals, and similar, though less severe, than the clinical state induced in animals fed diets substantially free of EFA, in which the lipid was provided as hydrogenated coconut oil. Analysis of plasma phospholipids in these cats showed they all had very low levels of LCP metabolites in plasma

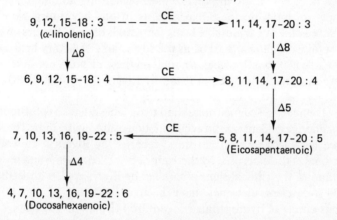

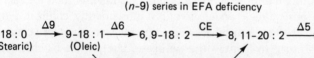

Linolenic acid (*n*–3) family

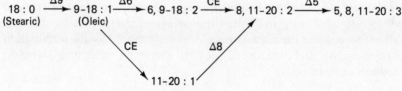

(*n*–9) series in EFA deficiency

§ CE is chain elongation of the fatty acid by the addition of a two-carbon fragment to the carboxyl end of the molecule.

*9, 12-18 : 2 is an 18-carbon fatty acid with double bonds between carbons 9 and 12 and between 12 and 13, numbering from carboxyl end of the molecule.

Δ6 indicates a desaturase enzyme capable of inserting double bonds between carbons 6 and 7.

Fig. 1. Linoleic acid (*n*−6) family, and (*n*−3) and (*n*−9) pathways.

phospholipids, and an absence of 5,8,11–20:3, which is normally associated with EFA deficiency. Animals which were fed linoleate and alpha-linolenate, had high levels of these fatty acids in tissue phospholipids. Rivers *et al.* concluded that their animals were EFA deficient because they lacked the ability to convert dietary EFA into their LCP metabolites, and postulated that this was due to a lack of Δ6- and Δ8-desaturases. From subsequent studies they concluded that the Δ5-desaturase was also absent in the cat (Hassam, Rivers & Crawford, 1977; Frankel & Rivers, 1978; Rivers & Frankel, 1980).

The nutritional significance of these observations is that, if correct, they would mean that it was essential for the cat to consume a diet containing preformed LCP, which are in practice diets containing animal lipid.

Attempts to duplicate these experiments in our laboratory were not entirely successful. We maintained cats for periods of up to 8 years on diets which contained safflower seed oil as the sole source of dietary lipid, i.e. in which the sole EFA was linoleate. Animals on these diets did not suffer from the severe symptoms reported in detail by Rivers & Frankel (1980), but appeared normal except for a slight dulling of the coat and generally an inability of females to produce more than two viable litters. At post-mortem the major change evident was an accumulation of lipid in the liver. The signs may have been less severe than those reported by Rivers & co-workers because some of the diets used by these authors lacked taurine and possibly other vitamins; the clinical signs associated by Rivers *et al.* with feeding a totally EFA-deficient diet may have been exacerbated by the known deleterious effects of hydrogenated coconut oil (Holman, 1971).

Nevertheless diets containing what would be adequate amounts of linoleic acid for other mammalian species clearly did not maintain the cat in good health in our studies, an observation subsequently confirmed by MacDonald *et al.* (1984). A number of experiments were therefore carried out in our laboratory to identify the nature of the precise biochemical deficiency, because the findings have clear implications for the formulation of experimental and commercial cat diets which are extensively based on products of plant origin.

Observations on EFA metabolism in the cat

When we fed cats diets containing purified gamma-linolenate (6,9,12–18:3; *n*-6) for 28 days, changes occurred in the fatty acid composition of erythrocyte phospholipids which suggested the rapid chain elongation of 18:3 *n*-6 to 20:3 *n*-6 and its subsequent conversion to arachidonic acid, 20:4 *n*-6, (Table 1). This led us to conclude that an active Δ5-desaturase was present in the cat (Sinclair *et al.*, 1979), a conclusion at variance with that drawn by Frankel & Rivers (1978), from their results obtained from gamma linolenic acid feeding. However, the failure of Frankel & Rivers to detect a Δ5-desaturase can be plausibly explained by a variety of

Table 1. *Concentration[a] of (n − 6) fatty acids in erythrocytes after feeding g methyl-gamma-linolenate to EFA deficient cats*

	Days of gamma-linolenate feeding		
Fatty acid	0	14	28
18:2 (n−6)	7.2	7.7	5.2
18:3 (n−6)	0.0	0.3	0.4
20:3 (n−6)	1.0	5.0	10.0
20:4 (n−6)	6.8	8.7	13.0

Note:
[a] mg fatty acid/100 ml erythrocytes.
Source: Data from Sinclair, Mclean & Monger (1979).

Table 2. *Concentration of polyunsaturated fatty acids in cat plasma total lipids (mg/100 ml plasma)*

Fatty acid	Normal diet	Linoleate-rich diet
18:2 (n−6)	50.2	54.3
20:2 (n−6)	0.2	0.7
Unknown	0.2	1.3
20:3 (n−6)	1.9	0.8
20:4 (n−6)	25.1	1.0
Total LCP (n−3)	21.3	0.3

Source: Data from Sinclair *et al.* (1979).

factors. Both the amount and duration of gamma-linolenic acid feeding was much lower in their experiments, alpha-linolenic acid was present in the diet they used and would have competed for the desaturase, and these authors based their conclusions upon analysis of plasma not erythrocytes, where changes will be less marked.

In separate experiments where cats were fed linoleate-rich diets for 10 months, analysis of the plasma lipids confirmed the observations of Rivers *et al.*, showing a marked decrease in arachidonate and both *n*-6 and *n*-3 LCP and an elevation of 18:2 *n*-6 and 20:2 *n*-6, together with a six-fold increase in an unknown 20 Carbon fatty acid (Table 2). In order to identify this unknown fatty acid, the phospholipids were extracted from the liver, and urea solubilisation was used to concentrate the polyunsaturated fatty acids (Table 3). This fraction was subjected to silver nitrate column chromatography, and a fraction rich in fatty acids with three double bonds and containing this unknown fatty acid was obtained (Table 3). To identify the position of the double bonds, the trienoic fraction was subjected to

Table 3. *Fatty acid composition of fractions from the liver of cats fed a linoleate-rich diet*

Fatty acid	Fatty acid composition (g/100 g)			
	Phospholipid fraction	Urea-soluble fraction	Trienoate fraction	Hydrogenation of trienoate fraction
16:0	14.8	2.1	0.3	1.0
16:1	5.9	4.4	—	—
18:0	16.2	1.6	0.3	13.8
18:1	33.5	18.6	0.6	0.4
18:2 (n−6)	22.4	46.8	6.5	0.6
18:3 (n−6)	Trace	0.5	3.4	—
18:3 (n−3)	0.2	0.8	4.5	—
20:0	0.1	—	—	74.8
20:1	0.6	0.2	—	2.3
20:2 (n−6)	0.8	1.1	0.1	—
20:3 (n−9)	Trace	0.1	0.7	—
Unknown	0.7	3.9	34.1	3.3
20:3 (n−6)	0.7	5.3	48.1	3.9
20:4 (n−6)	0.7	11.9	0.4	—

Source: Data from Sinclair *et al.* (1979).

ozonolysis and reduction and the products identified by GLC (Table 3). This enabled the unknown to be identified as 5,11,14–20:3 (Fig. 2, Sinclair *et al.*, 1979). Since levels of this fatty acid increased markedly after linoleate feeding, it seems likely that it was a metabolite of linoleic acid, produced by the action of the Δ5-desaturase on 11,14–20:2, which is the chain elongation product of linoleate, 18:2 *n*-6, as Figure 1 shows. These results also supported the idea of an active Δ5-desaturase.

In the same experiments, cats were fed either normal diets or linoleate-rich diets and dosed orally with either [1 – 14C]-linoleate or [2 – 14C]-eicosa – 8,11,14-trienoate (8,11,14–20:3; *n*-6). The methyl esters of the total liver lipid fatty acids were then subjected to silver nitrate-thin layer chromatography which separates the fatty acids according to the number of double bonds present (Table 4). The results showed that when radiolabelled linoleate was fed, most of the radioactivity was confined to the dienoic region of the plate, with minimal activity in any other region. The very slight amount of activity present in the trienoic band was probably due to the formation of 5,8,14–20:3 as described earlier, together with the production of some 20:3 *n*-6 produced by the Δ8-desaturation of 20:2 *n*-6, rather than as a result of any Δ6-desaturase activity. In contrast, there were significant amounts of radioactivity from labelled 8,11,14–20:3 found in the tetraenoate band, providing further evidence of a Δ5 desaturase and the production of arachidonate.

In a further series of feeding studies, [1–14C] – eicosa 11,14-dienoate

Fig. 2. Identification of the unknown fatty acid.

$CH_3-(CH_2)_4-CH=CH-CH_2-CH=CH-(CH_2)_4-CH=CH-(CH_2)_3-\underset{\overset{\|}{O}}{C}-O-CH_3$

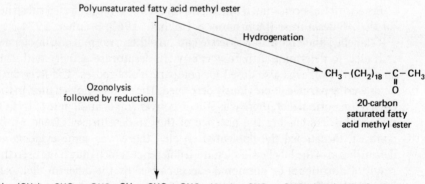

Polyunsaturated fatty acid methyl ester

Hydrogenation

$CH_3-(CH_2)_{18}-\underset{\overset{\|}{O}}{C}-CH_3$

20-carbon
saturated fatty
acid methyl ester

Ozonolysis
followed by reduction

$CH_3-(CH_2)_4-CHO + OHC-CH_2-CHO + OHC-(CH_2)_4-CHO + OHC-(CH_2)_3-\underset{\overset{\|}{O}}{C}-O-CH_3$

6-carbon aldehyde   Short chain dialdehydo-fragments   5-carbon aldehydo-ester

Structure identified as 5, 11, 14–20 : 3

Table 4. *Distribution of radioactivity from radiolabelled linoleic acid and eicosa-8,11,14-trienoic acid in cat liver fatty acids*[a]

AgNO3-TLC band	[1–14C]– 18:2		[2–14C]– 20:3	
	Normal diet	Linoleate-rich diet	Normal diet	Linoleate-rich diet
Saturated	200	98	5	73
Monoenoic	30	117	0	
Dienoic	7,603	17,497	17	105
Trienoic	85	293	4,748	22,167
Band between Δ3 and Δ 4	24	202	16	57
Tetraenoic	23	106	215	721
Penta and hexaenoic	36	25	14	90
Cholesterol	126	215	27	46

Note:
[a] cpm per fraction.
Source: Data from Sinclair *et al.* (1979).

(11,14–20:2; *n*-6) synthesised by the method of Sprecher (1977), was given to kittens fed a linoleate-rich diet. The fatty acids were isolated from the liver and the kidney, and the methyl esters subjected to silver nitrate thin layer chromatography. The results showed that there was substantial conversion of the radioactive 11,14–20:2 to trienoate and to a lesser extent to tetraenoate fatty acids. (Table 5). An analysis of the trienoate band showed it contained approximately 28% of 5,11,14–20:3 and 80% of 8,11,14–20:3. Reference to Figure 1 will make clear that the production of the latter indicates the existence of an active Δ8-desaturase in the cat.

Because it has been shown that alpha-linolenate is the preferred substrate for the Δ6-desaturase (Mohrhauer & Holman, 1963; Brenner, 1974), 1–14C-labelled linolenic acid was fed to cats on either a normal or a linolenate rich diet, to determine if there was any Δ6-desaturase activity with this substrate. A rat was also dosed for comparative purposes. The liver fatty acids were separated as previously described. The results showed that, in the cat fed the normal diet, there was still no conversion of linolenate to its LCP metabolites, confirming the absence of the Δ6-desaturase. (Table 6). In contrast, in cats fed the linoleate-rich diet, there was some evidence of desaturation of the labelled linolenate, a difference which may have been the result of nutritional or hormonal effects caused by the different diet. (see Jeffcoat, 1979). This desaturation would have been due to the presence of active Δ5- and Δ8-desaturases resulting in the production of 5,11,14,17–20:4 *n*-3 and 20:5 *n*-3. In addition some 22:6 *n*-3 may have been produced by Δ4-desaturation and the presence of these four fatty acids would have contributed to the radiolabel found in the tetra-, penta-, and hexa-enoic bands (see Fig. 1). The results of the rat experiment confirmed the validity

Table 5. *Distribution of radioactivity (dpm) in the fatty acids of cat liver and kidney after feeding [1-^{14}C]-eicosa-11,14-dienoate*

	Radioactivity	
AgNO$_3$-TLC band	Liver	Kidney
Dienoic	45,037	78,060
Band between Δ2 and Δ3	300	100
Trienoic	7,037	12,590
Tetraenoic	648	350

Source: Data from Monger (1986).

Table 6. *Distribution of radioactivity (dpm) in cat and rat liver fatty acids after feeding [1-^{14}C]-alpha linolenate*

	Cat		Rat
AgNO$_3$-TLC band	Normal diet	Linoleate-rich diet	Normal diet
Saturated and monoenoic	56	171	464
Trienoic	396	7,065	652
Band between Δ3 and Δ4	5	19	20
Tetraenoic	7	64	110
Penta- and hexaenoic	8	25	732

Source: Data from Monger (1986)

of the technique and the presence of a very active desaturase in this species.

Cats were maintained on either a normal diet or experimental semi-synthetic EFA deficient diet in which the only source of lipid was hydrogenated beef fat. After six months on the experimental diet the animals showed signs of EFA deficiency and the fatty acid composition of kidney lipids of the animals is shown in Table 7. In the deficient animals there appeared an unknown fatty acid which was isolated and characterised as 5,8,11–20:3, which is the classic indicator of EFA deficiency (Sinclair *et al.*, 1981). In the absence of a Δ6-desaturase this was produced by the action of the Δ8- and Δ5-desaturases, as outlined in Figure 1. Because the cat synthesised this fatty acid, it clearly had the ability to convert the dietary EFA into their LCP metabolites.

As a guide to determining the EFA status of the cat the profiles of liver fatty acids on different diets are shown in Table 8. The major features are the high levels of the *n*-6 family of fatty acids compared to the *n*-3 family even in the

Table 7. *Fatty acid composition (g/100 g) of kidney total lipids from normal and EFA deficient cats*

Fatty acid	Normal diet	EFA deficient diet
16:0	13.2	12.7
16:1	2.3	6.0
18:0	22.5	20.1
18:1 ($n-9$)	20.0	38.5
18:2 ($n-6$)	12.0	4.5
18:3 ($n-3$)	0.2	0.1
20:1 ($n-9$)	0.2	0.3
20:2 ($n-6$)	0.2	—
Unknown	—	2.3
5,11,14–20:3	—	0.3
20:3 ($n-6$)	0.4	0.6
20:4 ($n-6$)	14.3	4.0
20:5 ($n-3$)	1.6	0.6
20:6 ($n-3$)	3.0	0.1
Total LCP	29.4	12.3

Source: Data from Sinclair *et al.* (1981).

linolenate enriched experimental diets, and under all conditions there was an accumulation of saturated and monounsaturated fatty acids. The levels of 20:3 *n*-9 or the ratio of 20:3 *n*-9 to 20:4 *n*-6 are not reliable indicators of EFA status of the cat as they are in some other species, and therefore it is necessary to use features such as the absolute amount of EFA and their LCP metabolites to identify deficiencies in this species. (McLean & Sinclair, 1986).

Overall, this work confirmed the suggestion that the Δ6-desaturase was absent in the cat, but it demonstrated that the Δ5- and Δ8-desaturases were active and that therefore a pathway existed which allowed limited synthesis of LCP from dietary EFA. However, as noted earlier, while cats survived for long periods on semi-synthetic diets containing only linoleate, these animals were not normal and showed changes in their livers and decreased reproductive capacity. Presumptively, therefore, the capacity for desaturation via this alternative pathway is insufficient to meet the physiological needs for EFA in the cat, and a dietary requirement for LCP does in fact exist.

Dietary EFA requirements of the cat

Direct proof of the inadequacy of dietary EFA for maintaining normal physiology in the cat comes from three separate sets of observations: those of Rivers *et al.* and of Sinclair *et al.* already alluded to, and those of

Table 8. *Liver fatty acids in cats fed various diets[a]*

		Diet		
Fatty acid	Normal[b]	EFA[c] deficient	Linoleic[d] rich	Linoleic plus linolenic[e]
16:0	5,190			
18:0	6,450	} 1,860	} 12,169	} 9,195
20:0	10			
16:1	215	1,280	} 6,582	} 4,693
18:1 $(n-9)$	4,650	6,620		
18:2 $(n-6)$	3,010	130	15,292	4,910
20:2 $(n-6)$	20	105	135	75
20:3 $(n-6)$	90	25	30	60
20:4 $(n-6)$	2,220	90	630	710
18:3 $(n-3)$	20	200	30	2,070
20:5 $(n-3)$	130	30	5	65
22:6 $(n-3)$	425	5	5	70
20:3 $(n-9)$	20	50	5	5
5,11,14-20:3	30	5	150	0

Notes:

[a] Expressed as μg/g tissue.

[b] Commercial cat food (50% canned, 50% dry).

[c] Containing 29.5 g hydrogenated beef fat/100 g dry diet.

[d] Containing 20.5 g hydrogenated beef fat plus 3.85 g linoleic/100 g dry diet.

[e] Containing 20.5 g hydrogenated beef fat plus 1.07 g linoleic and 2.46 g linolenic/100 g dry diet.

Source: Data from Monger (1986).

MacDonald *et al.* (1984) in the USA. This latter group of authors were also unable to sustain normal reproductive function in cats maintained on diets that provided only linoleate as an EFA source.

This does not mean that linoleate is not essential for the cat. MacDonald *et al.* (1984) and Rivers & Frankel (1980) both studied the impact of dietary linoleate on skin permeability to water and concluded that, in the cat as in other mammal species, linoleic acid is essential for the regulation of skin permeability, which in desaturating species rises in EFA deficiency. Frankel was able to demonstrate that skin permeability can be reduced to very low levels in cats fed on diets providing only linoleate and linolenate in parallel with the elevation of linoleate in tissues of these animals. While it was not suggested that this reduced skin permeability to water was in any way pathological, it was associated with very low basal metabolic rates in these animals attributed by Rivers & Frankel to a reduced need for energy to meet the needs for evaporative water loss.

Frankel & Rivers (1978) reported that the clinical condition of their linoleate fed cats improved when evening primrose oil (providing gamma linolenic acid, 18:3 n-6) was added to the diet, but that the cats still were not normal. Skin and coat condition were improved and wound healing was normalised, female animals underwent oestrus, and sometimes became pregnant but no litters of viable young were delivered. The authors concluded that a mixture of linoleate and gamma linolenate was not able to meet the complete EFA requirements of the cat, though this failure to normalise clinical condition may be attributed to the low dose and duration of gamma-linolenic acid used by these authors (see above). Macdonald *et al.* fed supplements of arachidonic acid to cats fed linoleate-rich diets and were able to show that with these supplements at a level of 200 mg/kg diet (0.04% of the dietary energy) the cats were able to reproduce, with litters of viable kittens being delivered. These authors concluded that arachidonic acid is a dietary essential for reproduction in the female cat, at least at a level of 0.04% of the energy. However, when tuna oil was added to the diet, 0.04% of the energy as arachidonic acid was insufficient (MacDonald *et al.*, 1984), possibly because the n-3 LCP provided by the tuna oil were preferentially incorporated into tissue phospholipids, or because the tuna oil depressed the activity of the Δ5-desaturase (NRC, 1986). Since it is difficult to believe that the cat does not require n-3 LCP in the diet it seems likely that the arachidonic acid requirement in practical diets would be above 0.04%.

Conclusions

The metabolism by the cat of the 18 carbon EFA, linoleic acid and alpha-linolenic acid is unusual, because of undetectable activities of the Δ6-desaturase which is the rate-limiting enzyme in the metabolism in most mammals. LCP production does, however, occur in the cat by an alternative pathway involving the Δ5- and Δ8-desaturases which the cat, like other mammals, also possesses. However, rates of production are sufficiently slow that the cat exhibits a dietary requirement for arachidonic acid in addition to a requirement for linoleic acid. The role of alpha-linolenic acid and the n-3 LCP in cat nutrition has yet to be shown. There is every reason to think that it will require a dietary supply of at least some n-3 EFA but whether that will be met by alpha-linolenic acid or whether a n-3 LCP requirement will be shown to exist is at present unknown.

Acknowledgements

This work was supported by a grant from the Australian Research Grants Scheme. We thank Mrs K.J. Herbison and Mrs A.J. Killey for their assistance.

REFERENCES

Albert, D.H. & Coniglio, J.G. (1977). Metabolism of eicosa-11,14-dienoic acid in rat testes. Evidence for $\Delta8$-desaturase activity. *Biochemica & Biophysica Acta,* **489**, 390–6.

Brenner, R.R. (1974). The oxidative desaturation of unsaturated fatty acids in animals. *Molecular & Cellular Biochemistry,* **3**, 41–52.

Burr, G.O. & Burr M.M. (1929). A new deficiency disease produced by the rigid exclusion of fat from the diet. *Journal of Biological Chemistry,* **82**, 345–67.

Burr, G.O. & Burr, M.M. (1930). On the nature and role of the fatty acids essential in nutrition. *Journal of Biological Chemistry,* **86**, 587–621.

Frankel, T.L. & Rivers, J.P.W. (1978). The Nutritional and metabolic impact of γ-linolenic acid ($18:3\omega6$) on cats deprived of animal lipid. *British Journal of Nutrition,* **39**, 227–31.

Hassam, A.G., Rivers, J.P.W. & Crawford, M.A. (1977). The failure of the cat to desaturate linoleic acid: Its nutritional implications. *Nutrient Metabolism,* **21**, 321–8.

Holman, R.T. (1971). Essential fatty acid deficiency. *Progress in Chemistry of Fats and Other Lipids,* **9**, 275–348.

Jeffcoat, R. (1979). The biosynthesis of unsaturated fatty acids and its control in mammalian liver. *Essays in Biochemistry,* **15**, 1–36.

MacDonald, M.L., Anderson, B.C., Rogers, Q.R., Buffington, C.A. & Morris, J.G. (1984). Essential fatty acid requirements of cats: Pathology of essential fatty acid deficiency. *American Journal of Veterinary Research,* **45**, 1310–17.

MacDonald, M.L., Rogers, Q.R. & Morris, J.G. (1984). Nutrition of the domestic cat, a mammalian carnivore. *Annual Reviews of Nutrition,* **4**, 521–62.

McLean, J.G. & Sinclair, A.J. (1986). Assessment of essential fatty acid status. *Proceedings of the XIII International Congress of Nutrition,* 350–2.

Mead, J.F. (1986). Functions of the *n*-6 and *n*-3 polyunsaturated fatty acids. In *Proceedings of the XIII International Congress of Nutrition,* ed. T.G. Taylor & N.K. Jenkins, pp. 346–9. London: John Libbey.

Mohrhauer, H. & Holman, R.T. (1963). The effect of dose level of essential fatty acids upon fatty acid composition of the rat liver. *Journal of Lipid Research,* **4**, 151–9.

Monger, E.A. (1986). Polyunsaturated fatty acid metabolism in the cat. PhD Thesis, University of Melbourne, Australia.

Nakazawa, I., Mead, J.F. & Yonemoto, R.H. (1976). *In vitro* activity of the fatty acyl desaturases of human cancerous and noncancerous tissues. *Lipids,* **11**, 79–82.

Rivers, J.P.W. & Frankel, T.L. (1980) Fat in the diet of cats and dogs. In *Nutrition of the Dog and Cat,* ed. R.S. Anderson, pp. 67–99. Oxford: Pergamon Press.

Rivers, J.P.W., Sinclair, A.J. & Crawford, M.A. (1975). Inability of the cat to desaturate essential fatty acids. *Nature (London)*, **258**, 171–3.

Sinclair, A.J., McLean, J.G. & Monger, E.A. (1979). Metabolism of linoleic acid in the cat. *Lipids*, **14**, 932–6.

Sinclair, A.J., Slattery, W., McLean, J.G. & Monger, E.A. (1981). Essential fatty acid deficiency and evidence for arachidonate synthesis in the cat. *British Journal of Nutrition*, **46**, 93–6.

Sprecher, H. (1977). The organic synthesis of polyunsaturated fatty acids. In *Polyunsaturated fatty acids*, ed. Kunau, W. and Holman, R.T., pp. 69–79. Champaign: American Oil Chemists' Society.

Tinoco, J. (1982). Dietary requirements and functions of α-linolenic acid in animals. *Progress in Lipid Research*, **21**, 1–45.

21

Lipoprotein cholesterol distribution in experimentally induced canine cholestasis

JOHN E. BAUER, DENNIS J. MEYER, ROBERT L. GORING, C. HENRY BEAUCHAMP, JULIA JONES

The modification of serum lipids and lipoproteins by diet is a well-established phenomenon (Nelson, 1983). Liver disorders, including obstructive jaundice, are frequently accompanied by dramatic changes in both serum and tissue lipids because of the central role which the liver plays in lipid metabolism (Seidal, Alaupovic & Furman, 1969; Naito, 1984). Experimental bile duct ligation in dogs has previously revealed the occurrence of an abnormal low density lipoprotein, lipoprotein-X (LP-X), many features of which are identical to those of the human abnormal lipoprotein of cholestasis (Ritland & Bergan, 1975). The isolation and analysis of this lipoprotein in the dog has revealed a particle which contains approximately 5% protein, 59% phospholipid, 28% total cholesterol and 8% triglyceride while in the human it contains 6% protein, 66% phospholipid, 25% total cholesterol and 3% triglyceride (Seidel *et al.*, 1976). More generally, in both man and dog, although LP-X has a flotation density (d) similar to that of low density lipoprotein (LDL) ($1.019 < d < 1.063$ g/ml), it is characterised by both an unusually high proportion of plasma phospholipid and unesterified cholesterol and by a low protein content, comprising apo-peptides B and C and albumin (Naito, 1984).

Dogs which have been subjected to ligation and transsection of the common bile duct rapidly recover, continue to thrive and maintain vigour. Consequently, the use of this model in combination with subsequent dietary modification can lead to new, potentially useful, information on lipoprotein metabolism and cholestatic diseases and their possible treatment. In many cases of hepatic disease in the dog, dietary modifications are utilised as part of the treatment. The objectives of this study were to provide further information on the serum lipids and lipoproteins in experimental canine cholestasis and to obtain baseline data on the impact after feeding commercially available diets to dogs with surgically obstructed bile ducts.

343

Methods

Six mixed breed dogs of both sexes and varying age were maintained on a commercially available chow diet (Purina dog chow, Ralston Purina Co., St. Louis, MO, USA) prior to the experiment. Surgical ligation and transsection of the common bile ducts in four of the dogs were performed under general anaesthesia. The remaining two dogs were sham-operated and used as controls. After recovery, all dogs were again maintained on the commercial chow diet. The declared composition of this diet was: crude protein, $\geq$ 25%; crude fat, $\geq$ 9%; crude fibre, $\leq$ 4%; moisture, $\leq$ 12%; ash, $\leq$ 10%, and minerals, $\leq$ 2%. Source of dietary ingredients included: ground yellow corn, soybean meal, meat and bone meal, ground wheat, animal fat, ground beef pulp, dried milk product, calcium carbonate, fish meal, and added vitamins B_{12}, D_3, and E.

Serum samples were collected after an overnight fast immediately before surgery (day 0) and after surgery on days 3, 7, 10, 14, and 17. Serum total cholesterol, triglyceride, and total and direct bilirubin concentrations were determined by appropriate chemical and enzymatic methods (Anonymous, 1974; Neri & Frings, 1973; Walters & Gerarde, 1970). On samples collected on days 0 and 7 lipoprotein fractionation was performed by single-spin

Fig. 1. Serum bilirubin concentrations of dogs, mean ± SEM. Abbreviations used: C, control; L, ligated. Direct-reacting (or conjugated) bilirubin concentrations are indicated by the cross-hatched bars, indirect (or unconjugated) bilirubin by the open bars and total bilirubin concentrations by the sum of both cross-hatched and open bars. Standard errors are shown for total and direct-reacting bilirubin determinations only.

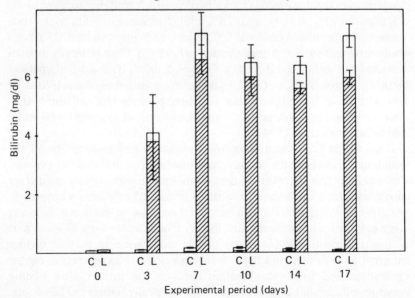

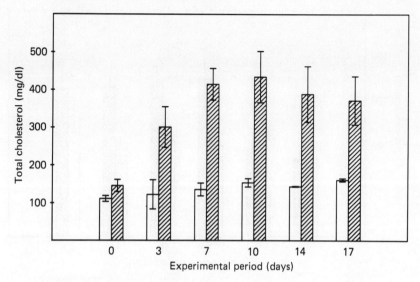

Fig. 2. Serum total cholesterol concentrations of dogs, mean ± SEM. Open bars, control; cross-hatched bars, ligated.

density gradient ultracentrifugation (Terpstra, Woodward & Sanchez-Muniz, 1981). Lipoprotein classes, prestained with Sudan black IV, were photographed and subsequently collected as six fractions. Total cholesterol and protein concentrations of collected lipoprotein fractions were measured (Lowry *et al.*, 1951).

Results
The time course of alterations in serum bilirubin concentrations can be found in Figure 1. Hyperbilirubinemia, characteristic of cholestasis, was evident by day 3 after surgery with a constant (presumably maximal) elevation 44 fold greater than controls observed from day 7 onwards. The majority of this increase was due to the direct-reacting (or conjugated) bilirubin which consistently accounted for 80–90% of the total serum bilirubin concentration in the experimental group. Changes in serum cholesterol and triglyceride concentrations after bile duct ligation are shown in Figures 2 and 3. Again, in the cholestatic animals, maximal elevation of these serum lipids were observed on day 7 and remained fairly constant thereafter. Overall, serum cholesterol concentrations were elevated approximately 4-fold while serum triglyceride concentrations were somewhat less than doubled.

Figure 4 shows the density gradients of prestained sera from the 6 animals on day 7. Density ranges of the lipoprotein fractions are given as a footnote to Table 1. The relative cholesterol distribution among the canine lipoprotein classes of both experimental and control dogs on days 0 and 7 are

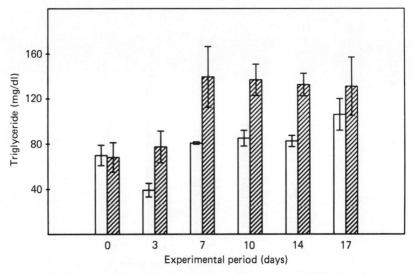

Fig. 3. Serum triglyceride concentrations of dogs, mean ± SEM. Open bars, control; cross-hatched bars, ligated.

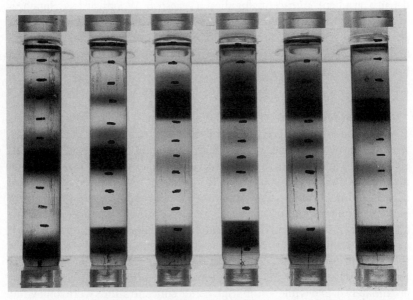

Fig. 4. Density gradient ultracentrifugation of dog serum. Samples prestained with Sudan Black IV before centrifugation. Left to right; 2 control and 4 bile duct ligated samples.

Table 1. *Per cent cholesterol distribution among canine lipoprotein classes.*
(Mean + SEM)

Density	Day 0		Day 7	
	Control	Ligated	Control	Ligated
VLDL	1.8 + 0.4	2.1 + 0.1	1.4 + 0.6	2.8 + 0.8
LDL1	4.9 + 0.2	13.4 + 4.0	8.5 + 5.7	42.1 + 6.0
LDL2	7.7 + 0.1	18.4 + 3.5	13.7 + 0.5	42.2 + 7.0
LDL3	14.6 + 5.0	11.4 + 0.7	22.0 + 3.8	6.8 + 0.4
HDL1	65.4 + 4.3	47.1 + 4.9	50.0 + 9.7	5.0 + 0.5
HDL2	1.4 + 0.1	5.1 + 2.0	3.5 + 0.9	0.3 + 0.1

Note:

Density ranges (g/ml) are: VLDL, $d < 1.0102$; LDLI, $1.0102 < d < 1.0210$; LDL2,
$1.0210 < d < 1.0425$; LDL3, $1.0425 < d < 1.0578$; HDLI, $1.0578 < d$
< 1.0981; HDL2, $1.0981 < d < 1.1395$.

shown in Table 1. Before surgery the relative percent cholesterol distribution was similar in the two groups. By day 7, cholesterol distributions were
dramatically altered resulting in a markedly higher percentage of the
cholesterol being found in the LDL_1 and LDL_2 fractions in the ligated
animals. Concomitant decreases in per cent total cholesterol were seen in
the LDL_3. HDL_1 and HDL_2 fractions. The absolute cholesterol concentrations
of lipoprotein classes in ligated dogs on day 7 were similarly affected as
Figure 5 shows. Absolute cholesterol concentrations were increased in the

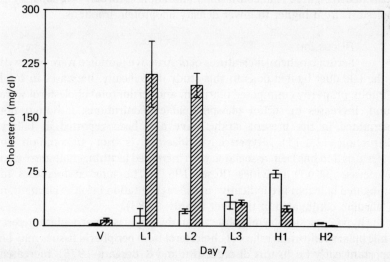

Fig. 5. Postsurgical (Day 7) lipoprotein class cholesterol concentrations of dogs,
mean ± SEM. Open bars, control; cross-hatched bars, ligated. For abbreviations
used see Table 1.

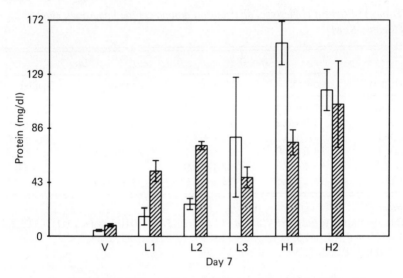

Fig. 6. Postsurgical (Day 7) lipoprotein class protein concentrations of dogs, mean ± SEM. Open bars, control; cross-hatched bars, ligated. For abbreviations used see Table 1.

LDL_1 and LDL_2 fractions with a major decrease observed in the HDL_1 fraction. No changes were observed in the cholesterol content of the other lipoprotein fractions. The protein concentrations of these lipoproteins on day 7 are shown in Figure 6. Increased protein concentrations were observed in both the LDL_1 and LDL_2 fractions, whereas, in the HDL_1 fraction it was somewhat decreased. Cholesterol to protein ratios for the lipoproteins, illustrated in Figure 7, indicate a shift in the cholesterol carrying capacity in cholestasis from higher to lower density lipoprotein fractions.

Discussion

Serum biochemical features of obstructive jaundice were observed in the bile duct ligated dogs in this study. Specifically, increases in total bilirubin, primarily conjugated bilirubin, and serum total cholesterol were found. Increases in serum phospholipid concentrations, although not determined in the present study, have also been reported in biliary obstruction (2,4,11). Hypertriglyceridaemia is not uncommon in cholestasis and has been associated with increased lecithin: cholesterol acyl transferase (LCAT) activities (Rose, 1981). The marked increases in conjugated bilirubin are indicative of the regurgitation into the circulation of bilirubin conjugated in the liver (Hardy, 1983).

The hypercholesterolaemia observed in this study is of particular interest. While biliary reflux and influx of cholesterol from peripheral tissues may be important early on in this disorder (Ritland & Bergan, 1975), increased cholesterol synthesis may also operate in biliary disease (Seidel *et al.*, 1969; Rose, 1981). Consequently, the rise in serum cholesterol and other lipids in this condition may exceed the capacity of the LCAT system and result in the formation of special lipoprotein particles as transport vehicles for the lipids. Evidence for these special lipoprotein particles has been found in the present study. The lipoprotein pattern observed was characterised by a decreased concentration of HDLs and an increased concentration of LDLs. Over 80% of the increased serum cholesterol was found in the LDL fractions. Similar findings have been documented in man, rat and dog and have been shown to be the result of the presence of an abnormal low density lipoprotein, LP-X (Naito, 1984; Felker, Hamilton & Havel, 1978). This lipoprotein has also been further characterised with respect to its chemical and physiochemical properties as well as its structural relationship of proteins to lipids (Mills, Seidel & Alaupovic, 1969; Hamilton *et al.*, 1971; Picard *et al.*, 1972). Its presence has been demonstrated to be exclusively limited to patients with cholestasis and obstructive jaundice (Seidel, Gretz & Ruppert, 1973; Ritland *et al.*, 1963).

Elevations in serum total cholesterol concentration in clinical cases of canine cholestasis may not be of sufficient magnitude to yield diagnostic or prognostic information. For example, the mean total serum cholesterol concentration of cholestatic dogs in the present work was 400 mg/dl, but we have observed values in apparently normal animals as high as 300 mg/dl (Bauer, unpublished observations). Consequently, the serum total choles-

Fig. 7. Postsurgical (Day 7) lipoprotein class cholesterol/protein ratios of dogs, mean ± SEM. Open bars, control; cross-hatched bars, ligated. For abbreviations used see Table 1.

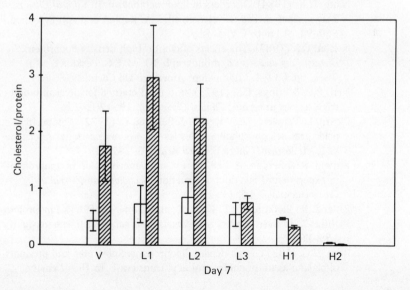

terol concentrations in many cases of liver disease may be of limited value. However, the use of the single spin density ultracentrifugal method as employed in the present work not only enables the rapid fractionation and quantitation of serum lipoproteins within the density range 1.006 to 1.140 g/ml but also an assessment of their cholesterol carrying capacity. It can thus be helpful in the evaluation of various therapeutic regimens, including dietary modification, in both spontaneous and experimentally induced liver diseases.

REFERENCES

Anonymous. (1974). Lipid and lipoprotein analysis. In *Manual of Laboratory Operations: Lipid Research Clinics Program*, vol. 1, DHEW publication No. (NIH) 72–628. Bethesda: NHLBI–NIH.

Felker, T.E., Hamilton, R.L. & Havel, R.J. (1978). Secretion of lipoprotein-X by perfused livers of rats with cholestasis. *Proceedings of the National Academy of Science, USA*, **75**, 3459–63.

Hamilton, R.L., Havel, R.J., Kane J.P., Blaurock, A.E. & Sata, T. (1971). Cholestasis: Lamellar structure of the abnormal human serum lipoprotein. *Science*, **172**, 475–8.

Hardy, R.M. (1983). Diseases of the liver. In *Textbook of Veterinary Internal Medicine. Disease of the Dog and Cat*, 2nd edn. ed. S.J. Ettinger, pp. 1372–434. Philadelphia: W.B. Saunders Company.

Lowry, O.H., Rosenbrough, N.J., Farr, A.L. & Randall, R.J. (1951). Protein measurement with the folin phenol reagent. *Journal of Biological Chemistry*, **193**, 265–75.

Mills, G.L., Seidel, D. & Alaupovic, P. (1969). Ultracentrifugal characterization of a lipoprotein occurring in obstructive jaundice. *Clinica Chimica Acta*, **26**, 239–44.

Naito, H.K. (1984). Disorders of lipid metabolism. In *Clinical Chemistry: Theory, Analysis and Correlation*, ed. L.A. Kaplan and A.J. Pesce, pp. 550–93. St Louis: C.V. Mosby.

Nelson, G.J. (1983). The effects of diet on high density lipoprotein. In *Dietary Fats and Health*, monograph 10, ed. E.G. Perkins & W.J. Visek, pp. 632–4. Champaign: American Oil Chemists' Society.

Neri, B.P. & Frings, C.S. (1973). Improved method for determination of triglycerides in serum. *Clinical Chemistry*, **19**, 1201–2.

Picard, J., Veissiere, D., Voyer, F. & Bereziat, G. (1972). Composition en acids gras des phospholipids des les lipoproteines seriques anormales de la cholestase. *Clinica Chimica Acta*, **36**, 247–50.

Ritland, S. & Bergan, A. (1975). Plasma concentration of lipoprotein-X in experimental bile duct obstruction. *Scandanavian Journal of Gastroenterology*, **10**, 17–23.

Ritland, S., Bloomhuff, J.P. Elgio, K. and Gjone, E. (1963). Lipoprotein-X (LP-X) in liver disease. *Scandinavian Journal of Gastroenterology*, **8**, 155–160.

Rose, H.G. (1981). High density lipoproteins: Substrates and products of plasma lecithin: cholesterol acyl transferase. In *High* Density

Lipoproteins, ed. C.E. Day, pp. 213–280. New York: Marcel Dekker Incorporated.

Seidel, D., Alaupovic, P. & Furman, R.J. (1969). A lipoprotein characterizing obstructive jaundice. I. Method for quantitative separation and identification of lipoproteins in jaundiced subjects. *Journal of Clinical Investigation*, **48**, 1211–23.

Seidel, D., Buff, H.U., Fauser, U. & Bleyl, U. (1976). On the metabolism of Lipoprotein-X (LP-X). *Clinica Chimica Acta*, **66**, 195–207.

Seidel, D., Gretz, H. & Ruppert, C. (1973). Significance of the LP-X test in differential diagnosis of jaundice. *Clinical Chemistry*, **19**, 86–91.

Terpstra, A.H.M., Woodward, C.J.H. & Sanchez-Muniz, F.J. (1981). Improved techniques for the separation of serum lipoproteins by density gradient ultracentrifugation: Visualization by prestaining and rapid separation of serum lipoproteins from small volumes of serum. *Analytical Biochemistry*, **111**, 149–57.

Walters, M. & Gerarde, H. (1970). An ultramicro method for determination of conjugated and total bilirubin in serum or plasma. *Microbiology Journal*, **15**, 231.

22

The role of fluid in the feline urological syndrome

C.J. GASKELL

The Feline Urological Syndrome (FUS, urolithiasis, urethral obstruction, cystitis) has been a problem in feline medicine for many years. In the United Kingdom an apparent increase in incidence was noted some 15 years ago, and there has been, and continues to be, much discussion of the underlying factors involved. Clinically, FUS may present as a cystitis, with haematuria and increased frequency of urination, or as urethral obstruction, with dysuria or anuria and the development of post-renal failure. The obstructing material is a variable mixture of magnesium ammonium phosphate crystals ($Mg\ NH_4\ PO_4 \cdot 6H_2O$), or struvite, and proteinaceous material, often with the consistency of gritty toothpaste. This material accumulates in the bladder and urethra, becoming moulded in the latter to form the obstructing plug at a narrow point. Osborne *et al.* (1984) have rightly warned against the use of FUS as a convenient label for all undiagnosed lower urinary tract disorders, but the syndrome is recognised clinically and epidemiologically as a distinct entity of an unknown aetiology. The absence of a known aetiology for FUS makes it a frustrating condition for clinicians to manage, although it is generally accepted that the syndrome is associated with a number of factors acting together; these factors, or their combination, may vary between individual cases. In an attempt to identify these factors, a number of epidemiological surveys were carried out in the United Kingdom, USA and Denmark in the 1970s (Willeberg, 1975; Reif *et al.*, 1977; Walker *et al.*, 1977).

Table 1 lists those features identified in the UK survey, based on some 430 cases of FUS and 600 control cats, (Walker *et al.*, 1977) as more likely to be found in a cat with FUS: other surveys were in broad agreement. A picture emerges of a sedentary fat-cat-on-the-mat type of animal: of the factors which might more easily be manipulated, that of diet has received the most attention. The epidemiological surveys gave no indication as to how diet, especially low-moisture or dry diets, might exert an effect, but various factors such as moisture content, ionic concentrations (particularly magnesium) and resultant urinary pH have all been suggested. It is logical, given

353

Table 1. *Factors epidemiologically associated with Feline Urological Syndrome (FUS) (Walker et al., 1977)*

A cat with FUS is more likely
to be > 2 years of age neutered, male
to take little exercise
to have less freedom to leave the house
to have a litter tray
to eat dry cat food
to drink less
to have had a previous episode of FUS
(to be lazy and overweight)

that struvite crystals are an integral part of the obstructing material, to assume that all of these factors are relevant as each may effect the concentration or solubility of the relevant urinary ions.

The epidemiological support for the hypothesis that a reduced fluid intake is important in FUS is based on the identification of both dry cat food and the tendency to drink less fluid as factors. Indeed, where both these factors occur together, the relative risk of FUS is enhanced (Willeberg, 1981). A number of different studies have been carried out to look at the effect of different diets and diet types on fluid intakes and urine outputs. While there has been some measure of agreement there has also been apparent conflict. Disagreement has been due in part to the large number of different dietary factors which may influence fluid balance and the failure of many studies to isolate individual components. Dry matter content of the diet is the most obvious factor, but formulation and nutrient composition, salt content, method of manufacture and amount of diet actually consumed also have a bearing on fluid intake and urinary excretion. Many of these dietary variables can be controlled by the use of semi-purified diets.

In our studies, such a diet was fed at different moisture levels (75%, 45% and 10%) to 16 adult, entire, specific-pathogen-free (SPF)-derived cats. Details of the diet and experimental design are published elsewhere (Gaskell, 1979). The results obtained are shown in Table 2. Total dry matter intake was similar on all diets: fluid intake differed significantly between the high-moisture (wet) and the low-moisture forms of the diet. On the wet diet, the majority (92%) of water intake was obligatory, being taken with the food: on the dry diet (10% moisture) the majority of the water intake (94%) was taken voluntarily and the cat did not drink water to the level of the fluid intake achieved on the wet food. On the diet of intermediate moisture level (45%), the amount of water derived from the food and from drinking were more evenly balanced (39:61), yet the mean value showed that the cats achieved a similar value for total fluid intake to that found for dry diet. These differences between diets can be represented using the ratio total water

Table 2. *Data from sixteen adult cats fed on semi-purified diet at different moisture contents*

Dietary moisture (%)	75	45	10
Food intake (g/day)	204	96	62
Dry matter intake (g/day)	51	53	56
Water in food (ml/day)	153	43	6
Water drunk /ml/day)	17	68	102
Total water intake /ml/day)	170	111	108
Daily water in food (%)	90	39	6
TWI:DMI	3.4	2.2	2.0
Urine volume (ml/day)	96	50	57
Urine osmolarity	1,213	2,100	2,075

intake to dry matter intake (TWI:DMI) and the values found here are in broad agreement with other studies using a variety of diets, mainly commercial, of different formulations and types. Ratio values for dry-type foods are generally between two and three, and for wet-type foods are above three (Table 3). These differences in fluid intake are reflected in the urinary parameters, cats on the dry diet producing small volumes of more concentrated urine (Table 2). It seems clear that fluid intake and urinary output may be influenced by the moisture content of the diet where other dietary variables are controlled. Although cats drank progressively more water in line with the decrease in diet moisture, the *total* water intake was independent of food moisture across the range of 10 to 45%. So it appears that, when given low moisture foods, the cats adjusted their total fluid balance to give a TWI:DMI ratio of around 2.0. It is difficult to correlate this observation with the development of FUS: none of the cats in this study developed the syndrome, nor do the majority of cats receiving dry food in the general population. It is logical, however, to assume that more concentrated urine represents an environment more conducive to the formation of struvite crystals and thus, presumably in the presence of other factors, to predispose to FUS. A similar, but perhaps more important point, arises from the clear observation that a large percentage of the fluid intake on dry diets is voluntary. The opportunity therefore exists for wide variations in intake to occur, and the mean values quoted in Table 2 do indeed conceal a considerable degree of between – and within – cat variation in daily fluid intake which is independent of dry matter consumption. In the home environment an individual pet cat's idiosyncratic tendencies to a low voluntary fluid intake may be compounded by a lack of sufficient fluid for drinking. It may be that voluntary or enforced episodes of reduced fluid intake, and the consequential rise in urinary concentration and solute load, provide an appropriate environment which, together with other factors,

Table 3. *Ratios of total water intake to dry matter intake for diets of different moisture content, from various studies (after Anderson, 1983)*

	A	B	C	D	E	F*
Dry food	2.33	2.3	2.3	2.3	2.5	2.0
	2.00			2.6	2.7	
	2.43					
Intermediate moisture food						2.2
High moisture (canned food)	3.45	5.5	3.9	3.9	4.9	3.4
	2.99	5.7				
	3.69					

Note:
* Data from Table 2.

may lead to clinical disease. That the potential for such episodes exists on dry cat food in a way that it does not on high moisture diets may well explain the epidemiological association of dry foods and FUS.

Such lines of argument support a role for fluid in the pathogenesis of Feline Urological Syndrome. As one of a number of factors, others of which may also be dietary in origin, changes in fluid balance may 'unmask' or 'find out' an individual cat already predisposed in other ways. To the clinician the importance of fluid intake (as with some other dietary factors), is that it is to some extent amenable to manipulation.

REFERENCES

Anderson, R.S. (1983). Fluid Balance and Diet *Proceedings of the 7th Kal Kan Symposium*, pp. 19–25. Vernon, California: Kal Kan Foods Inc.

Gaskell, C.J. (1979). Studies on the feline urological syndrome PhD Thesis, University of Bristol.

Osborne, C.A., Johnston, G.R., Polzin, D.J., Kruger, J.M., Bell, F.W., Poffenbarger, E.M., Feeney, D.A., Stevens, J.B. and McMenomy, M.F. (1984) Feline Urologic Syndrome: A Heterogeneous Phenomenon? *Journal of the American Animal Hospital Association*, **20**, 17–32.

Reif, J.S., Bovee, K., Gaskell, C.J., Batt, R.M. and Maguire, T.G. (1977). Feline Urethral Obstruction: a case control study. *Journal of American Veterinary Medical Association*, **170**, 1320–4.

Walker, A.D., Weaver, A.D., Anderson, R.S., Crighton, G., Gaskell, C.J. and Wilkinson, G.T. (1977). An epidemiological survey of the feline urological syndrome. *Journal of Small Animal Practice*, **19**, 283–301.

Willeberg, P. (1975). Diets and the Feline Urological Syndrome: a retrospective case-control study. *Nordisk Veterinaermedicin*, **27**, 15–19.

Willeberg, P. (1981). Epidemiology of the feline urological syndrome. *Advances in Veterinary Science and Comparative Medicine*, **25**, 311–44.

23

The role of diet in feline struvite urolithiasis syndrome

C.A. BUFFINGTON, N.E. COOK, Q.R. ROGERS AND J.G. MORRIS

Introduction

Feline Urological Syndrome (FUS), is a general term coined by Osbaldiston & Taussig (1970) to describe a variety of feline lower urinary tract diseases. The incidence of this syndrome in the cat population of the United States is approximately 0.85% (Lawler, Sjolin & Collins, 1985), and it accounts for up to 10% of veterinary clinic admissions of cats. The probability of reobstruction after an initial urethral blockage is high, and in one study the 6-month case fatality rate following initial obstruction was 36% (Walker et al., 1977). Lewis & Morris (1984a) have stated that 'FUS is by far the main health concern of cat owners'.

Although the descriptive term 'FUS' is relatively new, the occurrence of feline urolithiasis is not. For example, the Dresden Pathological Institute found evidence of the syndrome in 0.22% of cats examined between 1862 and 1897 (Hutyra, Marek & Manninger, 1938). Kirk describes the condition as 'very common' in his 1925 text, and numerous other reports of the problem have appeared over the years (Blount, 1931; Milks, 1935; Krabbe, 1949; Stansbury & Truesdail, 1955; Whitehead, 1964; Buffington et al., 1985). A wide variety of potential causative factors have been suggested, and the aetiology of the syndrome is currently thought to be multifactorial (Osborne et al., 1984c). Kirk (1925) reported that 'urinary deposits are believed to be induced by a too liberal diet of highly nutritious food, coupled with close confinement to the house'. He also stated that the disease was seen more frequently in Persians than in short-haired cats, usually occurred in adults 2–10 years of age, and that the possibility of a hereditary component existed. He discounted the effects of castration on the incidence of the disease because, 'The condition is not by any means infrequently encountered in entire males'. In 1931, Blount listed the 'chief factors which control the formation of a calculus (as): (1) The reaction of the urine; (2) presence of a nucleus; (3) excess of foods rich in salts and organic acids; (4) metabolic disturbances associated with abnormal excretion of salts'. Only dietary magnesium levels (Jackson, 1971), a Herpes virus (Fabricant,

357

Table 1. *Dietary factors reported to affect incidence of FUS*

1 'Ash'
2 Magnesium
3 Calcium and phosphate
4 Water
5 Fibre (via faecal H_2O loss)
6 Na^+ and K^+ and CL^- direct and via water
7 Protein (sulphur amino acids)
8 Urinary pH

1977), and multiple feedings (Willeberg, 1981) have been suggested as contributory factors additional to those in Blount's list. Vitamin A deficiency was reported to be a cause of phosphatic urinary calculi in rats in 1917 by Osborne and Mendel, but is not considered to be a causative or predisposing factor for feline urolithiasis (Lewis & Morris, 1984b). The importance of many of these factors has been stressed by various authors in a recent symposium (Osborne, 1984a); only the dietary factors listed in Table 1 will be discussed further here.

The final common denominator in nearly all cases of FUS is struvite urolithiasis. Struvite is magnesium ammonium phosphate hexahydrate, (Mg NH_4 PO_4 · $6H_2O$). This fact suggests that most cases of FUS would be more appropriately named FSUS, feline struvite urolithiasis syndrome. As shown in Table 2, struvite stones are the most common type found in the cat (and possibly other carnivores: Leoschke & Elvehjem, 1954; Long, 1984), which distinguishes the pattern of urolithiasis in the cat from that of the dog and human (Osborne *et al.*, 1984c; Bovee, 1984). Further, struvite crystals appear to be a primary cause of hematuria when present in the bladder, and comprise more than 90% of urethral obstructions in male cats (Osborne *et al.*, 1984b).

Physicochemical aspects of FSUS

Because struvite urolithiasis is such a prominent feature of FUS, a clear conceptual understanding of the determinants of struvite crystal formation is necessary for evaluation of the relative importance of various factors associated with the syndrome. Formation of any crystal depends on the activities of the solutes which aggregate to form the crystal, activity being defined as the concentration of solute which is free to react with other solutes in a solution, in this case urine (Finlayson, 1978). Activity is influenced by temperature, ionic strength, concentration of other solutes and pH of the urine. For simplicity, this discussion will assume a constant urine temperature of 38°C. The effect of ionic strength on struvite solubility has been discussed by Johnson (1959), but, to our knowledge, no reports on cat urine are available. These two factors will not therefore be discussed

Table 2. *Relative incidence of different types of urinary tract calculi in the cat, dog and human, as percentage total calculi found*

	Cat[1]	Dog[2]	Human
Struvite	87.5	60–69	15
Urate	2.7	5–10	8
Ca phosphate	2.4	1	6
Ca oxalate	2.4	10–15	33
Mixed calculi	3.6	6	34
Matrix	1.2	NR	NR
Cystine	0	3.5–27	3
Silica	0	35	1

Notes:
[1] Osborne *et al.* (1984c).
[2] Bovee, (1984).

further. The remaining variables affecting crystal formation, then, are concentration of the components of struvite (and other solutes) and pH of the urine. Although most work has been done with human urine there is no evidence that urine contains any specific inhibitors of struvite crystallisation (Robertson, Peacock & Nordin, 1968).

Urinary concentration of constituent ions

The effect of solute concentration on struvite crystallisation is shown in Figures 1 and 2. At a constant pH and concentration of other solutes, formation of crystals depends only on the product of the three constituent ion concentrations: $[Mg^{2+}] \times [NH_4^+] \times [PO_4^{3-}]$. For brevity this will be called MAP. In the zone of undersaturation (Figure 2), at low MAP, no crystallisation occurs and preformed crystals dissolve. The solubility

Fig. 1. Solutes affecting magnesium ammonium phosphate in urine.

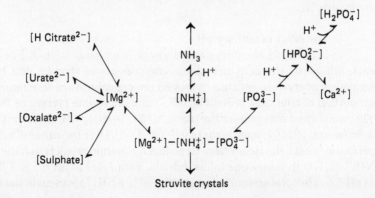

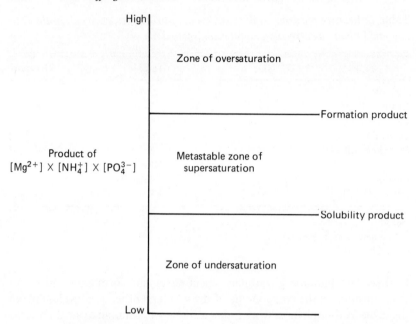

Fig. 2. Zones of urine saturation with magnesium ammonium phosphate
(Adapted from Osborne and Kruger, 1984d)

product, K_{sp}, is 1.15×10^{-13} (Elliot, Sharp & Lewis, 1959; Johnson, 1959;
Burns & Finlayson, 1982). As MAP rises, this is exceeded and the urine
becomes supersaturated with respect to struvite. In this zone, although
growth and aggregation of *preformed* crystals occurs, insufficient free energy
is available for spontaneous crystal formation. The free energy required for
crystallisation is attained at the formation product of struvite, k_{fp}. This is
2.5×10^{-13} (Robertson, Peacock & Nordin, 1968), approximately twice the
solubility product. Urine in this region is oversaturated with the components
of struvite; spontaneous crystallisation will occur, and preformed crystals
grow rapidly.

The effect of urinary pH

Urinary pH, the physiological range of which is 5.5 to 8.5 in adult
cats, influences the activities of the components of struvite and hence
formation of crystals, because hydrogen ion concentration determines the
proportion of ammonia present as NH_4^+ and phosphate present as PO_4^{3-}.
The equilibrium constant for the reaction $NH_4^+ \leftrightarrow NH_3 + H^+$ in urine (Bank
& Schwartz, 1960) is approximately 9.3×10^{-10}. Over the range of urinary
pH found in cats, the Henderson-Hasselbach equation predicts that the ratio
NH_4^+/NH_3 will change one thousand-fold, from 3.4:1 at pH 8.5 to 3400:1
at pH 5.5. The total amount of ammonia ($NH_3 + NH_4^+$) present in the urine

is also influenced by renal acid excretory mechanisms, varying directly with the acid load excreted (Halperin & Jungas, 1983). Chow *et al.* (1978) demonstrated that total urinary ammonia concentration approximately doubled, from 50 to 90 mM, as the urinary pH decreased from 7.2 to 6.1.

The total phosphorus concentration measured in the urine is the sum of all the ionic species of phosphorus; the proportions of urinary phosphorus present as H_3PO_4, $H_2PO_4^-$, HPO_4^{2-} and PO_4^{3-} are determined largely by the urinary pH. Elliot, Sharp & Lewis (1958) have reported the following equilibrium constants (at ionic strength 0.32 M) for phosphorus:

Reaction	Equilibrium constant
$H_3PO_4 \leftrightarrow H^+ + H_2PO_4^-$	$K_1 = 1.40 \times 10^{-2}$
$H_2PO_4^- \leftrightarrow H^+ + HPO_4^{2-}$	$K_2 = 2.37 \times 10^{-7}$
$HPO_4^{-2} \leftrightarrow H^+ + PO_4^{3-}$	$K_3 = 4.57 \times 10^{-12}$

The proportion of total urinary phosphorus present as PO_4^{3-} may be calculated by using the above equilibrium constants and the urinary pH in the following equation (Elliot *et al.*, 1958).

$$\frac{[PO_4^{3-}]}{[\text{Total phosphorus}]} = \frac{K_1 K_2 K_3}{[H^+]^2 K_1 + [H^+] K_1 K_2}$$

This equation predicts that at pH 8.5, the molar ratio of PO_4^{3-} to total phosphorus will be approximately 1:700. Moreover, a 1,000-fold increase in H^+ concentration, changing the pH from 8.5 to 5.5, will decrease the PO_4^{3-}:total phosphorus ratio to approximately 1:10,000,000, a 14,000-fold decrease in PO_4^{3-} concentration. Thus, lowering urinary pH causes a *decrease* in the proportion of total phosphorus present as PO_4^{3-} ten times as much as it *increases* NH_4^+ concentration.

Struvite crystals are rarely found in urine of pH less than 6.4, while calculi formation is encouraged at pH 7.0 or greater (Rich & Kirk, 1969). These pHs can therefore be considered to correspond approximately to the struvite solubility and formation products in cat urine. As the pH of the urine varies during the course of the day (Figure 3), MAP will also change, and struvite crystals may form, grow, or dissolve. Feeding large meals stimulates secretion of relatively large amounts of gastric acid, which results in an 'alkaline tide' in the blood followed by alkalinisation of the urine. This 'post-prandial alkaline tide' (Brunton, 1933) promotes struvite crystal formation. House cats do not eat large meals when food is continuously available, but generally eat 10–20 meals rather evenly spaced over the 24-hour period (MacDonald, Rogers & Morris, 1984). This *ad libitum* feeding pattern minimises the amount of gastric acid secretion for any individual meal, and attenuates the extent of post-prandial alkalinisation of the urine.

Urinary pH and other urinary solutes

The concentration of other solutes in the urine should also be determined to evaluate all components affecting struvite crystallisation because they can reduce the activity of magnesium, ammonium and phosphate. Computer programmes have been developed (Finlayson, 1977; Werness *et al.*, 1985) to predict the solubility of a variety of stones in urine from the concentration of a number of urine solutes. Variables in the programmes are urinary pH and concentrations of sodium, potassium, calcium, magnesium, ammonium, sulfate, phosphate, citrate, oxalate, urate, and chloride. These variables are used to calculate supersaturation of the urine with respect to the urolith of interest. The large number of chemical analyses required to provide programme inputs makes these programmes cumbersome to use, so Marshall & Robertson (1976) have provided a series of nomograms to predict supersaturation based on the results of fewer analyses. The nomogram for magnesium ammonium phosphate is presented in Figure 4; the coefficient of correlation between values for struvite supersaturation calculated by the nomogram or by their computer programme is 0.996. This nomogram may be used to demonstrate the relative importance of the determinants of struvite crystallisation. The activity product, the ultimate determinant of crystal

Fig. 3. Variation in urinary pH during the day with meal fed and *ad libitum* feeding conditions. The dashed lines indicate approximate pH of struvite solubility and formation products.

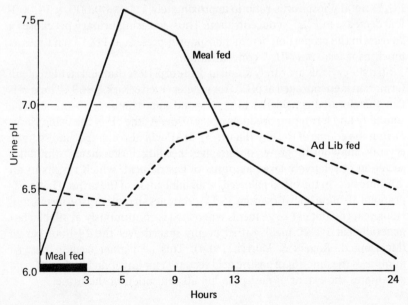

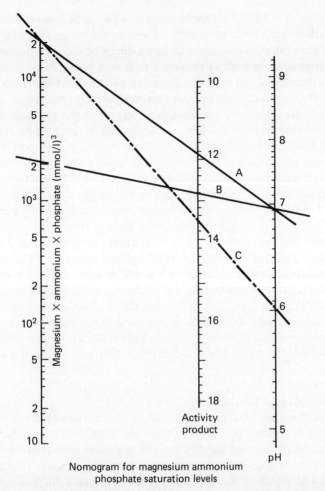

Nomogram for magnesium ammonium
phosphate saturation levels

Fig. 4. Nomogram for predicting struvite saturation in the urine. (Adapted from
Marshall and Robertson, 1976; used with permission.)

formation, is expressed in the nomogram as its negative logarithm, AP, just
as hydrogen ion concentration is expressed as pH, for ease of interpretation.
Lines A and B show that at pH 7, a 10-fold change in magnesium,
ammonium or phosphate concentration (from 2×10^3 to 2×10^4) changes
AP by approximately one unit (ten-fold in its absolute value). In contrast, a
ten-fold (line C) change in hydrogen ion concentration, from pH 7 to 6,
changes AP by 1.6 log units (i.e. a forty-fold change in its absolute value).
Thus, changing urinary hydrogen ion concentration ten-fold (i.e. by 1 pH
unit) is more effective at inhibiting struvite precipitation than a ten-fold
change in the concentrations of magnesium, ammonium or phosphate. A
ten-fold change in urinary magnesium, ammonium or phosphate is greater

than that which occurs when diets formulated especially to decrease the risk of FSUS, are substituted for normal diets. However, the range of urinary hydrogen ion concentrations which may be produced by dietary manipulation is much greater, as high as thousand-fold, or 3 pH units.

Despite the theoretical importance of urinary pH in determination of struvite solubility, its importance has not been consistently emphasised in veterinary literature. This may be because of reports of the presence of struvite crystals in the acidic urine of obstructed cats (Osbaldiston & Taussig, 1970; Rich & Kirk, 1969; Schecter, 1970). Such reports may in part be explained by the curve shown in Figure 3. This shows that urinary pH changes during the day, and periods of high urinary pH may allow sufficent time for net crystal accretion to occur. Moreover, although cats maintained on commercial diets may have a urine pH greater than 7.0 immediately after feeding, after 12 to 24 hours of food deprivation their urine is often acidic. An early event in the course of urinary tract obstruction is anorexia, so it is probable that a cat could be presented for treatment more than 12–24 hours after the previous meal, when its urinary pH would be acidic. The presence of crystals in acid urine indicates that either the urine is not sufficiently acidified for solubilisation of crystals, or that insufficient time has elapsed for their dissolution. A single determination showing acidic urine does not imply that the urine was sufficiently acidic to be undersaturated with respect to struvite before obstruction. Indeed, the presence of crystals proves it was not.

Urinary pH and infection

It is generally accepted that urinary tract infection is not associated with most cases of struvite urolithiasis in the cat (Lees, 1984; Schecter, 1970). By contrast, in dogs and humans struvite stones are often referred to as 'infection stones'. In dogs, 60–70% of naturally occurring struvite calculi are associated with infection (Bovee, 1984), while in humans the presence of urease-producing bacteria (usually Staphylococcus or Proteus), is considered a *sine qua non* of struvite stone formation (Griffith, 1978). Urease hydrolyses urea according to the following reaction:

$$NH_2 - \overset{\overset{\displaystyle O}{\|}}{C} - NH_2 \ \xrightarrow[H_2O]{urease} \ 2\ NH_3 + CO_2$$

$$2\ NH_3 + 2\ H_2O \leftrightarrow 2\ NH_4^+ + 2\ OH^- \ (pK = 9.2)$$

$$CO_2 + H_2O \leftrightarrow H_2CO_3 \leftrightarrow H^+ + HCO_3^- \ (pK = 6.3)$$

This results in the production of two ammonium ions per molecule which may be incorporated into the struvite crystal. The urine is also alkalinised, which further promotes crystallisation by increasing the conversion of HPO_4^{2-} to PO_4^{3-}.

Adding a urinary acidifier to the diet may reduce the incidence of struvite

Table 3. *Effect of dietary components on urinary pH*

Reaction	Effect on urinary pH
Carbohydrates $+ O_2 \rightarrow CO_2 + H_2O$	none
Neutral lipid $+ O_2 \rightarrow CO_2 + H_2O$	none
$CH_3COO^- + K^+$ (or Na^+) $+ O_2 \rightarrow CO_2 + H_2O + K^+ + HCO_3^-$	increases
Lipid $- PO_4^- + K^+ + O_2 \xrightarrow{H_2O} CO_2 + H_2O + K^+ + 2/3HPO_4^=$ $+ 2/3H^+$	decreases[1]
Protein $+ O_2 \rightarrow CO_2 + H_2O + $ urea $+ SO_4^= + 2H^+$ (per sulphur)	decreases
$MgO + H_2O \rightarrow 2OH^- + Mg^{++}$	increases[2]
$MgCl_2 + H_2CO_3 \rightarrow MgCO_3$(faeces) $+ 2HCl$ (absorbed)	decreases
$2NH_4Cl \rightarrow$ urea $+ 2H^+ + 2Cl^-$	decreases
$NaCl \leftrightarrow Na^+ + Cl^-$	none

Notes:
[1] The extent depends on the pH of the diet, which determines the proportion of monoester phosphorus ingested as $H_2PO_4^-$ and $HPO_4^=$; and on pH of urine.
[2] To the extent that the anion is absorbed in excess of the cation. Most of the magnesium is excreted in the faeces as carbonates.

uroliothiasis. This has been true for mink which are susceptible to struvite urolithiasis (Leoschke & Elvehjam, 1954), associated with infection in many clinical cases (Nielsen, 1956). In the early 1950s Leoschke & Elvehjam reported that feeding 1 g NH_4Cl per day to mink was effective in preventing urinary calculi. Dietary addition of phosphoric or sulphuric acids are currently used in the mink industry to prevent urolithiasis (NRC, 1982).

Effect of diet on FSUS

The diet is of primary importance in the aetiology of FSUS. Dietary ingredients affect the pH, mineral concentration and volume of the urine. The effect of diet on urinary pH and mineral composition depends on the chemical form of ingredients and the extent of their absorption and metabolism. Urinary volume is influenced by dietary moisture and ingredient composition. The effect of some individual dietary components on urinary pH is presented in Table 3. The constituents of foodstuffs which exert major effects on urinary pH are proteins and phospholipids which acidify the urine, and the salts of organic acids which are alkalinising (Harrington & Lemann, 1970). Acidification after protein ingestion occurs as a result of sulphate production during oxidation of sulphur-containing amino acids (Hunt, 1956), and after phospholipid consumption as a result of hydrolysis of the diester phosphate bonds (Harrington and Lemann, 1970). Salts of dietary organic acids which are metabolised are alkalinising because they are protonated during oxidation to carbon dioxide and water.

Table 4. *Studies supporting the role of magnesium in FSUS*

Author	Magnesium level %	Number of cats obstructing / Number of cats observed
Rich *et al.*, 1974	0.1	0/4
	0.25	0/8
	0.5	0/8
	0.75	2/8
	1.0	5/8
Chow *et al.*, 1978	0.75	7/11
Hamar *et al.*, 1976	0.75–1.0	6/8
Lewis *et al.*, 1978	0.75	6/18
	1.0	5/12
Kallfelz, Bressett & Wallace, 1980	1.0	7/16

Veterinarians have often ascribed the effects of dietary minerals on struvite crystallisation to the ash content of the diet. In 1953, Morris stated that, 'Opinion was quite unanimous that a low ash diet was essential for prevention and treatment of urinary calculi'. However, in 1956 Dickinson & Scott failed to produce urinary calculi in kittens by the addition of mineral salts derived from bone meal to the diet. They fed diets containing ash levels of 8.43, 21.86, and 30.07% for 10 weeks with no adverse results except for decreased growth rates observed in kittens fed the highest level of ash. Thus, while *components* of the ash are significant with respect to the pathogenesis of FSUS, ash *per se* has little meaning in this context, and its use should be discarded.

Dietary minerals which have been examined for their effects of FSUS include magnesium, calcium, phosphorus and sodium chloride. The importance of dietary magnesium to FSUS has been emphasised by a number of workers over the past 15 years. (Jackson, 1971; Rich *et al.*, 1974; Lewis *et al.*, 1978; Kallfelz, Bressett & Wallace, 1980; Lewis & Morris, 1984*a*). Results of some of these studies are summarised in Table 4. One conclusion drawn from these studies as recently as 1984 was that, 'although numerous factors . . . play a role in enhancing (struvite) urolith formation, *the key appears to be magnesium intake* and its subsequent effect on urine magnesium concentration. A high magnesium intake causes and low magnesium intake prevents FUS and all of its manifestations' (Lewis & Morris, 1983). It should be noted that the majority of crystals examined were reported in the original papers to be magnesium phosphate. Nevertheless, Lewis and Morris (1983) have recently stated that 'In all of these studies, the uroliths produced were magnesium ammonium phosphate with a small amount of calcium phosphate.' This apparent discrepancy was not explained by Lewis and Morris, neither do they comment on their observation that the mean dietary magnesium level in commercial diets is 0.16% a level of magnesium far below that necessary to induce obstruction in the experimental studies they review (Table 4).

In 1984, Taton, Hamar & Lewis reported a series of experiments designed to study the efficacy of urinary acidification in the prevention and treatment of FSUS. In their study two groups of 12 adult male cats were fed a high magnesium (0.37%) dry expanded diet. The diet of one of the groups was acidified with 1.5% ammonium chloride added before extrusion. The urinary pH of cats fed the acidic diet was found to be less than 6 (Fig. 5). In the control group (without added NH_4Cl), seven cats formed struvite calculi, and each suffered urethral obstruction twice. In the group fed the acidified diet, two cats obstructed once, in both cases shortly after being catheterised as paired controls for obstructed cats in the control group which had needed catheterisation. The diet of the seven control group cats which had obstructed was changed after the second obstruction. Three cats were placed on a commercial dry expanded diet containing 0.16% Mg, after which two reobstructed. These two, and the remaining four control cats, were all switched to the ammonium chloride-containing diet and had no further episodes of obstruction. Taton *et al.* reported that two of these six cats had radiographically visible calculi, which dissolved after three months of consumption of the ammonium chloride-containing diet. This report clearly demonstrated that struvite crystallisation could be prevented, and existing disease treated, even when a diet containing a 'calculogenic' level of magnesium was fed.

An experiment to study the effects of acid–base balance on dietary choice

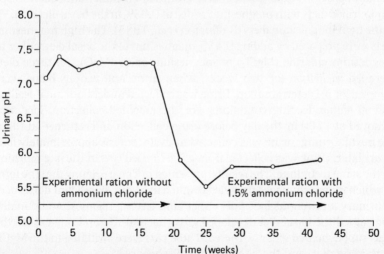

Fig. 5. Effect of ammonium chloride ingestion on urine pH in the cat. (Reproduced with permission from Taton *et al.*, 1984.)

of cats was in progress in our laboratory when Taton *et al.* reported their work. In our study (Cook, 1985), a group of male cats aged 2–7 months were fed purified diets which contained varying amounts of NH_4Cl or $NaHCO_3$ and $KHCO_3$ to manipulate systemic acid-base balance. The level of magnesium added to the purified diets was 0.045%. After 57 days of diet-induced urine alkalinity, one of the cats in a group with an average urinary pH of 8.2 suddenly died. Gross necropsy revealed that the bladder of this otherwise healthy cat was distended and severely haemorrhagic, containing a large amount of gritty yellowish material. The last centimetre of the penile urethra was obstructed. Crystallographic analysis of the material in both bladder and urethra indicated that its composition was solely struvite. By the end of 20 weeks, gross haematuria was evident in all six cats fed the basic diets. These cats were then placed on a diet containing 4% NH_4Cl for nine days; haematuria disappeared from all cats by the fourth day of feeding, and no further urinary tract problems were observed over the period studied. The cats were fed other purified and commercial dry diets for a period of approximately 15 months with no reccurrence of haematuria or other signs of FSUS.

Taken together, these two studies suggested that if urinary pH was less than 6.0, the levels of magnesium found in most commercial diets should be safe, while if urinary pH was high, even the level of dietary magnesium recommended by the National Research Council (1986) might still carry a risk of FSUS. When previous studies relating 'FUS' to dietary magnesium were reviewed in an effort to reconcile these new results with those previously reported, it was noticed that alkaline salts of magnesium (magnesium oxide or carbonate) had been used as magnesium sources. To determine if the form of magnesium added to the diet had influenced the results of previous experiments, three groups of 10 adult male cats were fed semipurified diets with magnesium added at 0.05% in the basal diet or 0.5% in the high magnesium diets (Buffington *et al.*, 1985). The high magnesium diets were prepared by adding 0.45% magnesium to the basal diet, either as magnesium chloride ($MgCl_2$) or magnesium oxide (MgO). All three diets were fed *ad libitum* for two weeks, when urine and venous blood were sampled for pH determination. Urine was collected under both meal-feeding and *ad libitum* feeding conditions. For the meal-fed collection, food was removed at 1700 hr the day before urine collection and returned at 0800 the next morning; urine was collected by cystocentesis approximately four hours later. Urine was collected from *ad libitum* fed cats in the late morning by the same technique. The results, presented in Table 5, show that the form in which magnesium was provided resulted in highly significant differences in urinary and venous blood pH under both feeding regimes. Additionally, one cat fed the basal diet developed gross haematuria and one cat fed the MgO supplemented diet obstructed. These cats were switched to the $MgCl_2$ containing diet, and the clinical signs disappeared.

Table 5. *Effect of source of dietary magnesium on urinary and blood pH*[1]

| Dietary Mg added | Source added | Ad libitum-fed | | Meal-fed[2] |
		Urine pH	Venous blood pH	Urine pH
		mean ± SE		
0.05	MgSO$_4$	6.9 ± 0.11[a,3]	7.35 ± 0.03	7.5 ± 0.12
0.50	MgO	7.7 ± 0.12[b]	7.35 ± 0.04	7.6 ± 0.12
0.50	MgCl$_2$	5.7 ± 0.11[c]	7.26 ± 0.03[a]	6.4 ± 0.12[a]

Notes:
[1] Food intake was 56–61 g per day.
[2] Food intake in 3 hours was at least half that of daily *ad lib* consumption.
[3] Column means with differing superscripts are significantly different ($p < 0.01$) ($n = 10$).
Source: Adapted from Buffington *et al.* (1985).

These results demonstrate that dietary magnesium concentration *per se* is much less important in the pathogenesis of FSUS than the form of the salt which is added to the diet. The influence of urine pH and dietary magnesium concentration in struvite formation is shown in Figure 6a, which was constructed using the nomogram in Figure 4. It was assumed that urinary phosphate and ammonium ion concentration decreased with increasing urinary pH (Mizgala & Quamme, 1985; Halperin & Jungas, 1983), that absorption of dietary magnesium was 8.5% (Sauer, Hamar & Lewis, 1985b) and that MAP in excess of the formation product was completely converted to struvite. Figure 6 shows that at a urinary pH less than 6, struvite production would be negligible at any level of magnesium found in commercial diets, but as urinary pH rises, the effect of magnesium becomes apparent, with struvite levels increasing more rapidly at higher dietary magnesium concentrations.

The effects of changes in dietary calcium and phosphorus levels on the incidence of feline urolithiasis have also been studied by Chow & colleagues (Chow, 1977). Their results, summarised in Table 6, show that dietary phosphorus had no influence on calculi production by high magnesium diets when calcium levels were low, but high phosphorus in the presence of high calcium appeared to exacerbate the effects of magnesium. Increasing dietary calcium and magnesium elevated urinary pH in these studies, while increasing dietary phosphate (as NaH$_2$PO$_4$) tended to decrease it, although urine was still alkaline. It is therefore likely that the calculogenic effect of the calcium and magnesium supplements resulted largely from their influence on urine pH.

The effect of dietary mineral levels on urinary mineral concentrations in these studies is shown in Table 7. Increasing dietary magnesium increased

(a)

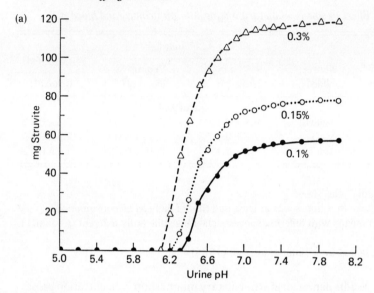

(b)

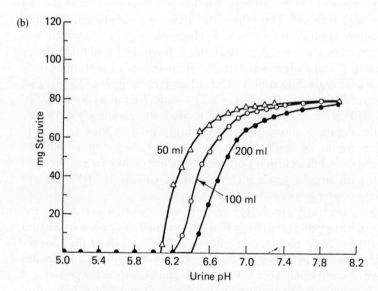

Fig. 6(a). Effect of pH and magnesium intake on struvite formation in urine. (100 ml); (b). Effect of pH and urine volume on struvite formation in urine. (Figures a and b were constructed assuming a 4 kg cat consumes 120 mg magnesium per day, of which 8.5% is absorbed and appears in the urine (Sauer *et al.*, 1985b). Urinary phosphate and ammonium concentrations were assumed to decrease from 105 mM at pH 5.4 to 40 mM at pH 8. Urinary magnesium was assumed to be unaffected by urinary pH. Saturation products were determined from the nomogram of Marshall & Robertson, 1976, Figure 3. All struvite in excess of the solubility product was assumed to precipitate in the bladder.)

Table 6. *Effect of dietary mineral levels on urinary pH and urethral obstruction in male cats*

Dietary mineral level (% dry matter)				
Mg	Ca	PO$_4$	Urinary pH[1]	Cats obstructing/ Total cats[2]
0.08	0.85	0.57	7.1	0/6
0.08	0.85	1.6	6.4	0/6
0.08	2.0	0.57	7.9	0/6
0.08	2.0	1.6	7.6	0/6
0.75–1.0	0.85	0.57	7.5	5/6
0.75–1.0	0.85	1.6	7.3	11/12
0.75–1.0	2.0	0.57	8.3	1/6
0.75–1.0	2.0	1.6	7.2	6/6

Notes:
[1] Relationship of time of urinary pH determination to feeding was not reported.
[2] Obstructing material was primarily magnesium phosphate.
Source: Adapted from Chow (1977).

its urinary excretion and decreased phosphorus excretion, and vice versa. Urinary calcium was relatively unaffected by the different treatments, except for an increased excretion on the high calcium, high magnesium, low phosphorus diet. Increasing dietary calcium decreased phosphorus excretion.

The effect of dietary salts on urinary pH is due to the presence of metabolisable anions and cations, and to differences in absorption of the cation and anion (Table 3). The absorption of magnesium and calcium from the gut is relatively low, but absorption of the accompanying anion can be high. Organic anions require a proton to be metabolised, which results in alkalinisation; first of the blood, and then the urine. Non-metabolisable anions absorbed in excess of their accompanying cation are acidifying, causing respiratory loss of bicarbonate as CO_2 (see example 2, Table 3). The kidney re-establishes systemic acid-base homeostasis by essentially reversing the process; the urine is acidified and bicarbonate is recovered. Because of these reactions, it is important that experimentally added minerals should have no effect on urinary pH to avoid introducing a confounding factor into a study (Finco, Barsanti & Crowell, 1985). If an acidifying salt of magnesium had been chosen as the source of added dietary magnesium in earlier experiments, magnesium might have been shown to *protect* cats from FSUS.

Differences in absorption between naturally occurring and added minerals must also be considered. For example, the availability of phosphorus from phosphorus salts may depend on the salt in question (ARC, 1965) and phosphorus bound to phytate in foodstuffs is generally poorly available

Table 7. *Effect of dietary mineral levels on urinary mineral concentration in male cats*

Dietary minerals % dry matter			Urinary minerals mM		
Mg	Ca	PO$_4$	Mg	Ca	PO$_4$
0.08	0.85	0.57	1.2	0.23	39.1
0.08	0.85	1.6	0.5	0.30	105.3
0.08	2.0	0.57	0.7	0.30	18.2
0.08	2.0	1.6	0.3	0.23	82.7
0.75–1.0	0.85	0.57	13.4	0.40	10.6
0.75–1.0	0.85	1.6	6.9	0.28	78.5
0.75–1.0	2.0	0.57	16.8	0.71	3.6
0.75	2.0	1.6	2.9	0.35	55.0

Source: Adapted from Chow (1977).

(NRC, 1986). A variety of other dietary factors also affect calcium and magnesium absorption, including: dietary levels of the minerals themselves, phytates, protein, vitamin D (Cobern, 1977) and phosphorus (Lewis *et al.*, 1978). Parathyroid hormone (PTH) levels may also influence absorption, which is important because PTH is influenced by acid–base status (Wachman & Bernstein, 1970; Wills, 1970; Hulter & Peterson, 1985).

Urine volume and water turnover are also potentially important determinants of FSUS risk because of their effects on the urinary concentration of magnesium, ammonium and phosphate. A 5-fold increase in the urine output of rats has been shown to both prevent formation and promote dissolution of struvite calculi, when urine pH was less than 7.5. At urinary pHs greater than 7.5, struvite crystals grew in the presence of this impressive diuresis (Grove *et al.*, 1950). The interaction in struvite production of urinary pH and urinary volume, over the range of daily urine volumes seen in cats, is presented in Figure 6b. The results parallel those in Figure 6a. At low urinary pH struvite does not form even at the low urine volumes, but as urinary pH rises, the influence of urine volume on struvite crystallisation becomes more significant.

Effects of dietary composition on water intake and excretion have been reported by a number of researchers (Hamar *et al.*, 1976; Thrall & Miller, 1976; Holme, 1977; Jackson & Tovey, 1977; Seefeldt & Chapman, 1979; Burger, Anderson & Holme, 1980; Anderson, 1982; Jenkins & Coulter, 1981; Lawler & Evans, 1984; Sauer, Hamar & Lewis, 1985*a*). The moisture level of the diet has variable effects on total water intake. Three studies (Jackson & Tovey 1978; Jenkins & Coulter, 1981; Sauer, Hamar & Lewis, 1985*a, b*) have reported a decrease in total water intake by animals fed dry diets, while two studies found no difference in water intake between dry and wet diets (Thrall & Miller, 1976; Seefeldt & Chapman, 1979). Dry expanded diets have been found to result in lower urinary and greater faecal output of water than canned foods (Jackson & Tovey, 1977; Sauer & Lewis, 1985a). Since this difference was not found for a non-expanded dry diet (Jackson & Tovey, 1977), the method of food processing may also affect urine volume.

Addition of sodium chloride to the diet is also known to increase total water intake in the cat (Hamar *et al.*, 1976; Jenkins & Coulter, 1981; Holme, 1977; Burger *et al.*, 1980; Anderson, 1982), although not to the same extent as the dog (Burger *et al.*, 1980). However, it should be noted that balance figures vary widely among individual cats; interindividual coefficients of variation of 50% are common. Further, one study (Lawler & Evans, 1984) reported no differences in water turnover between normal cats and those afflicted with FSUS. A detailed analysis of the influence of water on FSUS is presented by Gaskell in Chapter 22 of this book.

Sauer, Hamar & Lewis (1985*a,b*) have recently presented evidence that the increased risk of FSUS found among cats fed dry diets does not result from differences in total water turnover, but may be related to decreased energy density of the diet. The larger intakes of these foods needed to provide adequate energy intakes would increase the ingestion and subsequent urinary excretion of 'calculogenic minerals'. They concluded that, if urinary pH is not controlled, the amount of magnesium per kilocalorie of digestible energy should be minimised to reduce urinary magnesium excretion.

Strategies for prevention

We conclude from the available evidence that FSUS can largely be prevented by diet formulations that result in consistent maintenance of a urinary pH at or below 6.5. The high protein prey of feral cats results in urinary acidification due to oxidation and excretion of ingested sulphur amino acids as sulfate (Lemann & Relman, 1959), whereas the diet of herbivores contains an excess of the potassium salts of organic acids, the metabolism of which results in an alkaline urine. While we agree with Taton *et al.* (1984) that agents for urinary acidification should be administered with the food, we suggest that pet foods be formulated at the time of manufacture to assure an acid urinary pH. Addition of compounds which result in absorption of anions, e.g. chlorides or phosphates, in excess of balancing cations will increase urinary acid excretion, and lower urinary pH (Harrington & Lemann, 1970). Inclusion of ingredients which would maintain a urinary pH of less than 6.5 under *ad libitum* feeding conditions should prevent struvite crystal accretion and prevent the diet-induced component of FSUS. However, care must be taken against over-acidification. If more acid is consumed or produced than the animal is capable of excreting, systemic acidemia will occur, which may cause osteoporosis (Jaffe, Bodansky & Chandler, 1932; Kurtz *et al.*, 1983) and depletion of

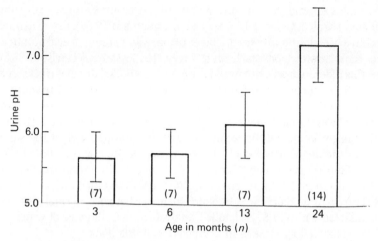

Fig. 7. Effect of age on urinary pH of cats fed a constant diet.

potassium. Our unpublished observations indicate that the cat, eating normal diets, is not capable of acidifying the urine below about pH 5.5. Urinary pH should be maintained above this level therefore, and at a pH which allows for 'normal' systemic acid–base status.

A change in commercial diet formulations could also influence other putative causes of FSUS. When diets are such that cats no longer produce urine in the metastable region of saturation with respect to struvite for long periods of the day, the risk associated with non-dietary factors should also be reduced. The onset of adulthood is one such factor (Willeburg, 1981). In fact, the increased risk associated with adulthood provides evidence to support the significance of urinary pH in FSUS. As shown in Figure 7, *ad libitum* urinary pH rises as a function of age in animals fed a constant diet. The increased urinary pH with advancing age is apparently a result of decreased bone accretion. As calcium and di- and monobasic hydrogen phosphate precipitate as calcium apatite, hydrogen ions are produced, which increases net acid excretion in the young (Chan, 1974). Skeletal growth in the cat is completed at about 16 months of age, the time when a sharp increase in the incidence of FSUS occurs (Willeberg, 1981).

A further advantage of proper diet formulation at the time of manufacture to control urine pH is avoidance of the possibility of toxicosis resulting from excessive ingestion of DL-methionine (Fau *et al.*, 1983; Cook, 1985) or food treated with DL-methionine and ammonium chloride (Brown & Fox, 1984). As a urinary acidifier, methionine has been recommended at a level of 1.5 g/cat/day (Lewis & Morris, 1984a); for a 3 kg cat eating 60 g of dry matter/day this would be 2.5% of the dry matter. This level has been shown to depress food intake and weight gain of kittens (Fau *et al.*, 1983). Therefore, methionine may not be safe to use as the only dietary component for urinary acidification.

Conclusions

Crystal formation in the urine is the result of complex physiochemical interactions, and is dependent on the free ion concentrations of the minerals involved (Finlayson, 1978). These interactions complicate exact prediction of struvite crystal formation at neutral to alkaline pH, but at low urinary pH crystal formation is impossible under normal circumstances, and dissolution occurs. Reformulation of commercial cat foods to include ingredients which acidify the urine, but do not jeopardise normal systemic acid–base homeostasis, is practicable and should result in a significant reduction in the incidence of feline struvite urolithiasis.

REFERENCES

Anderson, R.S. (1982). Water balance in the dog and cat. *Journal of Small Animal Practice*, **23**, 588–98.

Agricultural Research Council (1965). The nutrient requirements of farm livestock No. 2. Ruminants. Technical reviews and summaries p. 95 London HMSO.

Avioli, L.V. (1980). Calcium and phosphorus. In *Modern Nutrition in Health and Disease*, ed. R.S. Goodhart and M.E. Shils, pp. 294–309, Philadelphia, Lea and Febiger.

Bank, N. & Schwartz, W.B. (1960). Influence of certain urinary solutes on acidic dissociation constants of ammonium ion at 37°C. *Journal of Applied Physiology*, **15**, 125–7.

Blount, W.P. (1931). Urinary calculi. *Veterinary Journal*, **87**, 561–76.

Bovee, K.C. (1984). Urolithiasis in Canine Nephrology. Ed. K.C. Bovee, pp. 355–79, Philadelphia: Harwal.

Brown, J.E. & Fox, L.M. (1984). Ammonium chloride/methionine toxicity in kittens. *Feline Practice*, **14(2)**, 16–19.

Brunton, C.E. (1933). The acid output of the kidney and the so-called alkaline tide. *Physiological Reviews*, **13**, 372–97.

Buffington, C.A., Rogers, Q.R., Morris, J.G. & Cook, N.E. (1985). Feline Struvite urolithiasis: Magnesium effect depends on urinary pH. *Feline Practice*, **15(6)**, 29–33.

Burger, I.H., Anderson, R.S. & Holme, D.W. (1980). Nutritional factors affecting water balance in the dog and cat. In *Nutrition of the Dog and Cat*, ed. R.S. Anderson pp. 145–56, Oxford: Pergamon Press.

Burns, J.R. & Finlayson, B. (1982). Solubility product of magnesium ammonium hexahydrate at various temperatures. *Journal of Urology*, **128**, 426–7.

Chan, J.C.M. (1974). The influence of dietary intake on endogenous acid production. *Nutrition and Metabolism*, **16**, 1–9.

Chow, F.C. (1977). Dietary mineral effects on feline urolithiasis. *Proceedings of the Kal Kan Symposium for Treatment of Dog and Cat Diseases*, pp. 36–39, Vernon, California; Kal Kan Foods, Inc.

Chow, F.C., Taton, B.S., Lewis, L.D. & Hamar, D.W. (1978). Effect of dietary ammonium chloride, dl-methionine, sodium phosphate and ascorbic acid on urinary pH and electrolyte concentrations in male cats. *Feline Practice*, **8(4)**, 29–34.

Cook, N.E. (1985). The importance of urinary pH in the prevention of feline urologic syndrome. *Pet Food Industry*, **27**, March/April, 24–31.

Dickinson, C.D. & Scott, P.P. (1956). Failure to produce urinary calculi in kittens by the addition of mineral salts, derived from bone-meal, to the diet. *Veterinary Record*, **68**, 858–9.

Elliot, J.S., Sharp, R.F. & Lewis, L. (1958). The apparent dissociation constants of phosphoric acid at 38°C at ionic strengths from 0.1 to 0.5. *Journal of Physiological Chemistry*, **62**, 686–9.

Elliot, J.S., Sharp, R.F. & Lewis, L. (1959). The solubility of struvite in urine. *Journal of Urology*, **81(3)**, 366–8.

Fabricant, C.G. (1977). Herpesvirus-induced urolithiasis in specific-pathogen-free male cats. *American Journal of Veterinary Research*, **38**, 1837–42.

Fau, D., Smalley, D.A., Rogers, Q.R., *et al.* (1983). Effects of excess methionine in the kitten. *Federation Proceedings*, **42**, 542.

Finco, D.R., Barsanti, J.A. & Crowell, W.A. (1985). Characterization of magnesium-induced urinary disease in the cat and comparison with feline urologic syndrome. *American Journal of Veterinary Research*, **46**, 391–400.

Finlayson, B. (1977). Calcium stones: Some physical and clinical aspects. In *Calcium Metabolism in Renal Failure and Nephrolithiasis*, ed. D.S. David, pp. 337–82, New York: John Wiley & Sons.

Finlayson, B. (1978). Physiochemical aspects of urolithiasis. *Kidney International*, **13**, 344–60.

Griffith, D.P. (1978). Struvite stones. *Kidney International*, **13**, 372–382.

Grove, W.J., Vermeulen, M.D., Goetz, R. & Ragins, H.D. (1950). Experimental urolithiasis. II. The influence of urine volume upon calculi experimentally produced upon foreign bodies. *Journal of Urology*, **64**, 549–54.

Halperin, M.L. & Jungas, R.L. (1983). Metabolic production and renal disposal of hydrogen ions. *Kidney International*, **24**, 709–13.

Hamar, D., Chow, F.C., Dysart, M.I. & Rich, L.J. (1976). Effect of sodium chloride in prevention of experimentally produced phosphate uroliths in male cats. *Journal of American Animal Hospital Association*, **12**, 514–17.

Harrington, J.T. & Lemann, J. Jr. (1970). The metabolic production and disposal of acid and alkali. *Medical Clinics of North America*, **54**, 1543–54.

Holme, D.W. (1977). Research into the feline urological syndrome. *Proceedings of the Kal Kan Symposium for Treatment of Dog and Cat Disease*, 40–5, Vernon, California: Kal Kan Foods, Inc.

Hulter, N.M. & Peterson, J.C. (1985). Acid–base homeostasis during chronic PTH excess in humans. *Kidney International*, **28**, 187–92.

Hunt, J.N. (1956). The influence of dietary sulphur on the urinary output of acid in man. *Clinical Science*, **15**, 119–34.

Hutyra, F., Marek, J. & Manninger, R. (1938). *Diseases of the Kidney*. In *Special Pathology and Therapeutics of the Disease of Domestic Animals*, ed. J.R. Greig, J.R. Mohler & A. Eichhorn, p. 47, Vol. 3, Chicago: Eger.

Jackson, O.F. (1971). The treatment and subsequent prevention of struvite urolithiasis in cats. *Journal of Small Animal Practice*, **12**, 555–568.

Jackson, O.F. & Tovey, J.D. (1977). Water balance studies in domestic cats. *Feline Practice*, **7(4)**, 30–3.

Jaffe, H.C., Bodansky, A. & Chandler, J.P. (1932). Ammonium chloride decalcification, as modified by calcium intake: The relationship between generalized osteoporosis and osteitis fibrosa. *Journal of Experimental Medicine*, **56**, 823–4.

Jenkins, E.E. & Coulter, D.B. (1981). Effects of diet on water intake, food intake and faeces production in the cat. *Georgia Vet*, **33(6)**, 8–10.

Johnson, R.G. (1959). The solubility of magnesium ammonium phosphate hexahydrate at 38°C with considerations pertaining to the urine and the formation of urinary calculi. *Journal of Urology*, **81**, 681–90.

Kallfelz, F.A., Bressett, J.D. & Wallace, R.J. (1980). Urethral obstruction in random source and SPF male cats induced by high levels of dietary magnesium and phosphorus. *Feline Practice*, **10(4)**, 25–35.

Kirk, H. (1925). Urino-genital diseases. In *The Diseases of the Cat*, pp. 261–73, Chicago: Eger.

Krabbe, A. (1949). Urolithiasis in dogs and cats. *Veterinary Record*, **61**, 751–9.

Kurtz, I., Maher, J., Hulter, H.N., Schambelan, M. & Sebastian, A. (1983). Effect of diet on plasma acid–base composition in normal humans. *Kidney International*, **24**, 670–80.

Lawler, D.F. & Evans, R.H. (1984). Urinary tract disease in cats: Water balance studies, urolith and crystal analysis and necropsy findings. *Veterinary Clinics of North America*, **14**, 537–53.

Lawler, D.F., Sjolin, D.W. & Collins, J.E. (1985). Incidence rates of feline lower urinary tract disease in the United States. *Feline Practice*, **15(5)**, 13–16.

Lees, G. (1984). Epidemiology of naturally occurring feline bacterial urinary tract infections. *Veterinary Clinics of North America*, **14**, 471–79.

Lemann, J. & Relman, A.S. (1959). The relationship of sulphur metabolism to acid-base balance and electrolyte excretion: The effects of DL-methionine in normal man. *Journal of Clinical Investigation*, **38**, 2215–23.

Leoschke, W.L. & Elvehjem, C.A. (1954). Prevention of urinary calculi formation in mink by alteration of urinary pH. *Proceedings of the Society for Experimental Biological Medicine*, **85**, 42–4.

Lewis, L.D., Chow, F.C., Taton, G.F. & Hamar, D.W. (1978). Effect of various dietary mineral concentrations on the occurrence of feline urolithiasis. *Journal of the American Veterinary Medical Association,* **172**, 559–63.

Lewis, L.D. & Morris, M.L. Jr. (1983). Feline urologic syndrome (FUS). In *Small Animal Clinical Nutrition,* 2nd ed., ed. L.D. Lewis & M.L. Morris, Jr., pp. 9-1–42, Topeka: Mark Morris Associates.

Lewis, L.D. & Morris, M.L. Jr. (1984*a*). Feline urologic syndrome: Causes and clinical management. *Veterinary Medicine/Small Animal Clinician,* **79**, 323–37.

Lewis, L.D. & Morris, M.L. Jr. (1984*b*). Treatment and prevention of feline struvite urolithiasis. *Veterinary Clinics of North America,* **14**, 649–60.

Long, G.G. (1984). Urolithaisis in ranch foxes. *Journal of the American Veterinary Medical Association,* **185**, 1394–6.

Marshall, W. & Robertson, W.G. (1976). Nomograms for the estimation of the saturation of urine with calcium oxalate, calcium phosphate, magnesium ammonium phosphate, uric acid, sodium acid urate, ammonium acid urate and cystine. *Clinica Chimica Acta,* **72**, 253–60.

MacDonald, M.L., Rogers, Q.R. & Morris, J.G. (1984). Nutrition of the domestic cat, a mammalian carnivore. *Annual Review of Nutrition,* **4**, 521–62.

Milks, H.J. (1935). Urinary Calculi. *Cornell Veterinarian,* **25**, 153–161.

Mizgala, C.L. & Quamme G.A. (1985). Renal handling of phosphate. *Physiological Reviews,* **65**, 431–66.

Morris, M.L. (1953). Nutritive requirements of the cat: a preliminary report. *Veterinary Medicine,* **48**, 451–6.

National Research Council. (1986). *The Nutrient Requirements of Cats,* Washington, DC: National Academy Press.

National Research Council. (1982). *Nutrient Requirements of Mink and Foxes,* p. 19, Washington, DC: National Academy Press.

Nielsen, I.M. (1956). Urolithiasis in mink: Pathology, bacteriology and experimental production. *Journal of Urology,* **74**, 602–14.

Osbaldiston, G.W. & Taussig, R.A. (1970). Clinical report on 46 cases of feline urological syndrome. *Veterinary Medicine/Small Animal Clinician,* **65**, 461–8.

Osborne, C.A. (1984*a*). Symposium on disorders of the feline lower urinary tract. In *Veterinary Clinics of North America,* **14**, 407–717.

Osborne, C.A., Clinton, C.W., Brunkow, H.C., Frost, A.P. & Johnston, G.R. (1984b). Epidemiology of naturally occurring feline uroliths and urethral plugs. *Veterinary Clinics of North America,* **14**, 481–91.

Osborne, C.A., Johnston G.R., Polzin, D.J., Kruger, J.M., Pottenbauger, E.M., Bell, F.W., Feenmey, D.A., Gojal, S., Fletcher, T.F., Newman, J.A., Stevens, J.B., and McMenomy, M.F. (1984*c*). Redefinition of the feline urologic syndrome: Feline lower urinary tract disease with heterogenous causes. *Veterinary Clinics of North America,* **14**, 409–39.

Osborne, C.A. & Kruger, J.M. (1984d). Initiation and growth of uroliths. *Veterinary Clinics of North America,* **14(3)**, 439–545.

Osborne, T.A. & Mendel, L.B. (1917). The incidence of phosphatic urinary calculi in rats fed on experimental rations. *Journal of the American Medical Association*, **69**, 32–3.

Rich, L.J. & Kirk, R.W. (1969). The relationship of struvite crystals to urethral obstruction in cats. *Journal of the American Veterinary Medical Association*, **154**, 153–7.

Rich, L.J., Dysart, I., Chow, F.C. & Hamar, D. (1974). Urethral obstruction in male cats: Experimental production by addition of magnesium and phosphate to diet. *Feline Practice*, **4(5)**, 44–7.

Robertson, W.G., Peacock, M. & Nordin, B.E.C. (1968). Activity products in stone-forming and non-stone-forming urine. *Clinical Science*, **34**, 579–94.

Sauer, L.S., Hamar, D. & Lewis, L.D. (1985a). Effect of diet composition on water intake and excretion by the cat. *Feline Practice*, **15(4)**, 16–21.

Sauer, L.S., Hamar, D. & Lewis, L.D. (1985b). Effect of dietary mineral composition on urinary mineral composition and excretion by the cat. *Feline Practice*, **15(4)**, 10–15.

Schecter, R.D. (1970). The significance of bacteria in feline cystitis and urolithiasis. *Journal of the American Veterinary Medical Association*, **156**, 1567–73.

Seefeldt, S.L. & Chapman, T.E. (1979). Body water content and turnover in cats fed dry and canned rations. *American Journal of Veterinary Research*, **40**, 183–5.

Stansbury, R.L. & Truesdail, R.W. (1955). Occurrence of vesical calculi in cats receiving different diets. *North American Veterinarian*, **36**, 841–5.

Taton, G.W., Hamar, D.W. & Lewis, L.D. (1984). Urinary acidification in the prevention and treatment of feline struvite urolithiasis. *Journal of the American Veterinary Medical Association*, **184**, 437–43.

Thrall, B.E. & Miller, L.G. (1976). Water turnover in cats fed dry rations. *Feline Practice*, **6(1)**, 10–17.

Wachman, A. & Bernstein, D.S. (1970). Parathyroid hormone in metabolic acidosis: its role in pH homeostasis. *Clinical Orthopaedics and Related Research*, **69**, 252–63.

Walker, A.D., Weaver, A.D., Anderson, R.S., Crighton, G.W., Fennell, C., Gaskell, C.J., and Wilkinson, G.T. (1977). An epidemiological survey of the feline urological syndrome. *Journal of Small Animal Practice*, **18**, 283–301.

Werness, P.G., Brown, C.M., Smith, L.H.J., Finlayson, B. (1986). EQUIL2: A BASIC computer program for the calculation of urinary saturation. *Journal of Urology*, **134**, 1242–4.

Whitehead, J.E. (1964). Diseases of the urogenital system. In *Feline Medicine and Surgery*, ed. J.E. Catcott, pp. 270–6, Santa Barbara: American Veterinary Publishers.

Willeberg, P. (1975). Outdoor activity level as a factor in the feline urological syndrome. *Nordic Veterinary Medicine*, **27**, 523–4.

Willeberg, P. (1981). Epidemiology of the feline urological syndrome. *Advances in Veterinary Science and Comparative Medicine,* **25,** 311–44.

Willeberg, P. (1984). Epidemiology of naturally occurring feline urologic syndrome. *Veterinary Clinics of North America,* **14,** 455–69.

Wills, M.R. (1970). Fundamental physiological role of parathyroid hormone in acid-base homeostasis. *Lancet,* **2,** 802–4.

24

Canine and feline nutrition: future perspectives

J.P.W. RIVERS AND I.H. BURGER

Introduction

The reason for compiling this book was not merely to publish a symposium, but to record and focus upon the marked advances there have been, particularly in the last decade, in the understanding of the nutrition of cats and dogs. The extent of that progress is evident from the remarks made by Morris and Rogers in Chapter 5 as well as a comparison of this volume with its two intellectual predecessors. The symposium 'Canine and Feline Nutrient Requirements' (Graham-Jones, 1965), contains no mention of major concerns in the present volume, such as lipid metabolism or the metabolism of a variety of essential amino acids. Even comparison with the more recently edited *Nutrition of the Dog and Cat* (Anderson, 1980) shows marked advances that have occurred in the last decade, with topics presented there as provisional now being summarised comprehensively.

Is there a need for future research?

Our ability to compile such a record of progress makes it all the more necessary to consider where, if anywhere, the subject could go from here. We have no doubt that, although much effort has been expended on the climb, so that it is now possible to survey a whole terrain of knowledge, this does not mean that we have reached the summit.

Nevertheless it has become unfashionable to use the fact that there is much to learn as a justification for continuing with research. Scientists are expected to justify their curiosity by reference to practical gain, and the ascendancy of the cost-effectiveness criteria means it is usual these days, not merely to ask if scientific endeavour will result in some benefit, but to expect the answer to be framed in financial terms. Applied to work in this field such criteria would lead to the question: Is the potential gain to the pet animals, their owners, or the industry enough to justify a continuing programme of scientific work at the level which is reflected in the results in this book?

We have no doubt that retrospective application of such yardsticks would show that the work undertaken in the last 10 or 20 years could be more than

381

justified. There has never of course been any need to eliminate a body of malnutrition in pets fed reputable proprietary diets, and the mixture of commercially prepared foods fed has, as we noted in the introduction, often meant that pets get a better diet than their owners. What has been achieved by research is an increased confidence, which has set that dietetic rule of thumb on a firm footing and enabled pitfalls resulting from changes in diet to be avoided.

But of course such success in itself raises the spectre of diminishing returns. If so much has been discovered, it could be argued, might it be no longer worthwhile to continue to invest in the subject? Are the risks of nutritional disease now so low (and at the aggregate level the benefits from their avoidance therefore so slight) that research should be run down? Our view is that even if the analysis were correct, it should lead to the opposite solution. That is, it should be argued that, because the simple strides have been made and the obvious problems resolved, it is now necessary to begin to invest more heavily in research on the topic.

Increased expense is a general rule of science. Consider human medicine: an understanding of the cause of most infectious diseases could be achieved in the nineteenth century with relatively cheap and primitive science because such diseases presented scientifically simple problems. Cancer research, like research into cardiovascular disease has cost immeasurably more, and delivered less, because the problem is more complex. But in an age when death from infectious diseases is generally a thing of the past, cancer and cardiovascular disease are correspondingly more important.

In the present context we think three arguments must be considered. First, it would be remarkable indeed if, once the nutrient requirements of the dog and cat were established, there were no more consequences of diet for the health of pets. Secondly, it is almost a certainty that the complexities of such nutritional disease, whatever they are, will provide more intractable problems than simple deficiencies. Finally, even though such disease may be a *statistically* minor phenomenon, it will seem no less significant to the owner, or of course the pet. Hence the demand for more knowledge will not lessen as the benefit fades, rather increased demands will be made on scientific research to deliver the pet owner's millennium.

Thus we think that, if nothing else, enlightened self-interest by the pet food industry will make it likely that research into pet animal nutrition will expand rather than diminish in the future. We have no doubt that the bulk of that funding will come from industry. For reasons that are difficult to accept, virtually all government funding given for the nutrition of domestic animals currently goes into production of livestock rather than the maintenance of companion animals, and we cannot see any signs of a change in these priorities.

We cannot, of course, say with certainty quite in which direction future studies will go. Some were adequately pointed out by contributors to the book, but four others of particular interest seem to us to be sufficiently likely or important to focus upon here.

Future areas of research

Nutritional requirements and deficiencies

Clearly the most immediately obvious research areas are those clarifying the central thrust of recent work by continuing with the study of the intakes of nutrients required to prevent deficiency disease. Already the pet food industry compounds diets out of natural products such that cats and dogs can be fed for prolonged periods without any nutrient deficiency developing. But semi-purified diets could not be made with the same confidence, simply because the complete list of nutrient requirements for the dog and cat are not known. Most of the essential nutrients for cats and dogs are known, but clearly some have yet to be established. One obvious example is the exact nature of the essential fatty acid requirements, and particularly the question of whether the *n*-3 fatty acids are required, and, if so, which members of the homologous series. Another area of ignorance about dogs and cats is the nature of their trace element requirements.

Even for the established nutrients, the exact requirements have still in many cases to be quantified, and the metabolic basis for such requirements, where they differ from the normal mammalian pattern, has still to be found. The importance of research in this latter area cannot be understated, and we feel it must continue. Ultimately, only an understanding of the nature of nutrient requirements will provide a secure basis for their quantification.

The need for such research seems likely to be particularly evident in the case of the cat and the idiosyncratic nutrient requirements to which reference has already been made throughout this book. This is made all the more important because the nutrient requirements of the feline are not simply a reflection of the loss of certain enzyme capacities due to gene deletion; synthetic capacity is often present and the requirements appear to exist for complex reasons. It is clear that the cat does possess the enzymic capacity to desaturate linoleic acid to di-homo gamma-linolenic acid, (Chapter 20), to convert methionine to taurine (Chapter 10) or tryptophan to niacin (Chapter 11), or to synthesise arginine (Chapter 10). The nutrient requirements of the animal arise from a failure to regulate these enzymes appropriately in relation to diet, something which presents much more complex problems than simple gene loss.

Besides these reductionist explorations, it is important that studies of nutrient need do not remain focused on the mean requirement, but more clearly characterise the extent and nature of variability. Two obvious aspects of this call out for attention: the study of *predictable* variability, and the quantification of *unexplained* variation.

The former, on which work has to some extent begun, involves the study of the impact on nutrient need of environment and definable aspects of physiological status. The goal of such work is the classification of nutrient needs by age, sex, and reproductive status, size and activity, yielding the expansion and strengthening of the sort of estimates that are contained in Appendix 2. It will have been evident from the discussion in this book, and particularly Chapters 3 and 4 that much of this definition is still relatively crude and primitive, and that much refinement is needed. We have drawn attention to the paucity of information on the impact of body size variation in nutrition in Chapter 6, though necessarily our comments were confined to protein and energy requirements. Nothing systematic is known of the allometry of micronutrient requirements in the dog.

Another aspect of predictable variability that has yet to be resolved is the age variation in essential amino acid requirements in cats and dogs. Moreover, the very brevity of the discussion of the impact of environmental factors like temperature on energy and protein requirements illustrates that work in this field has effectively yet to begin.

Even once such defineable sources of variability are clarified, the important issue of unexplained variability will remain. It is now generally accepted that variability exists between ostensibly similar animals in a similar environment, and it is possible that a given animal exhibits temporal variations in nutrient need. In human nutrition this problem has been found to complicate the interpretation of nutrient need (see, for example FAO/WHO/UNU, 1985).

It may be enough in pet animal nutrition merely to conclude, for example, that nutrient requirements should include a generous allowance for this variability, that is, that they should be aimed at the animals with the highest observed requirement, not at the population mean. But other problems may exist: for example, the correlation of the unexplained variation of individual nutrient needs with each other may complicate this. It would not be enough, for example, to aim to feed abundant amounts of EFA and vitamin E if, as is possible, the higher intake of EFA elevated the vitamin E requirement. An understanding of the variation of one nutrient requirement with another, the scientific clarification of that over-used term 'a balanced diet', is clearly required. Also called for in this context is research to ascertain whether variability in requirement is in anyway reflected in variability in food choice.

Nutrition and disease

Earlier we suggested that, apart from simple deficiencies, nutrition may play a role in diseases of dogs and cats. Except for a few obvious examples, like obesity, or bone and joint disease, we can bring no evidence to support our view, though nevertheless, we regard it as sufficiently important to make further study worthwhile. Investigation of these putative relationships may require large scale studies, but, unless they are done, it is premature to promote the view that specialist foods are necessary to maintain normal healthy animals in optimal condition.

As with human degenerative disease, nutritional epidemiology seems likely to be the area which will yield immediate results here, allowing subsequent metabolic study. The veterinary profession can, and does, undertake quite sophisticated studies of morbidity and mortality in pets. Investigating the correlation between such problems and nutritional status is simplified by the fact that pet animals, unlike their owners, live on defined diets. Product standardisation, and brand loyalty, makes it possible to specify to a remarkable degree the dietary history of animals over a considerable period of time, so that, in theory at least, disease incidence could be related to diet. The major objection to such studies is that aetiology may be hidden by the very lack of variability of diet, but this aside, the project seems so worthwhile embarking on that we cannot but believe that pet animal epidemiology will soon become a rapidly growing area of nutritional research. It may even provide clues to clarify some of the imponderables of human nutrition.

In this context it is worth noting that historically the dog, more than any other animal, has provided insights into nutritional problems in humans, as Stewart noted in an unduly neglected review (Stewart, 1957). Much of this work was experimental, which, whatever one might feel about its ethical aspects, provides a clear demonstration of the potential relevance of dog diseases to human ones. Pavlov's work is one obvious example (Pavlov, 1910), as is Goldberger's study of canine black tongue which proved to be the canine equivalent to pellagra (Goldberger & Wheeler, 1928; Goldberger *et al.*, 1928). Schaumann's work on the nutritional value of heated foods (cited by Vedder, 1913), Cowgill's work on vitamin B1 (Cowgill, 1921, 1934), Mellanby's work on canine rickets and osteomalacia (Mellanby, 1921), and a variety of studies on vitamin A deficiency and bone diseases (see Stewart, 1957) are other examples. Some experimental studies are bizarre, such as Andrews persuading Polynesian mothers whose infants had died of infantile beriberi to suckle puppies, which in turn developed the disease (see Stewart, 1957). Sometimes, though, the insight did not involve experimental work, and the analogy between the dog's diseases and human diseases has been a valuable guide in the development of ideas about human disease. Stewart noted that it was a knowledge that dogs develop diabetes spontaneously that persuaded Best to study them in his work, and Whipple whose observation that dogs eating raw liver could withstand regular blood loss persuaded Minot to study Vitamin B12 deficiency in these animals for details see Stewart, 1957).

The interaction of nutrition and disease is not unidirectional, and, just as diet may affect the development of disease, so non-nutritional disease will affect nutritional needs and nutritional status. Little is known in practical

Table 1. *Tendency to obesity among some breeds of dogs. (Mean prevalence of obesity among 6,826 dogs of 49 breeds visiting veterinary practices was 29.3%)*

Breeds with high prevalence of obesity				Breeds with low prevalence of obesity			
Breed	% Obese	n	Sig.*	Breed	% Obese	n	Sig.*
Cairn terrier	49	160	d	Miniature poodle	24	341	a
Dachshund	49	88	c	Yorkshire terrier	20	393	d
Bull terrier	48	23	a	Pekingese	18	86	b
Bassett hound	47	43	c	Dobermann	17	124	c
Beagle	43	66	b	Great Dane	16	46	a
Corgi	42	54	a	Staffordshire bull terrier	15	73	b
Cocker spaniel	41	88	d	German shepherd dog	15	643	d
Labrador	40	937	d	Whippet	12	45	b
Cavalier King Charles spaniel	38	157	b	Weimeraner	9	23	a
Jack Russell terrier	35	369	b	Lurcher	5	26	b
				Greyhound	5	99	d

Notes:
* Statistical significance tested by Chi-square, and shown as follows: $a = 0.1 > p > 0.05$; $b = 0.05 > p > 0.01$; $c = 0.01 > p > 0.001$; $d = 0.001 > p > 0.0001$.

Source: Edney & Smith (1986).

terms of the impact of the normal disease load of dogs and cats on their nutrient needs or metabolism, and studies are clearly called for.

Overnutrition

The most obvious nutritional disease of excess that affects pets is that of obesity, although extant data on prevalence (see Table 1) leave unexplained the question of why this is a particular problem in dogs but not interestingly, cats, and the related question of why breed variation exists in dogs.

The disease is in part the outcome of a marked ability of some dogs to exhibit hyperphagia, but possibly the reported absence of dietary induced thermogenesis in beagles, a particularly obesity prone breed, may also be of some importance, and opens up interesting areas for research. It is a mistake, of course, to describe obesity purely in terms of excess energy stored, since it also reflects a difference in the partitioning of the energy retained. In both laboratory animals and farm animals it has been demonstrated that there are marked strain or breed differences in the propensity to deposit fat or lean tissue at the same level of energy intake, and it seems reasonable to suggest that such a phenomenon may apply in dogs. It is notable in laboratory animals that the dietary composition, particularly the protein:energy ratio of the diet may affect the composition of gain (Coyer, Rivers & Millward, 1987). Such observations suggest that it is possible that breed differences in the incidence of obesity might be modulated by variations in the protein:energy ratio of the diet, with each breed having a different optimum. This surely is fertile ground for study.

Finally, and less speculatively, two obvious candidates as aetiological factors are still understudied: the impact of energy expenditure due to activity and the effect of environmental temperature on energy expenditure.

Besides obesity, the other consequence of high food intakes is rapid growth. In farm animal production rapid growth is almost always regarded as desirable, since it is economically the best strategy, while in human nutrition it has taken on an almost mystical status, with largely uncritical acceptance of the idea that maximal growth is optimal growth. This view while it lacks a scientific basis, may well also dictate the way that owners feed their pets.

The proposal (see Chapters 17 and 18) that too rapid growth in giant breeds may predispose to joint disease makes it necessary for the implications of growth rate to be considered. Many other lacunae exist, among them the extent to which overfeeding in the growing animal predisposes to adult obesity (an ideal area for a veterinary epidemiological study), the extent to which growth rate influences carcass composition and conformation, the impact of dietary quality on the nature of growth and the energy cost of growth.

In this context it is worth noting that much could be done to guide owners on feeding regimes if growth standards were established.

Table 2. *The relative toxicity of some xenobiotics for cats as compared to other mammals*

	Lethal dose (as mg/kg weight)			
	Cat	Rabbit	Rat	Guinea pig
Phenol	80	—	—	450–550
o-Cresol	55	450–500	650	350–400
m-Cresol	180	500–600	900	300–400
p-Cresol	80	300–400	500	200–300
2-Nitrophenol	600	1,700	—	—
4-Nitrophenol	197	600	—	—
1-Naphthol	100–150	9,000	—	—
2-Naphthol	100–150	3,800	—	—

Note:
Results are for subcutaneously administered substances except for data on naphthols which refer to oral doses. Results are lethal doses (i.e. the amount which kills an animal), except for phenol and 4-nitrophenol where the minimal lethal dose was determined from studies on groups of animals. A dash indicates no data.
Source: Data of Spector (1956), cited from Hirom, Idle & Millburn (1976).

Toxicology

Diets are not of course merely vehicles for providing nutrients. They are mixtures of foods, which are themselves complex mixtures of organic compounds, some nutritionally valuable, some irrelevant and others deleterious. The ability to tolerate some toxins, and to otherwise detoxify the diet consumed ranks not far behind Claude Bernard's 'constancy of the milieu interieur' as a condition of *la vie libre*.

A study of detoxification mechanisms in the dog and cat would be justified by this alone, but more potent reasons for their investigation exist, since evidence makes it clear that there are clear differences between the cat and the dog and other mammals in their ability to deal with xenobiotics.

The source of most organic toxins is the plant kingdom, so that animals tend to be more tolerant of dietary toxins the more important plants are in their diet. Thus, not surprisingly, the cat shows an extremely poor ability to detoxify, as if indeed it evolved making the assumption that the liver of its prey would have done its detoxifying for it! This is why the cat is easily poisoned by a variety of toxins, as some older and unpleasant experiments summarised in Table 2, make clear. In practical nutritional terms this sensitivity is reflected in the sensitivity of the cat to certain food additives notably benzoic acid and propylene glycol (NRC, 1986).

Much is known of the differences between the cat and dog and other mammals in their ability to detoxify xenobiotics. The cat appears to be especially poor at detoxifying phenolic compounds, because of a relatively inactive glucuronic acid conjugation pathway. As Table 3 shows, glucuronide conjugation is active in most mammals, including the dog, but appears relatively inactive in all Felidae. However some metabolites (e.g. bilirubin) and xenobiotics (e.g. phenylphthalein) are conjugated with glucuronic acid by the cat and the low activity overall is probably due to a lack of activity of the enzyme responsible for one mechanism of glucuronic acid conjugation, UDP-glucuronyltransferase (Hirom, Idle & Millburn, 1976).

The cat preferentially conjugates most phenols with sulphate, which pathway has a lower capacity than the glucuronide pathway, and consequently the rate of clearance of these compounds is extremely slow. This is why the cat is sensitive to aspirin and to benzoic acid at quite low levels. There is some evidence that in the rat the sulphate moiety in detoxification comes from methionine, in which case it is possible that the requirement for detoxification may explain the high sulphur amino acid requirement of the cat noted by NRC (1986).

Most xenobiotics with carboxyl groups are conjugated with glycine by both the dog and cat, as indeed by most mammals. Nevertheless, as Table 4 shows, for the cat the arylacetic acids, like the bile acids, are conjugated with taurine. Given the low rate of production of taurine by the species, its use in detoxification mechanisms is again of direct significance in considering requirements for the compound. In its use of taurine for arylacetic acid conjugation the cat is not unique amongst carnivora: after 1-naphthylacetic acid dosage for example, the percentages of the 24 h excretion as glycine and taurine conjugates for the dog are 56% and 25%, very similar to those for the cat (59% and 39%). Interestingly, for the ferret the proportions are 6% and 63%, while, for the lion, they are 94% and 0%. Thus while the propensity to use taurine as a conjugating agent is widespread amongst the Carnivora, there is considerable species variation, and no systematic pattern is discernible between the Canids and the Felids.

As might be expected, since it is evolutionarily a scavenger and generally nutritionally omnivorous, the dog is more able than the cat to deal with xenobiotics. Nevertheless, as the above might suggest, it is by no means normal in its ability to handle toxicants, as for example, the evidence on the toxicity of chocolate to the dog testifies. Interestingly, though many cases of chocolate poisoning in the dog have been recorded (see, for example, Glauberg & Blumenthal, 1983) and are due to the theobromine in this material, there are no case reports of the same phenomenon in cats, although as the NRC committee (1986) point out, a consumption of 40–50 g of cocoa powder would provide a dose equal to the LD50. Even chocolate would provide this at an intake of about 500 g.

An important aspect of toxicity, particularly given the fact that diet compounders are expected to incorporate safety factors for nutrient

Table 3. *The excretion and pattern of conjugation of several xenobiotics in cats, dogs, rats and pigs*

	Cat			Dog			Rat			Pig		
	A	B	C	A	B	C	A	B	C	A	B	C
Phenol	63	<1	88	—	18	82	95	44	55	97	94	6
1-Naphthol	91	1	98	—	—	—	59	47	53	81	66	32
2-Naphthol	93	3	98	—	—	—	86	52	48	84	94	6
Morphine	88	0	71	—	—	—	72	57	0	—	—	—
Phenacetin	56	3	86	—	—	—	58	24	72	32	55	11
Phenolphthalein	27	60	40	—	—	—	100	98	0	—	—	—
Benzoic acid	29	0	[100]	94	18	[82]	100	<1	[99]	—	—	—
1-Naphthylacetic acid	63	0	[98]	—	6	[81]	64	51	[24]	—	—	—
Phenylacetic acid	100	0	[91]	—	—	—	95	0	[100]	—	—	—
4-Chlorophenylacetic acid	100	0	[94]	—	—	—	95	0	[100]	—	—	—
4-Nitrophenylacetic acid	55	0	[20]	—	—	—	89	0	[69]	—	—	—
Indolyl-3-acetic acid	93	0	[75]	—	—	—	76	0	[67]	—	—	—
Methylphenylacetic acid	58	41	[38]	—	—	—	81	79	[1]	—	—	—
Diphenylacetic acid	40	76	[0]	—	—	—	100	95	[0]	—	—	—

Notes:

A: % of administered dose excreted in 24 hours.

B: % of 24 h excretion as glucuronide.

C: % of 24 h excretion as sulphate [or amino acid acid conjugate].

Source: Data for cats, dogs and rats are from Hirom, Idle and Millburn (1976); data for pig are from Capel, Millburn & Williams (1974). Compounds were administered by intraperitoneal injection.

Table 4. *Conjugation of carboxylic acids by the cat as compared to the rat*

| | Cat | | Rat | |
| | % of 24 h excretion conjugated with | | | |
	Glycine	Taurine	Glycine	Taurine
Benzoic acid	100	0	99	0
Phenylacetic acid	90	1	99	1
4-Chlorophenylacetic acid	92	2	97	0
4-Nitrophenylacetic acid	4	16	69	0
Indolyl-3-acetic acid	75	0	57	0
1-Naphthylacetic acid	59	39	23	< 1
Methylphenylacetic acid	15	23	< 1	0
Diphenylacetic acid	0	0	0	0

requirements, is the toxicity of nutrients. The toxicity of the fat soluble vitamins A & D for mammals is now well established and unlikely to be overlooked in dogs or cats which appear to be capable of developing both toxicities (see NRC 1985, 1986).

Though other nutrients are less toxic, they are by no means non-toxic, and overdosage with other vitamins (water soluble as well as fat soluble) and minerals has been shown to occur in various mammals. There is a marked disparity on the evidence for this between cats and dogs. In the former, there are no reliable records for hypervitaminoses apart from A and D, while the only toxicity recorded for a nutritionally essential mineral is for magnesium, something which itself requires careful reconsideration as is given by Buffington *et al.* in Chapter 23 of this book. In the dog, by contrast, the 1985 NAS/NRC committee noted that adverse effects for overdosing have been recorded for vitamin A, D, E, and K (at least as K_1), as well as for niacin (fed as nicotinic acid). Thiamin and vitamin B_{12} have both been shown to be toxic when injected and toxicity of an oral dose may be inferred. Adverse responses to overdosing with iodine, fluorine, iron and copper have also been noted, though no adverse effects of excess magnesium are known.

It is unlikely that this disparity simply indicates a greater tolerance of high doses by the cat (though the species does appear to be relatively tolerant of high doses of iodine and selenium). More probably, it reflects the paucity of evidence on the topic. Similarly, we are inclinded to suspect that the absence of reports of the toxic effects of pyridoxine, in either dog or cat, likewise reflects the poor state of our knowledge.

While there is no necessity for the development of large bodies of information on LD50 for compounds for the dog and cat, it is clear that close study of the peculiarities of the way the species deal with foreign substances in their diet may be, in the long run, a method of avoiding considerable problems.

Conclusion

This chapter will hopefully have justified our conviction that pet nutrition is likely to continue to be an expanding area of scientific interest. To return to the metaphor with which we began, it is clear to us that while cat and dog nutrition has made important progress, this book is in no sense a report from the summit, but, we hope, an informative message from a good vantage point. Because we have every confidence in the future of the climb, we anticipate, albeit with mixed feelings, the day when our book will be superseded.

REFERENCES

Anderson, R.S. (1980). *Nutrition of the Dog and Cat*. Oxford: Pergamon Press.

Capel, I.D., Millburn, P., & Williams, R.T. (1974). The conjugation of 1- and 2-naphthols and other phenols in the cat and pig. *Xenobiotica*, **4**, 601–15.

Cowgill, G.R. (1921). A contribution to the study of the relation between vitamin B and the nutrition of the dog. *American Journal of Physiology*, **57**, 420–31.

Cowgill, G.R. (1934). *The vitamin B Requirement of Man*. New Haven: Yale University Press.

Coyer, P.A., Rivers, J.P.W. & Millward, D.J. (1987). The effect of dietary protein and energy restriction on heat production and growth costs in the young rat. *British Journal of Nutrition*, **58**, 73–85.

Edney, A.T.B. & Smith, P.M. (1986). Study of obesity in dogs visiting veterinary practices in the United Kingdom. *Veterinary Record*, **100**, 391–6.

FAO/WHO/UNU (1985). *Energy and Protein Requirements*. World Health Organisation Report Series no 522. Geneva: World Health Organisation.

Glauberg, A., & Blumenthal, H.P. (1983). Chocolate poisoning in the dog. *Journal of the American Animal Hospital Association*, **19**, 246–8.

Goldberger, J. & Wheeler, G.A. (1928). Experimental blacktongue of dogs and its relation to pellagra. *Public Health Reports*, **43**, 172–239.

Goldberger, J., Wheeler, G.A., Lillie, R.D. & Rogers, L.M. (1928). A further study of experimental blacktongue with special reference to the blacktongue preventative in yeast. *Public Health Reports*, **43**, 657–94.

Graham-Jones, O. (1965). *Canine and Feline Nutritional Requirements*. Oxford: Pergamon Press.

Hirom, P.C., Idle, J.R. & Millburn, P. (1976). Comparative aspects of the biosynthesis and excretion of xenobiotic conjugates by non-primate mammals. In *Drug Metabolism-from Microbe to Man* ed. D.V. Parke & R.L. Smith, London: Taylor & Francis.

Mellanby, E. (1921). *Experimental Rickets*. Medical Research Council Special Report Series, no 61. London: Her Majesty's Stationery Office.

National Research Council (1985). *Nutrient Requirements of Dogs*. Washington DC: National Academy of Sciences.

National Research Council (1986). *Nutrient Requirements of Cats*. Washington DC: National Academy of Sciences.

Pavlov, I.D. (1910). *The Work of the Digestive Glands*. London: Thompson.

Spector, W.S. (1956). *Handbook of Toxicology (Volume 1)*. Philadelphia: W.B. Saunders and Co.

Stewart, R.J.C. (1957). The dog's place in medical research. *Journal of the Animal Technicians' Association*, **8**, September issue.

Vedder, E.B. (1913). *Beri-Beri*. London: Bale Sons and Danielssohn.

Whipple, G.H. & Robscheit, F.S. (1925). Blood regeneration in severe anaemia II. *American Journal of Physiology*, **72**, 408–18.

Appendix 1: Abstracts

Plasma urea concentration in healthy dogs in relation to protein intake, time after feeding and protein source

E. KIENZLE

Determination of plasma urea concentration is of considerable importance in the diagnosis of renal diseases. Nevertheless, mild ureamia is not necessarily indicative of early renal failure because plasma urea is known to be markedly affected by factors such as sampling time and protein intake. In order to quantify the effects of these variables on plasma urea level we have evaluated the results of several investigations conducted at the Hanover Veterinary School (see references below). In all of these studies normal healthy adult dogs were used with a bodyweight range from 8 to 25 kg. Protein intakes were 3 or 7 g/kg bodyweight/day. The results are summarised in the two figures. Figure 1 shows the effects of different types of protein on the increase in plasma urea values after feeding. There was some variation in response especially in the values obtained four or more hours postprandial. However in a different series of studies (Fig. 2) little effect of protein source was observed. From these results we conclude that the most important variables to be considered in the interpretation of plasma urea levels are protein intake and time of sampling. Protein source did not appear to be a key factor under the conditions of our investigations.

REFERENCES

Apel, U. (1981). *Untersuchungen über Akzeptanz und Verträglichkeit proteinarmer Rationen für Hunde.* Hannover, Tierärztliche Hochschule, Dissertation.

Barthold, K. (1985). *Einfluss geringer Dosen von Anti- und Chemobiotika auf Kotparameter (Ammoniak, flüchtige Fettsauren, Trockensubstanz) beim Hund.* Hannover, Tierärztliche Hochschule, Dissertation.

Elbers, H. (1985). *Praecolonale Verdaulichkeit und Gesamtverdaulichkeit verschiedener Proteine beim Hund (mit und ohne Antibiose).* Hannover, Tierärztliche Hochschule, Dissertation.

Junker, S. (1985). *Untersuchungen über den Einfluss einer reduzierten Keimflora im Darm des Hundes auf die Verdaulichkeit verschiedener Eiweisse.* Hannover, Tierärztliche Hochschule, Dissertation.

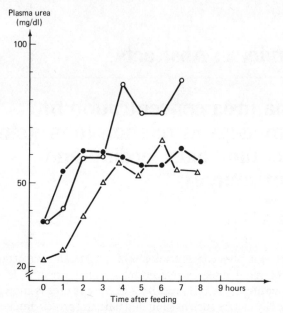

Fig. 1. Postprandial changes in plasma urea level in dogs fed different proteins –
protein intake 7 g/kg bodyweight; one meal per day.
△—△ raw lung ($n=6$)
●—● tinned dog food ($n=2$)
○—○ casein diet ($n=2$)

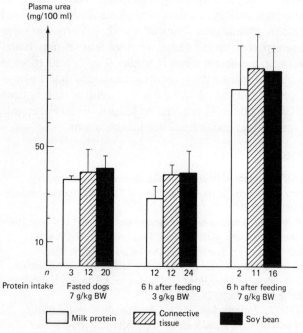

Fig. 2. Influence of protein source on plasma urea level in fed and fasted dogs
with different protein intakes.

Appendix 1: Abstracts

Lack of effect of age on digestibility of protein, fat and dry matter in Beagle dogs.*

C.A. BUFFINGTON, J.E. BRANAM, & G.C. DUNN

Three groups of Beagle dogs aged 2 to 3 ($n = 4$), 8 to 10 ($n = 5$), and 16 to 17 years ($n = 6$) were used to study the effects of age on digestive function and response of plasma amino acids to feeding. All dogs were housed in metabolism cages and fed a kibble-based diet (60%) supplemented with meat (20%), vitamins (1%) and blood meal (0.5%) in amounts which maintained bodyweight. After acclimatisation to the metabolism cages, 4-day quantitative urine and faecal collections were made. On the fourth day, venous blood was collected before, and three hours after, feeding for amino acid analysis. Mean nutrient digestibilities ($\pm$ SE) for young, mature and old dogs were: dry matter 71.3 ± 2.0, 71.5 ± 0.9, 69.5 ± 1.8; nitrogen 77.6 ± 1.4, 78.5 ± 1.2, 73.4 ± 1.8; fat 90.8 ± 1.0, 91.1 ± 0.4, 86.4 ± 2.2, respectively. No significant differences were found between groups. Age-related effects on prefeeding plasma amino acid levels included lower ($p < 0.01$) levels of arginine, citrulline, glutamate, isoleucine and valine in young dogs, and higher ($p < 0.01$) levels of aspartate, leucine, lysine, methionine, and histidine in old dogs. Three hours after eating, young dogs had lower ($p < 0.01$) levels of arginine and citrulline, while histidine, lysine and methionine were lower ($p < 0.01$) in old dogs than in other groups. No other differences in plasma amino acid levels between groups were found. We conclude that age related changes in nutrient digestibility and assimilation are modest in healthy dogs, with few differences observable between animals of greatly different ages.

* Supported by a grant from Allen Products Co., Inc., Allentown, PA.

Appendix 2

Nutrient requirements of the dog and cat

J.P.W. RIVERS & I.H. BURGER

Currently, standards for diets for dogs and cats are usually based upon reports produced by the US National Research Council (NRC, 1985, 1986). These review the evidence on different nutrients and make estimates of nutrient requirements. For both dog and cat these requirement estimates assume full bio-availability of nutrients and will need to be corrected for actual nutrient availability in any diet.

This appendix provides a summary of this data on nutrient requirements in a form that is of use in diet design and evaluation, and allows comparison of dogs and cats.

Energy requirements (Table 1)

Estimates of energy requirements, as metabolisable energy (ME) intakes for different physiological groups are presented in Table 1. Producing such estimates is complicated for both species: for dogs by difficulties in coping with the wide bodyweight variation in the species (see Chapter 6) and for cats by the fact that traditional estimates of ME, based on Atwater's factors may be overestimates by as much as 25% (Kendall, Burger & Smith, 1985).

For dogs, NRC (1985) provides two scales of estimates, one, in Table 4 of the report, which is scaled according to $W^{0.67}$, and the other in the text of the report, expressed as a ratio to adult maintenance requirements, which are also estimated in the report to be closely approximated to $W^{0.979}$. Thus the overall scaling has the same allometry. However, when expressed as ratios to maintenance these provide different estimates which are essentially similar for different ages and physiological states.

The impact of differences in age and physiological status on requirements for the cat (taken from NRC, 1986) is essentially similar to the dog, although the actual adult maintenance requirement is smaller for bodyweight (see note c to Table 1 below). The significance of this and any explanation are unknown.

399

Table 1. *Energy requirements for cats and dogs in various physiological states (as multiples of adult maintenance requirement)*

	Dog[a]	Dog[b]	Cat[c]
Early weaning	2.8		
Late weaning	2.6		
Early growth	2.4	2.0	
30% grown			3.1
50% grown	1.6		1.5
60% grown		1.6	
80% grown		1.2	
90% grown			1.1
Maintaining adult	1.0	1.0	1.0
Early gestation	1.0	1.0	
Late gestation	1.6	1.6	
Total gestation	1.2	1.2	1.2
Lactation[d] (large litters)	3.9	3.0	2.7

Notes:

[a] NRC (1985) Table 4, but scaled according to $W^{0.67}$ at all ages. Adult maintenance values given as $132–159$ kcal/$W^{0.67}$/day. An adult value of $145W^{0.67}$ assumed here for calculation.

[b] NRC (1985) from text. Values are cited as multiples of adult maintenance values for the breed which, on the basis of maintenance used by NRC, indicates an allometry of 0.88, and an adult value of 100 kcal/$W^{0.88}$/day.

[c] From NRC (1986) Table 1. Values for all except normal adults correct by 0.8 as recommended by NRC, to allow for overestimation of ME values. Adult value is for an inactive cat, given as 70 kcal/kgW/day. Adult weights are 4 kg (males), 3 kg (females).

Scaling these values allometrically gives:

	$W^{0.88}$	$W^{0.67}$
Inactive male	83	111
Active male	94	126
Inactive female	80	101
Active female	91	115

[d] Requirements for lactation vary with litter size and duration of lactation. These are for large litters in 1–3 weeks of lactation.

Nutrient standards for dog and cat diets

Nutrient requirements can be expressed in a variety of ways: per animal, per unit bodyweight or per kg of diet are common methods. The use of nutrient densities, that is amounts of nutrient per unit of energy, is one of the most versatile since it allows the evaluation of diets of different compositions, as well as providing insights into the differences and similarities between species. This system is used in Table 2, with nutrient densities calculated from the data provided by NRC (1985, 1986) and expressed as weights per 1000 kcal ME.

NRC (1986) only gives nutrient requirements for kittens, which are shown in column 1 of Table 2 below. Columns 2 and 3 give nutrient densities suitable for growing and adult dogs. These are calculated from the requirement data listed by NRC (1985) in their Table 4. Note that estimates of energy requirements and nutrient requirement of dogs vary in such a way that the same nutrient density is required by both age groups for most nutrients with the exception of protein and the indispensable amino acids. NRC (1986) gives no requirement for maintenance in the adult cat but notes that the kitten requirement is adequate.

Columns 4 and 5 are estimates of nutrient densities which are required in a diet which is suitable for all age groups including reproduction. This includes estimates of what the NRC committees termed prudent intakes of nutrients which are not actually necessary in the diet of normal healthy animals. These data are intended as a standard for diet design, not for evaluation.

Amino acid patterns for protein in dog and cat diets

In evaluating the quality of dietary proteins it is often necessary to have amino acid requirements per unit protein. These values, which might serve as provisional scoring patterns, are shown in Table 3.

It will be clear from the table that a protein which has an adequate pattern for growing dogs will be adequate for cats and dogs of all physiological states.

Note that in addition to amino acids a cat diet should contain the amino sulphonic acid taurine.

REFERENCES

Kendall, P.T., Burger, I.H. & Smith, P.M. (1985), Methods of estimation of the metabolisable energy content of cat foods. *Feline Practice*, **15(2)**, 38–44.
NRC (1985) and NRC (1986) – see page 393.

Table 2. Nutrient requirements for the dog and cat expressed as nutrient densities

	Amounts required per 1000 kcal ME			All ages[e]	
	Growing cat[a]	Growing dog[b]	Maintaining[c] adult dog[d]	Dog[e,f]	Cat[e,g]
Protein[h] (g)	48.0	24.5	20.0	25	48
Fat (g)	⌡	13.5	13.5	13.5	⌡
Essential amino acids					
Arginine[j] (g)	2.0	1.4	0.3	1.4	2.0
Histidine[j] (g)	0.6	0.5	0.3	0.5	0.6
Isoleucine[j] (g)	1.0	1.0	0.7	1.0	1.0
Leucine[j] (g)	2.4	1.6	1.1	1.6	2.4
Lysine[j] (g)	1.6	1.4	0.7	1.4	1.6
Methionine & Cystine[j,k] (g)	1.5	1.1	0.4	1.1	1.5
Phenylalanine + Tyrosine[j,l] (g)	1.7	2.0	1.2	2.0	1.7
Threonine[j] (g)	1.4	1.3	0.6	1.3	1.4
Tryptophan[j] (g)	0.3	0.4	0.2	0.4	0.3
Valine[j] (g)	1.2	1.1	0.8	1.1	1.2
Total indispensable amino acids[j] (g)	13.7	11.8	6.3	11.8	13.7
Dispensable amino acids[j] (g)	13.7	17.1	17.1		
Accessory food factors					
Vitamin A (retinol)[m,n] (mcg)	200	300	300	300	360
Vitamin D (cholecalciferol)[m,p] (mcg)	2.5	2.8	2.7	2.8	2.5
Vitamin E (α-tocopherol)[m,q] (mcg)	6.0	6.1	6.7	6.7	6.0
Vitamin K (phylloquinone)[r] (mcg)	20	220	300	270	20

Vitamin B$_1$ (Thiamin) (mg)	0.3	0.3	0.3	1.0
Riboflavin (mg)	0.7	0.7	0.7	0.8
Vitamin B$_6$ (Pyridoxine) (mg)	0.3	0.3	0.3	0.8
Niacin (mg)	3.0	3.0	3.0	8.0
Pantothenic acid (mg)	2.7	3.0	2.6	1.0
Folic acid (mcg)	54	54	545	160
Vitamin B$_{12}$ (Cyanocobalamin) (mcg)	7.0	7.0	7.0	4.0
Biotin (mg)	[s]	[s]	0.3	14[s]
Linoleic acid (g)	2.7	2.7	2.7	1.0
Arachidonic acid (mg)	—[t]	—[t]	—[t]	40
Taurine (mg)	—[t]	—[t]	—[t]	100
Choline (mg)	340	340	340	480
Inositol[u] (mg)	—	—	—	40
Minerals				
Calcium (g)	1.6	1.6	1.6	1.6
Phosphorus (g)	1.2	1.2	1.2	1.2
Potassium (g)	1.2	1.2	1.2	0.8[w]
Sodium (g)	0.15	0.15	0.15	0.1
Chloride (g)	0.25	0.25	0.25	0.4
Magnesium (mg)	110	110	110	80
Iron (mg)	9	9	9	16
Copper[x] (mg)	0.8	0.8	0.8	1.0
Manganese (mg)	1.4	1.4	1.4	1.0
Zinc (mg)	10	10	10	10
Iodine (mcg)	160	160	160	70
Selenium (mcg)	30[v]	30[v]	30[v]	20

Notes to Table 2

Notes:

[a] NRC (1986) estimate based upon data for 10–20 week old kittens. NRC suggest that a diet meeting these levels would be adequate for all physiological states, except for reproduction when an additional 360 mcg retinol (1200 IU vitamin A) and 100 mg of taurine would be required per 1000 kcal ME.

[b] NRC (1985) estimate based upon a 3 kg beagle puppy consuming 600 kcal ME per day.

[c] There are no specific requirements for the maintaining adult cat given by NRC (1986), but it is stated in a footnote that levels for the growing kitten would be more than adequate for adults, and that 'protein and methionine can be reduced to 140 and 3 g/kg diet [i.e. 28 and 0.6 g/1000 kcal ME] respectively. It is likely that the minimum requirement for all other nutrients are lower for maintenance than for the growing kitten'.

[d] NRC (1985) estimate in their Table 1 which is based upon an 'average 10 kg bodyweight dog consuming 742 kcal ME/day'. It is assumed by NRC that these ratios are independent of size, though in fact the allometry of most nutrient requirements has not been established in the dog.

[e] Diets providing these levels should according to NRC estimates be adequate for all ages and physiological states.

[f] From text and Table 1 of NRC (1985).

[g] From NRC (1986) Table 2, with levels of vitamin A and taurine increased to those required for reproduction as recommended.

[h] For cat, protein requirements are $N \times 6.25$. (NRC, 1986). For dog, the NRC (1985) report provides only estimates of dispensable (DAA) and indispensable amino acids (IDAA). A protein requirement was calculated from these as the sum of the weights of the amino acid residues, assuming that, for the DAA, the residues made up 86% of the total amino acid weight. The N:protein ratio for the amino acid mixture suggested by the NRC would vary with the exact DAA fed. If they were in the proportions provided by mixed animal proteins it could be assumed to not differ materially from 6.25. The amounts of protein required assume that the protein meets all the essential amino acid requirements.

[i] NRC (1986) do not specify any total fat requirement for cats apart from the requirements for linoleic and arachidonic acids. However, they note that fat is also essential as a carrier for fat soluble vitamins and to enhance the palatability of the diet.

[j] These are as amino acids not amino acid residues. Values have been rounded to the nearest 0.1 g.

[k] For the cat, at least 0.8 g must be provided as methionine.

[l] For the cat, at least 0.8 g must be provided as phenylalanine.

[m] Values can be converted to international units (IU) using the following factors:
Retinol 1 mcg = 3.33 IU
Cholecalciferol 1 mcg = 40 IU
Tocopherol 1 mcg = 1 IU

[n] Carotene is not a source of vitamin A activity for the cat, but it is well utilised by the dog.

^p Vitamin D_2 (ergocalciferol) and vitamin D_3 (cholecalciferol) are well utilised by both species.

^q Values are for a moderate to low fat diet. Requirement may be elevated 5 times by feeding high levels of polyunsaturated fatty acids (PUFA). A vitamin E intake of at least 0.5 mg/g PUFA is recommended as a target for dog diets.

^r There is some evidence that menadione may be less well utilised than phylloquinone by the cat and dog.

^s No dietary requirement has been demonstrated in animals fed diets compounded from natural ingredients, although a metabolic requirement exists.

^t No dietary requirement is believed to exist although a metabolic requirement exists.

^u Dietary essentiality of inositol has not been studied. Diets fed to cats invariably contain at least 30–40 mg inositol /1000 kcal ME.

^v If vitamin E intake is limited a value of 80 is recommended.

^w Potassium requirements may be elevated on high protein diets.

^x Provided the dietary mineral concentrations are not excessive.

Table 3. *Amounts of indispensable (essential) amino acids required (as grams per kg protein)*[a]

	Growing cat	Growing dog	Adult dog
Arginine	42	56	14
Histidine	13	20	15
Isoleucine	21	40	33
Leucine	50	65	57
Lysine	53	57	34
Methionine and cystine[b]	31	43	21
Phenylalanine and tyrosine[c]	35	80	59
Threonine	29	52	30
Tryptophan	6	17	9
Valine	25	43	41
Dispensable amino acids	885	697	859

Notes:

[a] Values are for free amino acids, not amino acid residues. They therefore total more than the weight of the protein.

[b] For cat diets, at least 16.7 g of this must be as methionine. There is no specific cystine requirement for either species.

[c] For cat diets, at least 16.7 g must be fed as phenylalanine. There is no specific requirement for tyrosine by either species.

Calculated from data tabulated/presented by NRC 1985 and NRC 1986 as $N \times 6.25$. The DAA requirements are calculated from this by subtraction of the IDAA residues from crude protein and assuming that the ratio of residues to free DAA is 0.86.

Index

407